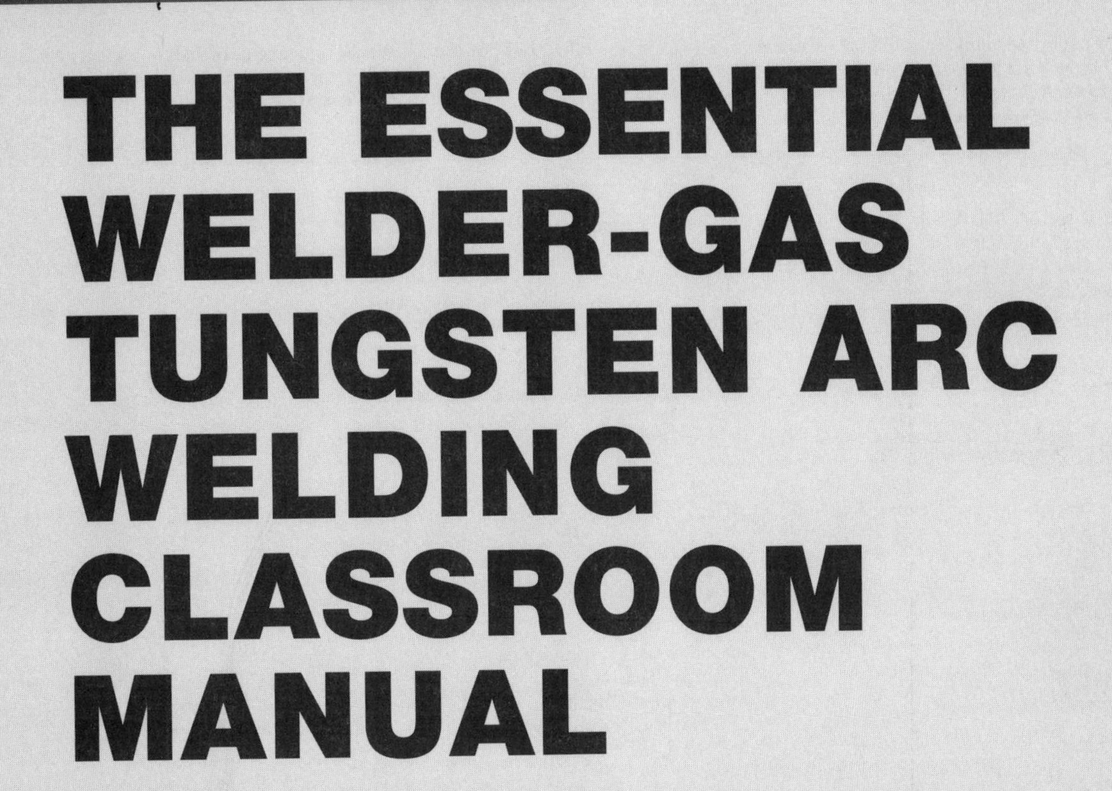

THE ESSENTIAL WELDER-GAS TUNGSTEN ARC WELDING CLASSROOM MANUAL

Larry Jeffus

Africa • Australia • Canada • Denmark • Japan • Mexico • New Zealand • Philippines • Puerto Rico • Singapore • Spain • United Kingdom • United States

NOTICE TO THE READER

Publisher does not warrant or guarantee any of the products described herein or perform any independent analysis in connection with any of the product information contained herein. Publisher does not assume, and expressly disclaims, any obligation to obtain and include information other than that provided to it by the manufacturer.

The reader is expressly warned to consider and adopt all safety precautions that might be indicated by the activities herein and to avoid all potential hazards. By following the instructions contained herein, the reader willingly assumes all risks in connection with such instructions.

The Publisher makes no representation or warranties of any kind, including but not limited to, the warranties of fitness for particular purpose or merchantability, nor are any such represenations implied with respect to the material set forth herein, and the publisher takes no responsibility with respect to such material. The publisher shall not be liable for any special, consequential, or exemplary damages resulting, in whole or part, from the readers' use of, or reliance upon, this material.

Delmar Staff
Business Unit Director: Alar Elken
Executive Editor: Sandy Clark
Acquisitions Editor: Vern Anthony
Editorial Assistant: Bridget Morrison
Executive Marketing Manager: Maura Theriault
Channel Manager: Mona Caron
Marketing Coordinator: Kasey Young
Executive Production Manager: Mary Ellen Black
Project Editor: Barbara L. Diaz
Art Director: Cheri Plasse

Printed in the United States of America
4 5 6 7 8 9 10 XXX 05 04 03

For more information, contact Delmar, 3 Columbia Circle, PO Box 15015, Albany NY 12212-5015; or find us on the World Wide Web at *http://www. delmar.com*

Asia
Thomson Learning
60 Albert Street, #15-01
Albert Complex
Singapore 189969

Australia/New Zealand
Nelson/Thomson Learning
102 Dodds Street
South Melbourne, Victoria 3205
Australia

Canada
Nelson/Thomson Learning
1120 Birchmount Road
Scarborough, Ontario
Canada M1K 5G4

International Headquarters
Thomson Learning
International Division
290 Harbor Drive, 2nd Floor
Stamford, CT 06902-7477
USA

Japan
Thomson Learning
Palaceside Building 5F
1-1-1 Hitotsubashi, Chiyoda-ku
Tokyo 100 0003
Japan

Latin America
Thomson Learning
Seneca, 53
Colonia Polanco
11560 Mexico D.F. Mexico

Spain:
Thomson Learning
Calle Magallanes, 25
28015-Madrid
Espana

UK/Europe/Middle East
Thomson Learning
Berkshire House
168-173 High Holborn
London
WC1V 7AA United Kingdom

Thomas Nelson & Sons Ltd.
Nelson House
Mayfield Road
Walton-on-Thames
KT 12 5PL United Kingdom

Library of Congress Cataloging-in-Publication Data

Jeffus, Larry.
The essential welder: gas tungsten arc welding classroom manual /
Larry Jeffus.
p. cm.
ISBN 0–8273–7614–6 (alk. paper)
1. Gas tungsten arc welding Handbooks, manuals, ect. I. Title.
TK4660.J395 1999
671.5'212—dc21

99–6466
CIP

CONTENTS

PREFACE

The gas tungsten arc welding (GTAW) process can be used to join almost any metal. It can be used in all positions and on a wide variety of thicknesses from a few thousands of a inch up to several inches. In addition to this wide range, the welds produced are clean, slag free, and in most cases require no postweld cleanup. All in all, this is a very versatile process.

In this textbook we focus on the essentials of this GTAW process. Once you master these skills, further developments in your welding abilities are practically unlimited with this process. GTAWelding as been used on everything from nuclear submarines to spacecraft and from mining equipment to aircraft. Everyone who works on such equipment had to develop their skills in the same manner as you will as you go through this text and lab guide.

One complaint about GTAW is that it is slow in comparison to other processes such as MIG or flux-cored wire welding. However, with practice, welding speeds can be improved, and often any question about speed is far outweighed by the precise control of the welding that is possible. In critical cases, being fast is not as important as being accurate, especially when lives are at stake, and no other manual welding process provides a comparable level of weld metal control.

Because many of the skills you will develop are easily transferred from one metal to another, we start with less expensive metals, such as mild or low-carbon steel; after you have developed the basics, you will move on to the more expensive materials, such as stainless steel and aluminum.

Acknowledgments

I would like to express my sincere thanks and appreciation to some of the people that helped make this book possible: Carol Jeffus for proofreading, Tina Ivey and Jessica Aga for their typing and organizing of material, and Wendy Jeffus and Amy Jeffus for their moral support.

Dedication

This book is dedicated to three very special people: my wife, Carol, and my daughters, Wendy and Amy.

CHAPTER 1

INTRODUCTION TO GAS TUNGSTEN ARC WELDING

INTRODUCTION

The gas tungsten arc welding (GTAW) process has been called by several names over the years. In the late 1930s when the process was developed, helium shielding gas was used with an arc to produce welds, hence, the name Heliarc®. The term Heliarc® is a registered trademark used by the Linde Company on its GTAW equipment. During the late 1940s and early 1950s argon gas was introduced as an alternate shielding gas. In the 1960s argon became the primary shielding gas, and the term *tungsten inert gas* welding (TIG) became the most commonly used term. In the 1970s, as the American Welding Society began standardizing welding terms, the process was given the official name of *gas tungsten arc welding* (GTAW). This book will use the term as well as all other current American Welding Society terminology.

The process has long been held in high regard by welders. GTA welding has from its very beginning been considered by most people to be the best welding process. And those welders who could heliarc were looked up to by both welders and nonwelders.

Stories from the early days of GTA welding relate that heliarc welders would not allow others to watch the welding process. The reason that others may have been excluded and the welding done after the shop was closed may have been simply that weldners needed to reduce the possibility of drafts that could easily blow the light helium shielding gas away from the molten weld pool. Such limitations helped to promote the air of mystery that surrounded the early heliarc welders.

The use of the term *heliarc* or even TIG within the welding community is often considered a sign of a lack of knowledge; therefore both terms have fallen out of favor within the welding industry. However, the special place heliarc has with the nonwelding public is demonstrated today by the commercial use of the term. Advertisements in telephone books and signs on welding shops proudly proclaim their ability to "Heliarc." Welders sometimes hear "Yes, but can you heliarc?" I have personally been asked, "I know that you can TIG weld, but can you heliarc?" My response has been, "No, but I can do GTA welding."

PROCESS OVERVIEW

The American Welding Society defines GTAW as "an arc-welding process that uses an arc between a tungsten electrode (nonconsumable) and the weld pool. The process is used with shielding gas and without the application of pressure" (Figure 1-1). The arc produces the heat needed to melt the work. The shielding gas keeps oxygen in the air away from the molten weld pool and the hot tungsten. There may or may not be filler metal added to the molten weld pool during the process.

The most commonly used welding currents are direct current electrode negative (DCEN) and alternating current (AC) (Figure 1-2). In some unique applications direct current electrode positive (DCEP) may be used.

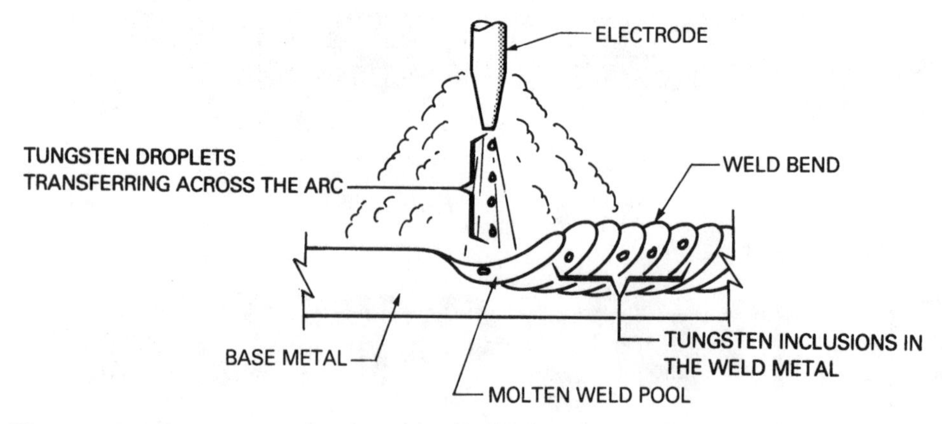

Figure 1-1 Some tungsten will erode from the electrode and can be transferred across the arc and become trapped in the weld deposit.

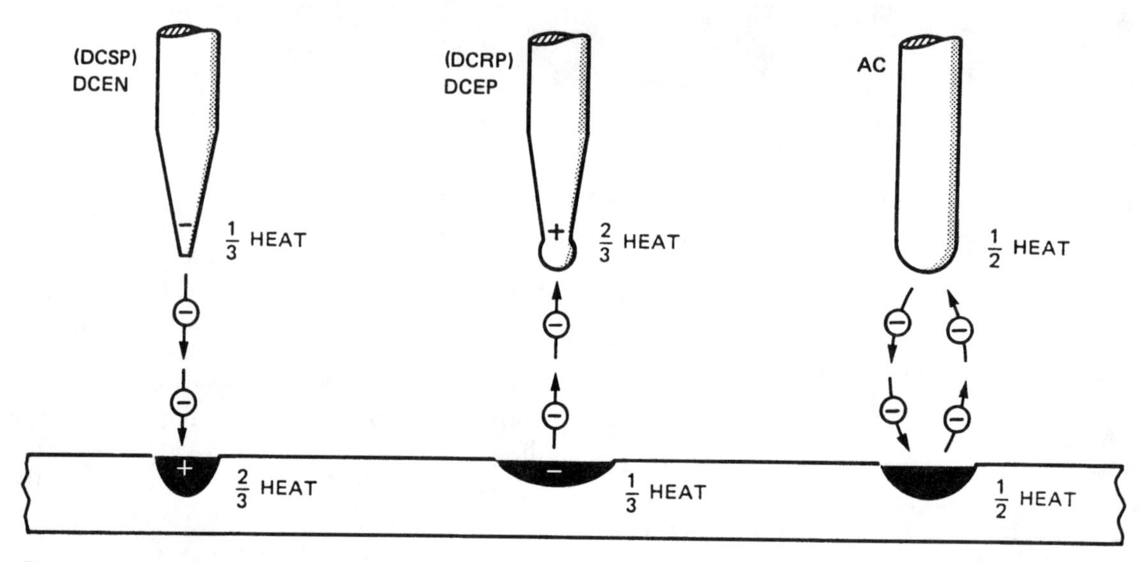

Figure 1-2 Heat distribution between the tungsten electrode and the work with each type of welding current.

Tungsten is used for the electrode because of its high melting temperature and good electrical characteristics. Argon gas is the most commonly used shielding gas because it provides very good coverage of the weld zone and is less expensive than helium.

The addition of filler metal in the form of rods or wire provides both the weld reinforcement and changes the weld chemistry to provide the desired weld physical properties.

WELD CHARACTERISTICS

GTAW is considered to be one of the most versatile of all welding processes. It can be used to make high-quality welds in almost any metal, in any welding position, and on almost any thickness of sheet, plate, or pipe. For example, Ernest Levert, a welding engineer working on the thermal control units (heat rejection panels) for the international space station, has welders producing manual GTA welds on stainless steel and inconel tubing that ranges in diameter from 1/8″ to 1″ (6 mm to 25 mm) with a wall thickness of 0.028″ (0.66 mm) (Figure 1-3). Tens of thousands of these tiny welds have been made successfully in the production of these panels at the Lockheed Martin Vought Systems' facility in Grand Prairie, Texas.

The GTA welding process has a number of advantages over most other welding processes. It is a very clean and versatile process. Because it leaves no slag such as that found on most other welding process, there is no need for postweld cleanup, unlike with most other welding processes. The versatility of the process is demonstrated by its ability to be used in welds with almost any metal of any thickness

Figure 1-3 Large structure that could be fabricated in space someday. (Courtesy of NASA)

range in any position. Before gas tungsten arc welding was developed, metals such as aluminum and magnesium were very difficult or impossible to weld. Welds that were produced in these metals had very poor mechanical properties because they were very porous and had very poor corrosion resistance.

The first major change that occurred with the new process known as heliarc was the development of an inexpensive source of argon. Argon is a by-product of the production of oxygen. Working with argon as a shielding gas, technicians developed the process, changing from DC electrode positive to DC electrode negative. Today DC electrode negative is the most commonly used current in the GTA process because of its arc stability and high heat input to the workpiece. With this change, gas tungsten arc welding became the most popular method of joining metals considered to be difficult to weld such as aluminum, magnesium, titanium, and some grades of stainless steel.

Major drawbacks to an even wider use of gas tungsten arc welding is its expense. Manual GTA welds require a great deal of time, which results in higher total cost for the product. Advances in automation with the GTAW process have reduced these costs somewhat; however, it is still considered to be a relatively expensive process as compared to other joining methods. These expenses are often outweighed by the high-quality welds that are produced. There are some welds that often can be produced only by using this process. Most GTA welds are produced in materials that are less than 1.25″ (6 mm).

The eye–hand coordination required to make GTA welds is very similar to that required to produce oxy-fuel welds. GTA welding is often easier to learn if a person can gas weld, although gas welding is not a prerequisite for learning GTA welding skills.

Weld Contamination

Because gas tungsten arc welding does not have a flux, the weld can be easily contaminated from a number of sources. The major sources of potential contamination are as follows:

- Filler metal. The weld may be contaminated by oil, dirt, grease, or oxides that are on the filler metal.
- Shielding gas. The weld may be contaminated by air or oxygen in the shielding gas moisture picked up by the shielding gas due to a poor fitting in the torch water leakage.
- Base metal. The weld may be contaminated by surface oxides or inclusions that are on or in the base metal itself and that are released during welding.
- Welders' hands. The weld may be contaminated by oils and dirt from the welders' hands as they handle filler metals or the plate.

Heat Input

Manual GTA welding has a relatively slow travel speed, which results in a higher heat input to the weldment. The higher heat input can be an advantage or a disadvantage, depending on the material being welded, the thickness of the material being welded, the weld configuration, and the shape of the weldment (Figure 1-4).

Thin sections of steel are highly susceptible to distortion or warping as a result of the slow travel speed and high input from the GTA process. The slower travel rate and higher heat input required on thicker sections of the same stainless steel tend to reduce the formation of martensitic grain structure and the brittle phase of stainless steel, thus reducing its tendency for cracking. Metals such as aluminum, which have high thermal conductivity, may pull the heat away from the weld so rapidly that the entire part becomes warm or hot during the welding process. This can be a particular problem if such high temperatures damage the nylon bushings, bearings, or finishes that may already exist on the aluminum part. Heating up the entire aluminum part during the weld may result in distortion. Such distortion can be controlled for the aluminum by holding the parts in place during the entire welding cycle with jigs or fixtures.

Arc and Stability

A number of factors can result in an erratic arc during the GTA welding process. Some of these factors are as follows:

- Magnetic interference (arc blow)
- Surface contamination
- Impurities in the shielding gas
- Imperfections in the shielding gas nozzle (chips or cracks)
- Metal oxides on the welded surface
- Improper welding current or voltage
- Improper electrode size
- Incorrect welding manipulation
- Improper electrode end shape

Magnetic interference, referred to as *arc blow,* is the result of a phenomenon caused when electricity flows through a conductor. As the electrons flow through a material, magnetic fields are created at right angles to that current flow. These magnetic fields may exert minute forces on the arc, causing it to move in an erratic manner sometimes referred to as arc wandering (Figure 1-5). Arc blow is most easily corrected by changing from direct (DC) welding current to AC. However, this remedy is not normally available for GTA welding due to the unfavorable heat distribution that exists with AC (Figure 1-6).

A major factor in the creation of these undesirable magnetic fields can be the location of the ground, or work, connection. The majority of the current flows directly from the arc to these connections. Changing their location may eliminate or disrupt the magnetic fields that are causing the arc blow.

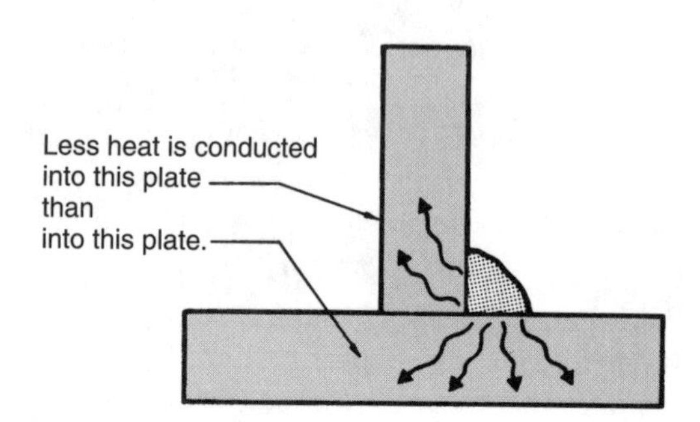

Figure 1-4 The heat is not distributed evenly in a tee joint. This will result in one plate melting faster than the other.

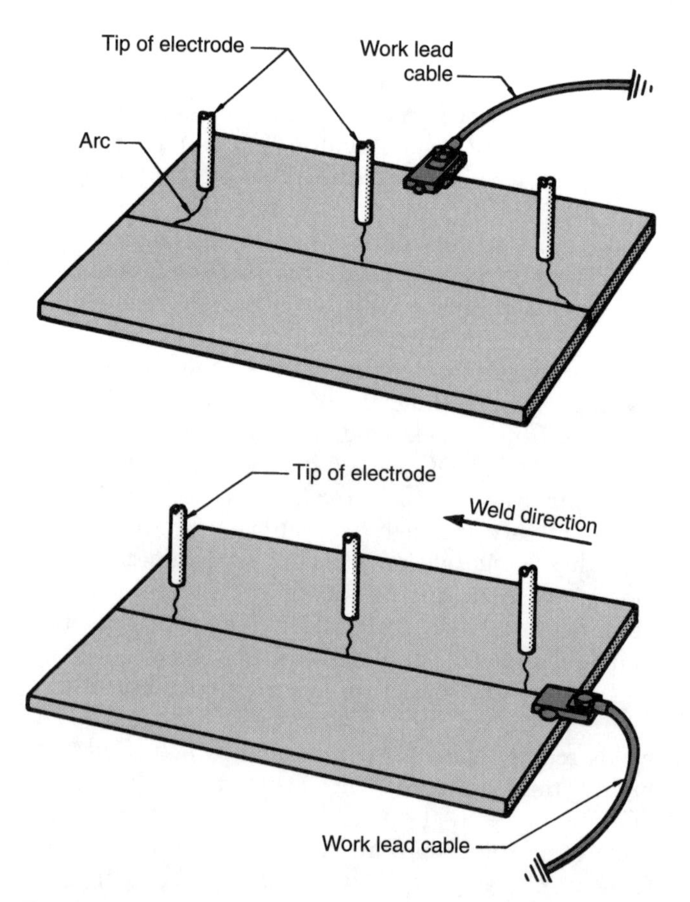

Figure 1-5 Arc blow can sometimes be controlled by changing the location of the work lead.

Additional parts surrounding the groove may also contribute to the formation of magnetic fields (Figure 1-7). Such a problem is likely to occur when trying to weld in the interior of a corner (Figure 1-8). Eliminating this joint configuration wherever possible will aid in reducing this problem. However, this is not possible for most welding applications. The best alternative is usually to use a slightly higher current and a slightly shorter arc and to carry the weld through the corner. Not stopping or starting in the actual corner (in addition to reducing arc blow) will make the corner weld stronger and less likely to have imperfections (Figure 1-9).

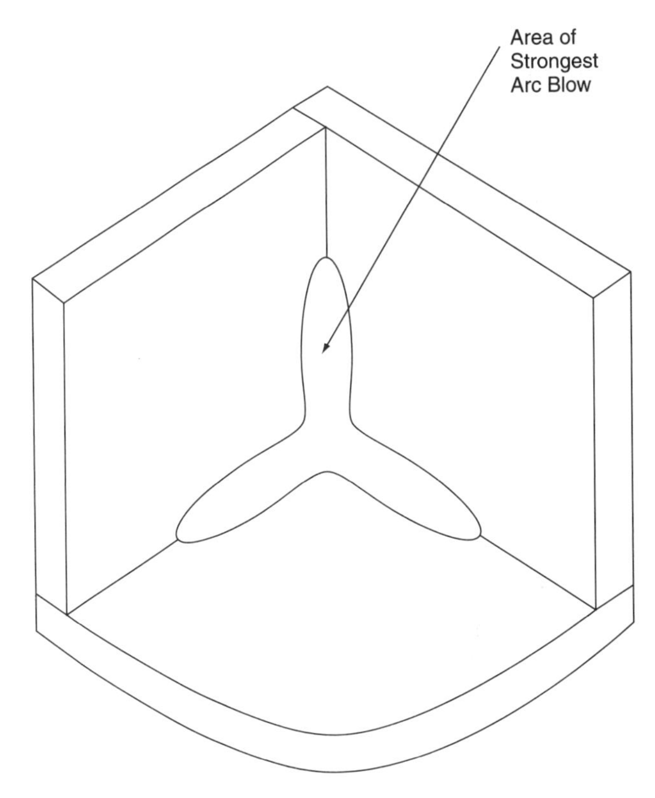

Figure 1-8 Interior corners have the strongest arc blow.

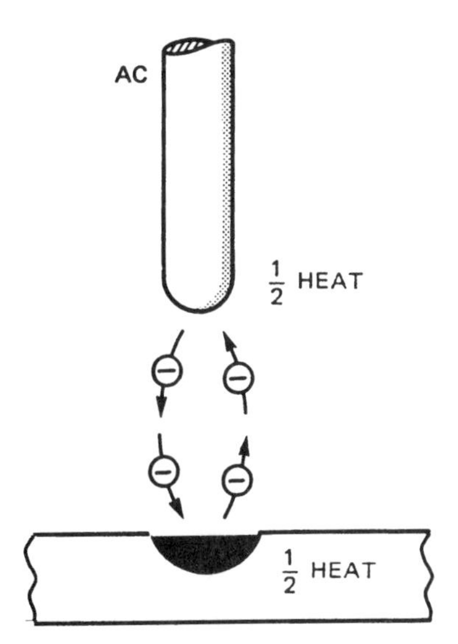

Figure 1-6 Heat distribution for AC.

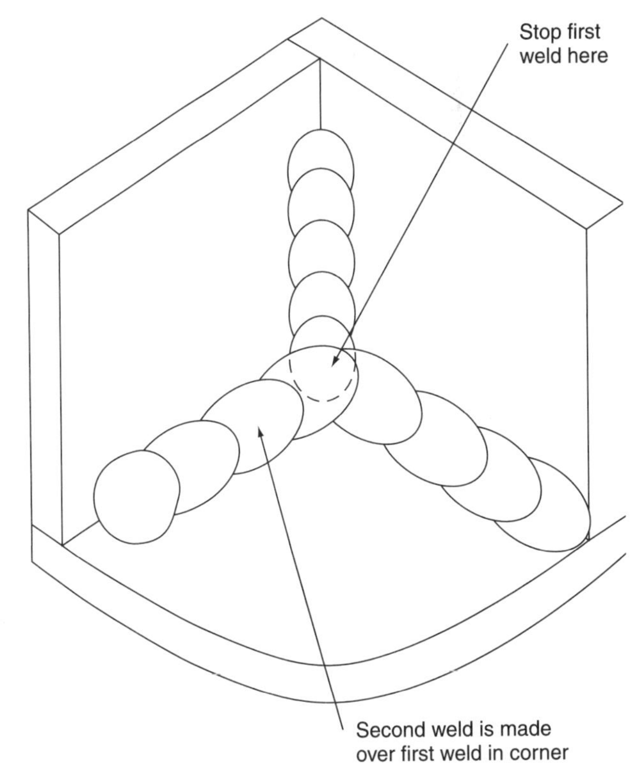

Figure 1-9 Not stopping or starting in a corner will help reduce arc blow problems.

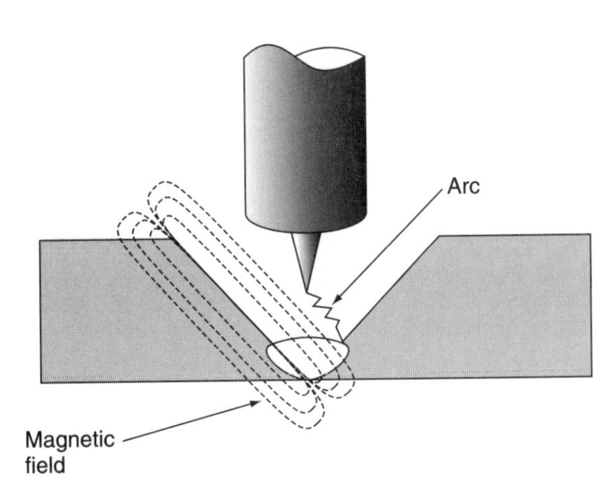

Figure 1-7 Uneven magnetic field causes arc to be blown off center.

Surface contaminates such as heavy oxides or, in the case of aluminum, anodized surfaces can also result in an erratic arc. Such surface conditions prevent the arc from stabilizing as they force it to move about in search of a good spot. If the molten weld pool is not sufficiently large to dislodge surface impurities, the arc may jump over them. This skipping to a new position will result in a weld discontinuity (Figure 1-10).

PLATE WELDING POSITIONS

The flat welding position is ideal for most joints. It allows a larger molten weld pool to be controlled. Usually the larger the weld pool, the faster the weld joint can be completed. It is not always possible to position a part so that all the welds can be made in the flat position. When weld positioning is not possible, changes in the weld joint design may help. For example, the bevel joint is often the best joint for horizontal welding (Figure 1-11).

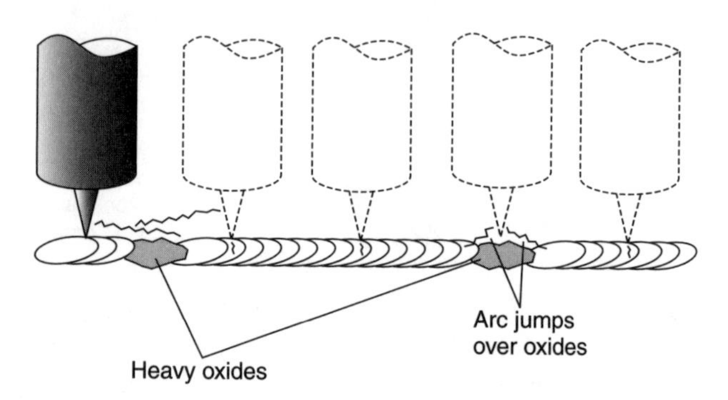

Figure 1-10 Heavy oxides on base metal can cause the arc to jump, resulting in weld discontinuities.

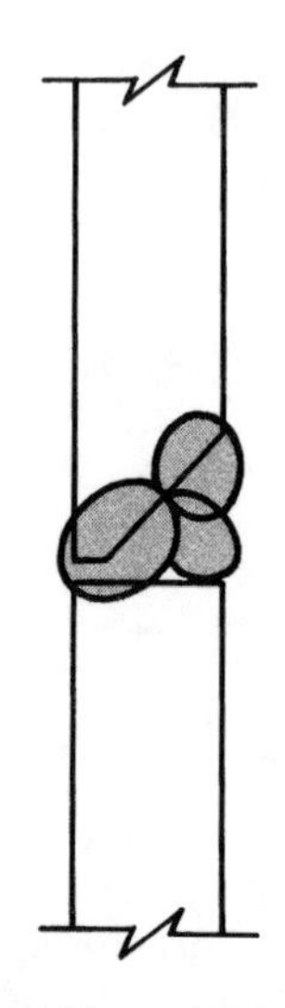

Figure 1-11 Horizontal bevel joint.

The American Welding Society has divided plate welding into four basic positions for groove (G) and fillet (F) welds as follows:

- Flat 1G or 1F: Welding is performed from the upper side of the joint, and the face of the weld is approximately horizontal (Figure 1-12).

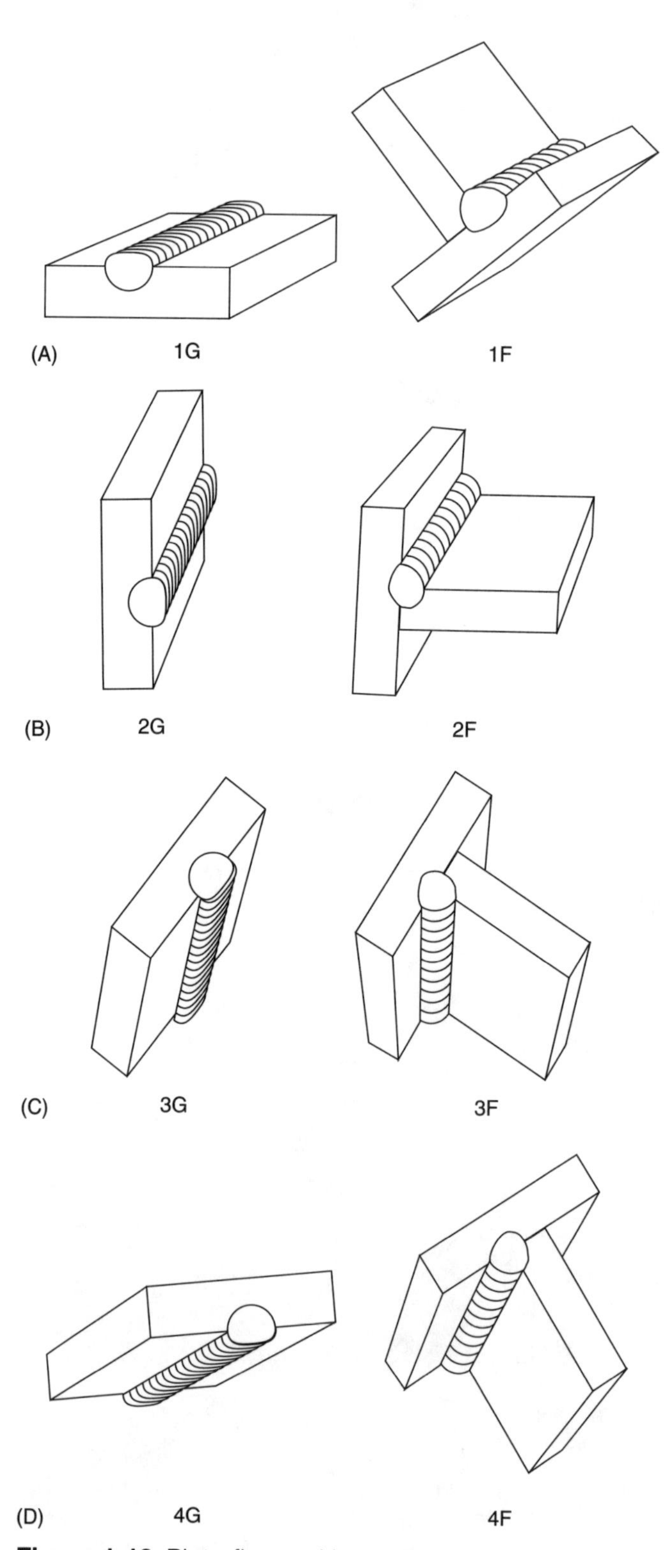

Figure 1-12 Plate flat position.

- Horizontal 2G or 2F: The axis of the weld is approximately horizontal, but the type of the weld dictates the complete definition. For a fillet weld, welding is performed on the upper side of an approximately vertical surface. For a groove weld, the face of the weld lies in an approximately vertical plane (Figure 1-13).

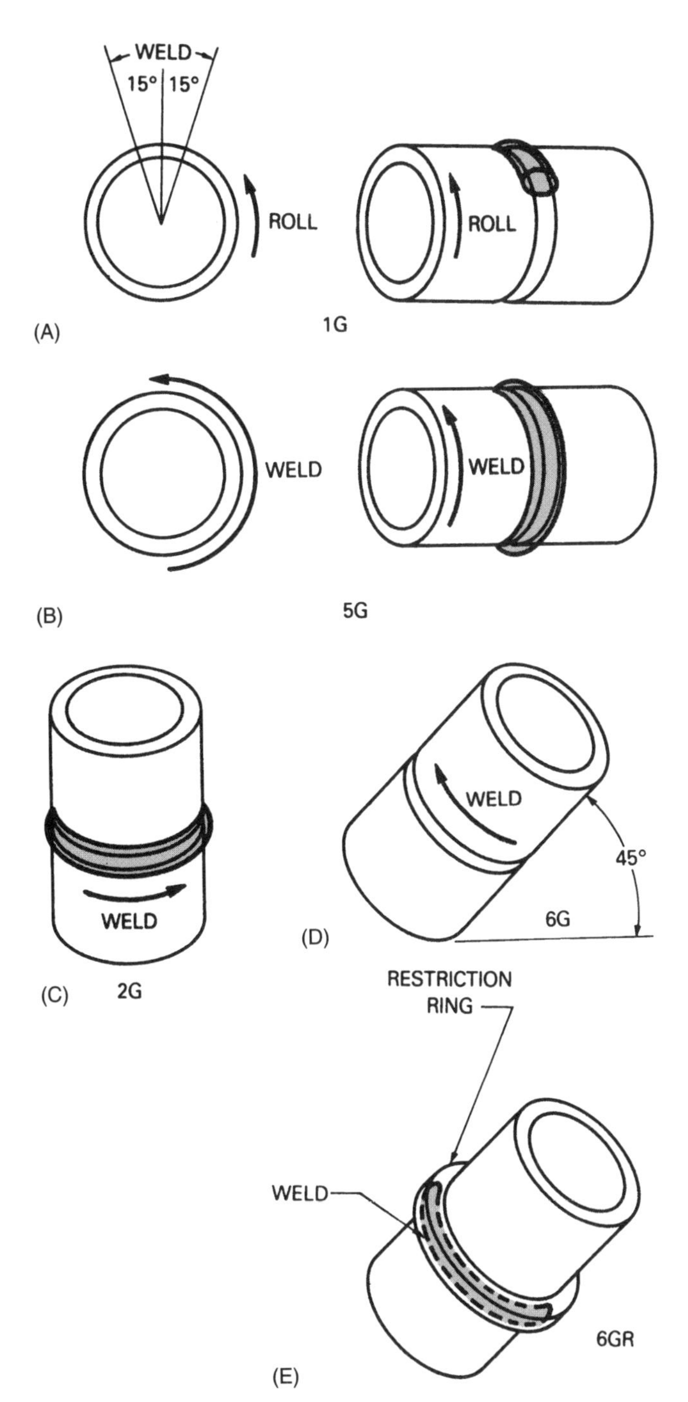

Figure 1-13 Plate horizontal position.

- Vertical 3G or 3F: The axis of the weld is approximately vertical.
- Overhead 4G or 4F: Welding is performed from the underside of the joint.

Pipe Welding Positions

The American Welding Society Pipe has divided pipe welding into the following five basic positions:

- Horizontal rolled 1G: The pipe is rolled either continuously or intermittently so that the weld is performed within 0° to 15° of the top of the pipe.
- Horizontal fixed 5G: The pipe is parallel to the horizon, and the weld is made vertically around the pipe.
- Vertical 2G: The pipe is vertical to the horizon, and the weld is made horizontally around the pipe.
- Inclined 6G: The pipe is fixed in a 45° inclined angle, and the weld is made around the pipe.
- Inclined with a restriction ring 6GR: The pipe is fixed in a 45° inclined angle, and a restricting ring is placed around the pipe below the weld groove.

SUMMARY

The gas tungsten arc welding process normally requires more training time, manual dexterity, and welder concentration than do other welding processes such as gas metal arc, shielded metal arc, and flux core arc welding. GTAW lends itself very well to all weld joint geometries and overlays for plate, sheet, pipe, tubing, and so on. It is, however, most practical for welding sections 3/8″ (9 mm) or less in thickness or for pipe and tubing 1″–6″ in diameter. Thicker sections can be welded but generally are not economically feasible. A combination of GTA welding in conjunction with other processes such as shielded metal arc, gas metal arc, or flux core arc welding can be advantageous when doing pipe. Gas tungsten arc provides a very smooth, uniform root pass with good root surface characteristics that can provide a basis for a more economical application of process gas metal arc, shielded metal arc, or flux core to complete the process.

REVIEW QUESTIONS

1. What are the most commonly used welding currents?
2. Why is tungsten used for the electrode?

3. What is the most commonly used shielding gas and why?
4. What are some of the advantages of GTAW process?
5. Which current is used today in the GTA process and why?
6. What are some major sources for potential contamination?
7. Why is slow travel speed bad for thin sheets of metal?
8. What may happen if you heat up an entire section of aluminum?
9. What factors may result in an erratic arc?
10. How is arc blow corrected?
11. What is a major factor in creating undesirable magnetic fields?
12. Why is the flat welding position ideal?
13. Which joint is often best for horizontal welding?
14. What are the four basic positions for groove and fillet welds?
15. In which position is the pipe rolled either continuously or intermittently so that the weld is performed within 0° to 15° of the top of the pipe?
16. When the pipe is vertical to the horizon and the weld is made horizontally around the pipe, what type of pipe welding is it?
17. When the pipe is fixed in a 45° inclined angle and a restricting ring is placed around the pipe below the weld, what type of groove is it?

CHAPTER 2

SAFETY

INTRODUCTION

Preventing accidents is everyone's job. A work place's safety is dependent on everyone acting in a responsible, safe manner. The safety information included in this text is intended as a guide. There is no substitute for caution and common sense.

Job safety is an attitude that can and must be developed to prevent injury to yourself or others. Every job has its potential safety hazards. Welding is no different. Numerous potential hazards to your safety and that of others exist during any welding and fabrication. If you are unsure of the possible dangers associated with a new welding job or situation, before work begins you must identify those dangers and find out how to avoid any possible injury to yourself and others or damage to equipment and materials.

A safe job is no accident. It takes work to make the job safe. It is important that each person works to keep the job safe.

You must approach the job with your safety in mind, which is your own responsibility. This text may not cover all dangers that may arise. You can get specific safety information from your supervisor, the shop safety officer, or other workers.

You should also read any safety booklets supplied with the equipment before starting any project.

CAUTION

High-frequency welding current used for some GTA welding processes may present a life-threatening hazard to persons who have heart pacemakers or other such medical implants. Check with your physician before entering an area where GTA welding may occur.

Fabrication Safety

Fabrication may present some potential safety problems not normally encountered in practice-type welding. Larger fabricates may need to be welded on outside the usual enclosed practice welding booths. In addition, several welders may be working on the same weldment at the same time. Because of the involvement of others, everyone must be conscious of co-workers to ensure that injuries do not occur from such hazards as arcs and hot sparks.

You may be working in an area that has welding cables, hoses, grinders, clamps, tools, and other fabrication-required items lying around on the floor. Cables, hoses, extension cords, and so on must be flat on the floor and should be covered if they are in a walkway to prevent accidental tripping. Hand tools, clamps, grinders, drills, and so forth must not be left in the walkways and should be kept up off the floor where they may become damaged from welding sparks. Despite all the potential hazards in welding, welders have been found to be as healthy as workers employed in other industrial occupations.

BURNS

Burns are one of the most common and painful injuries that can occur in the welding shop. You may be burned by exposure to ultraviolet light rays as well as by contact with hot welding material. Most welding shop burns are so small they may go unnoticed by an experienced welder. Burns are divided into three classifications, depending upon the degree of severity. The three classifications include first-degree, second-degree, and third-degree burns. The size of the burn has no bearing on its classification. For example, a slight sunburn (first-degree) may cover a large area of the body. Although very painful, it still is a first-degree burn. In contrast, a very small burn caused by a hot spark may be classified as a third-degree burn, but because it is so small, it may be painful for only a few moments.

All injuries including burns should receive proper first aid or medical attention. Burns may become infected as a result of the dead tissue caused by the burn. Proper first aid and medical attention can prevent such infections.

First-Degree Burns

A first-degree burn occurs when the unbroken surface of the skin becomes reddish in color, tender, and painful. A first-degree burn is painful because of the damage to the nerve ending just below the skin surface. The first step in treating a first-degree burn is to immediately put the burned area under cold water (not iced) or apply cold-water compresses (clean towel, washcloth, or handkerchief soaked in cold water) until the pain subsides. Then cover the area with sterile bandages or a clean cloth. Do not apply butter or grease. Do not apply any other home remedies or medications without a doctor's recommendation (Figure 2-1).

Second-Degree Burns

A second-degree burn occurs when the surface of the skin is severely damaged. Blisters form and may break. Again, the most important first step is to put the area under cold water (not iced) or apply cold-water compresses until the pain decreases. Gently pat the area dry with a clean towel and cover the area with a sterile bandage or clean cloth to prevent infection. Seek medical attention.

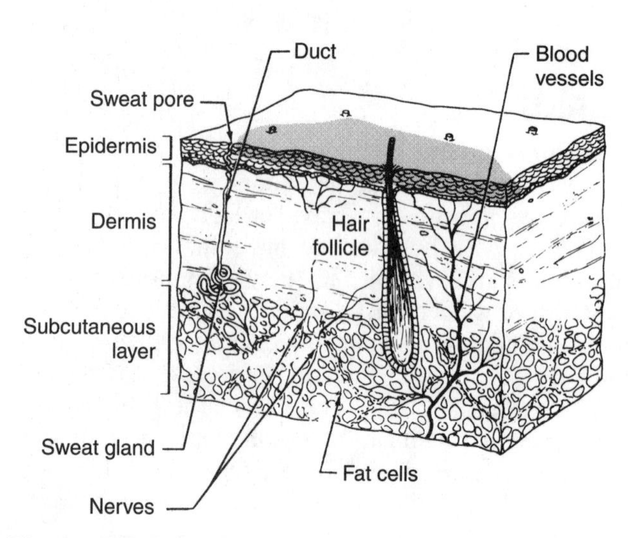

Figure 2-1 First-degree burn — Only the skin surface (epidermis) is affected.

CAUTION

> If the burns are around the mouth or nose or involve singed nasal hair, breathing problems may develop as a result of lung damage caused by inhaling hot vapors. Seek immediate medical attention.

Do not apply ointments, sprays, antiseptics, or home remedies. Note: In an emergency most cold drinks—water or iced tea—can be poured on a burn. This reduces the skin temperature as quickly as possible to reduce tissue damage (Figure 2-2).

Third-Degree Burns

A third-degree burn occurs when the surface of the skin and possibly the tissue below the skin turns white or black. Even though this is the most severe type of burn, initially there may be little pain present because nerve endings have been destroyed. If the victim is on fire, smother the flames with a blanket, rug, or jacket. During first-aid treatment:

- Do not remove any clothes that may be stuck to the burn.
- Do not put ice water or ice on the burns; this could intensify the shock reaction.
- Do not apply ointments, sprays, antiseptics, or home remedies to burns.

First aid things to do are as follows:

- Place a cold cloth or cool (not iced) water on burns of the face, hands, or feet to cool the burned areas.
- Cover the burned area with thick, sterile, non-fluffy dressings.
- Call for an ambulance immediately if necessary.

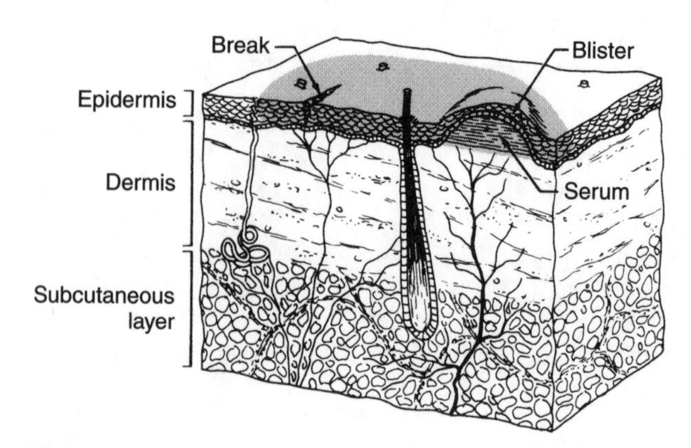

Figure 2-2 Second-degree burn — The epidermal layer is damaged, forming blisters or shallow breaks.

People with even small third-degree burns need to consult a doctor. See Figure 2-3.

Burns Caused by Light

The three types of light are ultraviolet, infrared, and visible. Ultraviolet and infrared are types of light that can cause burns that are invisible to the unaided human eye. During GTA welding, all three types of light are present. GTAW produces a great deal of ultraviolet light that will burn exposed skin. No gaseous cloud is formed from the burning on an electrode such as shielded metal arc or stick welding to absorb any of this light. Therefore burns can be severe and occur quickly.

The light from the welding process can cause burns whether it comes directly from the arc or is reflected from walls, ceilings, floors, or other large surfaces. To prevent burns from any welding light proper, protective equipment and clothing must be worn. To reduce the danger from reflected light, the welding area, if possible, should be painted flat black. Flat black reduces the reflected light by absorbing more of it than any other color can absorb.

When the welding is to be done on a fabrication or job site or in a large shop or other area that cannot be painted, portable welding curtains should be placed to contain the welding light (Figure 2-4). Special portable welding curtains may be either transparent or opaque. Transparent welding curtains are made of a special high-temperature, flame-resistant plastic that prevents the harmful light from passing through but allows you to see large shapes through them.

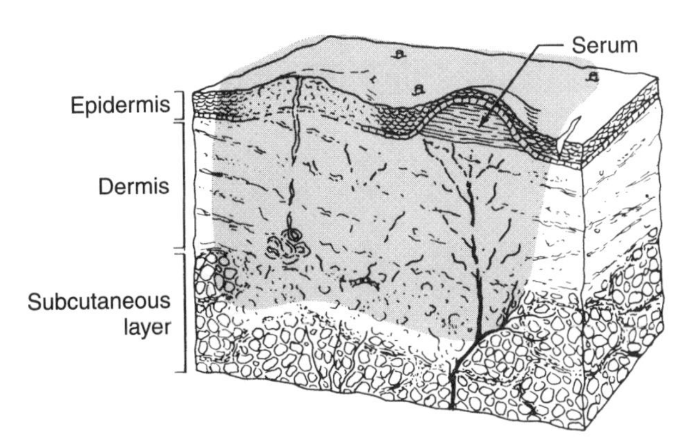

Figure 2-3 Third-degree burn — The epidermis, dermis, and subcutaneous layers of tissue are destroyed.

CAUTION

Welding curtains must always be used to protect other workers in an area that might be exposed to the welding light.

Ultraviolet light

Ultraviolet light (UV) is the most dangerous form of light. UV light can cause first-degree and second-degree burns to a welder's eyes or to exposed skin. Because welders cannot see or feel UV light while being exposed to it, they must keep themselves properly protected when in the area of any arc welding. The closer any unprotected skin or eyes are to the arc and the higher the current, the quicker it will burn. The UV light is so intense during some GTA welding that a welder's eyes can receive a flash burn within seconds, and exposed skin can be burned within minutes. UV light can pass through loosely woven clothing, thin clothing, light-colored clothing, and damaged or poorly maintained arc-welding helmets.

Infrared light

Because infrared light is the light wave that is felt as heat, a person immediately feels the heat from this type of light exposure. Burns can easily be avoided for this reason.

Figure 2-4 Portable welding curtains. (Courtesy Frommelt Safety Products)

Visible light

Visible light is the light that we see. It is produced in varying quantities and colors during welding. Too much visible light may cause temporary night blindness (poor eyesight under low light levels). Too little visible light may cause eye strain. Visible light alone is not hazardous and will not cause burns, but all visible light produced by GTA welding contains both harmful types of light unless it passes through a welding lens.

Whether burns are caused by UV light or hot material, you can avoid them by wearing proper protective clothing and other equipment.

FACE, EYE, AND EAR PROTECTION

Face and Eye Protection

Some form of eye protection must be worn in the shop at all times. Your eyes may be protected by safety glasses (Figure 2-5), goggles, or a full-face shield. For protection when working in brightly lit areas or outdoors, some welders wear flash glasses, which are special, lightly tinted, safety glasses. This type of safety glasses provides protection from both reflected light and flying debris.

Figure 2-5 Safety glasses with side shields.

Proper eye protection is important because eye damage caused by excessive exposure to arc light may not be noticed until well after welding is completed. Most welding light injuries occur without warning of pain, much like a sunburn that is not felt until the following day. Therefore welders must take appropriate precautions in selecting filters that are suitable for the specific GTA welding being performed (Table 2-1). Selecting the correct shade lens is important because both extremes of too light or too dark can cause eye strain. Because most GTA welding is performed with low currents that produce small molten weld pools, good visibility is essential. New welders, however, often select too dark a lens assuming it will give them better protection. A shade lens that is too dark can result in eye strain in the same manner as if one were trying to read in a poorly lit room. Any approved arc welding lenses will filter out all the harmful UV light. Select a filter lens that lets you see comfortably.

UV light can burn the eye by injuring either the white of the eye or the retina, which is the back of the eye. Burns on the retina are not painful but may cause some loss of eyesight. The whites of the eyes can also be burned by UV light. The whites of the eyes are very sensitive, and burns are very painful. When the eye is burned, it feels as though there is something in the eye. Without a professional examination, however, it is impossible to tell if there is a foreign object in the eye. Because there may actually be something in the eye and because of the high risk of infection, home remedies or other medicine should never be used for

TABLE 2-1

RECOMMENDED EXTENSION CORD SIZED FOR USE WITH PORTABLE ELECTRIC TOOLS

	Nameplate Ampere Rating															
CORD LENGTH	**0 TO 5**	**6**	**8**	**8**	**9**	**10**	**11**	**12**	**13**	**14**	**15**	**17**	**17**	**18**	**19**	**20**
25'	18	18	18	18	18	18	16	16	16	14	14	14	14	14	12	12
50'	18	18	18	18	18	18	16	16	16	14	14	14	14	14	12	12
75'	18	18	18	18	18	18	16	16	16	14	14	14	14	14	12	12
100'	18	18	18	16	16	16	16	16	14	14	14	14	14	14	12	12
125'	18	18	16	16	16	14	14	14	14	14	14	12	12	12	12	12
150'	18	16	16	16	14	14	14	14	14	12	12	12	12	12	12	12

Note: Wire sizes shown are AWG (American Wire Gauge) based on a line voltage of 120.

eye burns. Anytime you receive an eye injury, you should seek medical attention. Welders must check their helmets daily for potential problems that may occur from normal use or accidents. Small, undetectable leaks of UV light in an arc-welding helmet can cause a welder's eyes to itch or feel sore after a day of welding. To prevent these leaks, make sure the lens gasket is installed correctly (Figure 2-6). The outer clear lens may be either glass or plastic, but the inside clear lens must be plastic. Each time the clear lenses are replaced, check the filter lens for cracks by twisting it between your fingers (Figure 2-7). Worn or cracked spots on a helmet must be repaired. Some welders use tape as a temporary repair until the helmet can be replaced or permanently repaired.

Safety glasses with side shields are adequate for general use, but if heavy grinding, chipping, or overhead work is being done, goggles or a full-face shield should be worn in addition to safety glasses (Figure 2-8). Safety glasses, either clear or tinted, are best for general protection because they can be worn under an arc-welding helmet.

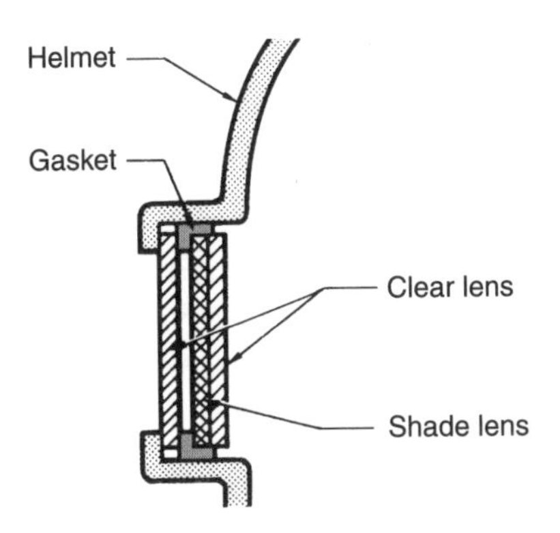

Figure 2-6 The correct placement of the gasket around the shade lens is important because it can stop ultraviolet light from bouncing around the lens assembly.

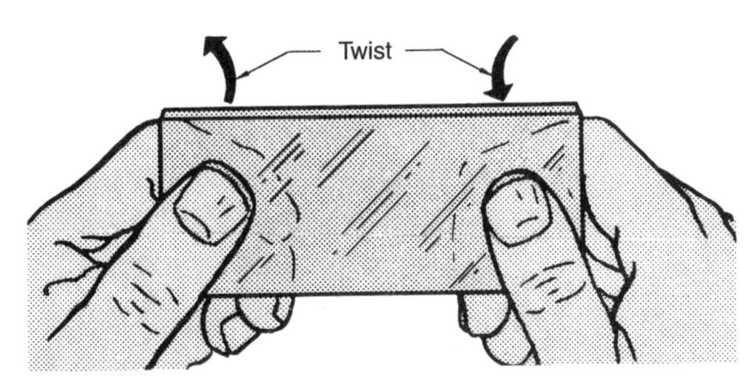

Figure 2-7 To check the shade lens for possible cracks, gently twist it.

Ear Protection

The welding environment can be very noisy. The sound level is at times high enough to cause ear pain and some loss of hearing if the welder's ears are unprotected. In addition to protecting the ears from sound, ear protection can prevent hot sparks from dropping into an open ear, causing severe burns.

One type of ear protection is earmuffs that cover the outer ear completely (Figure 2-9). Another type of

Figure 2-8 Full face shield. (Courtesy of Jackson Products)

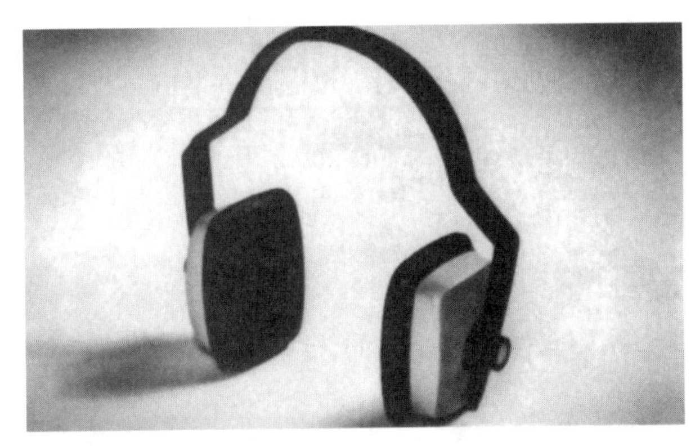

Figure 2-9 Earmuffs provide complete ear protection and can be worn under a welding helmet. (Courtesy of Mine Safety Appliances Company)

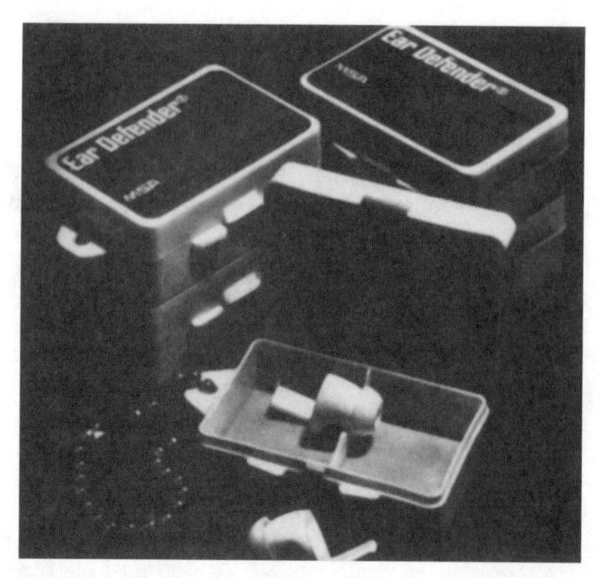

Figure 2-10 Earplugs used as protection from noise only. (Courtesy of Mine Safety Appliances Company)

protection is ear plugs that fit into the ear canal (Figure 2-10). Both of these protect a person's hearing, but only the earmuffs protect the outer ear from burns.

CAUTION

Damage to your hearing caused by high sound levels may not be detected until later in life, and any resulting permanent loss in hearing is nonrecoverable. It will not get better with time. Also, each time you are exposed to high levels of sound, the more damaged your hearing will become.

MATERIALS SPECIFICATIONS DATA SHEETS (MSDSs)

All manufacturers of potentially hazardous materials must provide to the users of their products detailed information regarding possible hazards resulting from the use of their products. These materials specifications data sheets are often called MSDSs and they must be provided to anyone using the product or anyone working in the area where the products are in use. Often companies will post these sheets on a bulletin board or put them in a convenient place near the work area.

RESPIRATORY PROTECTION

All GTA welding produces undesirable fumes and gases. Production of these by-products cannot be avoided. They are created when the temperature of metals is raised above their boiling or decomposing temperatures.

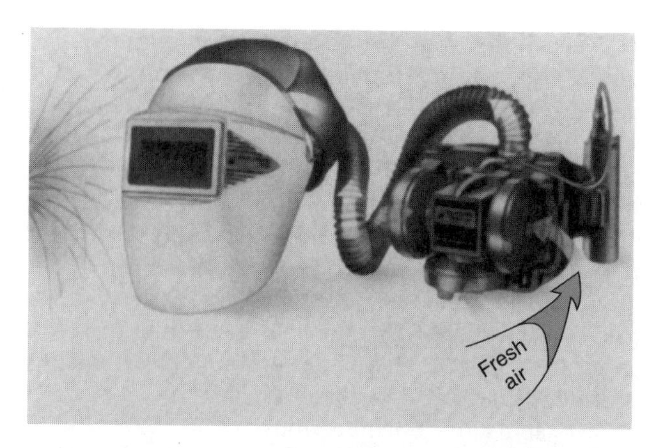

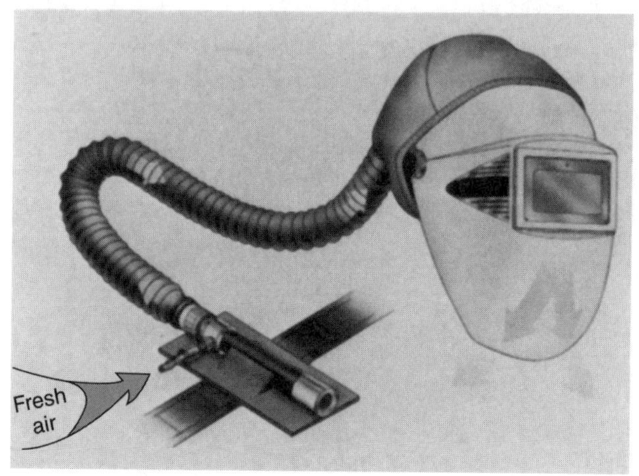

Figure 2-11 Filtered fresh air is forced into the welder's breathing area. The air can come from a belt-mounted respirator or through a hose for a remote control. (Courtesy of Hornell Speedglas, Inc. [R.H. Blake Inc.])

Never take chances. You should never inhale fumes of any type, regardless of their source. The best way to avoid problems is to provide adequate ventilation. If this is not possible, you should use breathing protection. Protective devices for use in poorly ventilated or confined areas are shown in Figure 2-11 and Figure 2-12.

CAUTION

Welding or cutting must never be performed on drums, barrels, tanks, vessels, or other containers until they have been emptied and cleaned thoroughly, eliminating all flammable materials and all substances (such as detergents, solvents, greases, tars, or acids) that might produce flammable, toxic, or explosive vapors when heated.

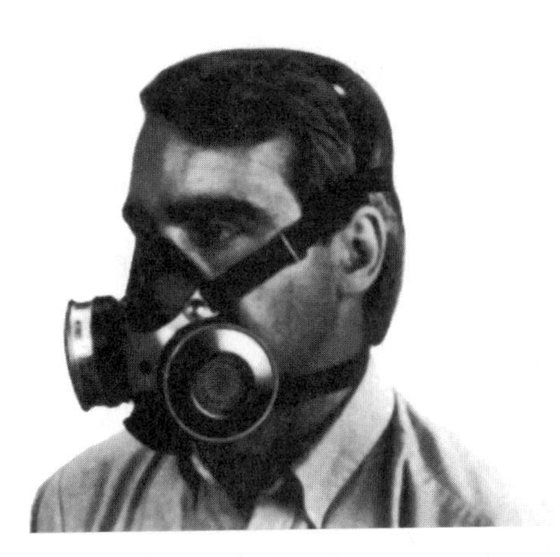

Figure 2-12 Typical respirator for contaminated environments. The filters can be selected for specific types of contaminants. (Courtesy of Mine Safety Appliances Company)

Although all GTA welding requires that the welding metal be thoroughly cleaned, some materials remaining outside the weld zone can cause problems. Some of these materials that can cause respiratory problems are used as paints, coating, or plating on metals to prevent rust or corrosion. In some cases other potentially hazardous materials may be used as alloys in metals to give them special properties.

Clean the surface thoroughly before welding any metal that has been painted or has any grease, oil, or chemicals on the surface. This cleaning may be done by grinding, sandblasting, or applying an approved solvent.

Lead

Most paints that contain lead have been removed from the market. But industrially some special purposes, such as marine or ship applications, still use these lead-based paints. Old machinery and farm equipment often have lead-based paints on them.

Inhalation and ingestion of fumes from lead oxide and other lead compounds will cause lead poisoning. Symptoms include a metallic taste in the mouth, loss of appetite, nausea, abdominal cramps, and insomnia. In time, anemia and a general weakness, chiefly in the muscles of the wrists, develop.

Cadmium

Cadmium is a plating material that is used to prevent iron or steel from rusting. Cadmium is often used on bolts, nuts, hinges, and other hardware items, and it gives the surface a yellowish-gold appearance.

Acute exposures to high concentrations of cadmium fumes can produce severe lung irritation. Long-term exposure to low levels of cadmium in the air can result in emphysema (a disease affecting the ability of the lungs to absorb oxygen) and can damage the kidneys.

Zinc

Zinc—often in the form of galvanized rust proofing—may be found on steel pipes, sheet metal, bolts, nuts, and many other types of hardware. Thin zinc plating may appear as a shiny metallic patchwork with a crystalline pattern. Thicker hot-dipped zinc appears rough and may appear dull.

Zinc is also used in large quantities in the manufacture of brass and brazing rods. Inhalation of zinc oxide fumes can occur when welding or cutting on these materials. Exposure to these fumes causes metal fume fever, whose symptoms are very similar to those of common influenza.

Chrome

When using stainless steels, concern has been expressed about the possibility that some of the chromium compounds may cause lung cancer.

CAUTION

Extreme caution must be taken to avoid the fumes produced when welding is done on coated or used metal. Any chemicals that might be on the metal will become mixed with the welding fumes, creating an extremely hazardous combination. All metal must be cleaned before welding to avoid this potential problem.

Ozone

Ozone (O_3) is a form of oxygen that is produced in the air by the UV light radiation in the vicinity of a welding arc. Proper ventilation or respirators are necessary when you are welding in confined spaces.

Ozone is very irritating to mucous (sinus) membranes, with excessive exposure producing pulmonary edema. Other effects of exposure to ozone include headache, chest pain, and dryness in the respiratory tract.

Shielding Gas

Care also must be taken to avoid the concentration of argon, carbon dioxide, or other gases into a confined working space such as when welding in tanks. The collection of gases in a work area can go unnoticed by the welders. Concentrated gases can cause asphyxiation if they replace the oxygen in the air.

VENTILATION

Welding areas should be well ventilated. Excessive fumes, ozone, or smoke may collect in the welding area, and ventilation must be provided to remove them. Natural ventilation such as are provided by large open spaces is best, but forced ventilation may be required. Areas that have 10,000 cubic feet (283 cubic meters) or more per welder or that have ceilings 16 feet (4.9 meters) high or higher may not require forced ventilation unless fumes or smoke begin to collect (Figure 2-13).

Forced Ventilation

Small enclosed shops or shops with a large number of welders require forced ventilation, which can be general (for example, exhaust fans located in the wall or ceiling) or localized, using fixed or flexible exhaust pickups. General room ventilation must be at a rate of 2,000 cubic feet per minute (CFM) (56 cubic meters per minute) or more per person welding. Localized exhaust pickups must have a draft strong enough to provide 100 linear feet per minute (30.5 meters per minute) velocity of welding fumes away from the welder. Any system of ventilation should draw the fumes or smoke away before they rise past the level of the welder's face.

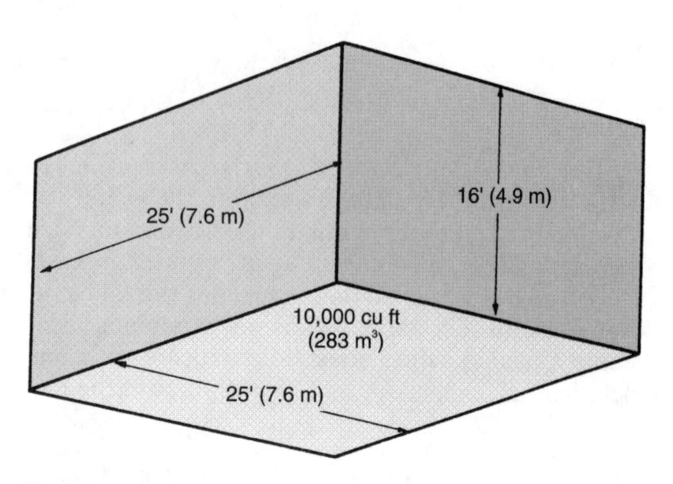

Figure 2-13 A room with a ceiling 16′ (4.9 m) high may not require forced ventilation for one welder.

Forced ventilation is always required when welding on metals that contain zinc, lead, beryllium, cadmium, mercury, copper, Austenitic manganese, or other materials that give off dangerous fumes.

Local, state, or federal regulations may require that welding fumes be treated to remove hazardous components before they are released into the atmosphere.

ELECTRICAL SAFETY

Injuries and even death can be caused by electric shock unless proper precautions are taken. Most welding and fabrication operations involve electrical equipment in addition to the GTA welding power supplies, such as grinders, remote controls, and drills. Most welders and other electrical equipment in a welding shop are powered by alternating current (AC). Power is supplied with voltages ranging from 115 volts to 460 volts. Shocks are almost never fatal; however, deaths have occurred even when working with equipment operating at less than 80 volts AC.

Most electric shocks in the welding industry result from accidental contact with bare or poorly insulated conductors and do not occur from contact with welding electrode holders.

Electrical connections must be tight. Terminals for welding leads and power cables must be shielded from accidental contact by personnel or by metal objects. Cables must be used within their current-carrying and duty-cycle capacities; otherwise, they will overheat and break down the insulation rapidly. Cable connectors for lengthening leads must be insulated. Cables must be checked periodically to ensure that they have not become frayed; if they have, they must be replaced immediately.

CAUTION

Welding cables must never be spliced within 10 feet (3 m) of the electrode holder.

Welders should not allow the metal parts of electrodes or electrode holders to touch their skin or wet coverings on their bodies. Dry gloves in good condition must always be worn. Rubber-soled shoes are advisable. Precautions against accidental contact with bare conducting surfaces must be taken when the welder is required to work in cramped kneeling, sitting, or prone positions. Insulated mats or dry wooden boards provide desirable protection from the earth.

Electrical resistance is lowered, and the chance of shock is highest in the presence of water or moisture. Welders must take special precautions when working under damp or wet conditions, including perspiration. Any GTA welding equipment that has coolant water leaks must not be operated until the leaks are repaired. Figure 2-14 shows a typical warning label attached to welding equipment. Grounding of any welding system is important, but it is especially important when working with a GTA system that uses high frequency for starting or welding.

The workpiece being welded, the frame or chassis of the welding machine, and all other electrically powered machines as well as any other metal items within 10 feet must be connected to a good electrical ground. The work lead from the welding power supply is not an electrical ground and is not sufficient. A separate wire is required to reduce static hazards caused by the high-frequency power.

Welding circuits must be turned off when the work station is left unattended. It must be turned off and locked or tagged to prevent electrocution when working on the welder, welding leads, torches, or other parts (Figure 2-15).

GENERAL WORK CLOTHING

Because of the high heat and temperatures produced during welding, you should choose general work clothing that will minimize the possibility of getting burned.

Wool clothing (100% wool) is the best choice but difficult to find. All cotton (100% cotton) clothing is a good second choice and is the most popular material used. Synthetic materials, including nylon, rayon, and polyester, should not be worn because they can easily burn, producing a hot sticky ash, and some produce poisonous gases. Because these materials become sticky, burns can be more severe. The clothing must also be a dark color, thick, and tightly woven to stop UV light from passing through.

Here are some guidelines for selecting work clothing:

- Shirts must be long sleeved to protect the arms, have a high buttoned collar to protect the neck (Figure 2-16), be long enough to tuck into the pants to protect the waist, and have flaps on the pockets to keep sparks out (or have no pockets).

WARNING

READ AND UNDERSTAND THIS WARNING — PROTECT YOURSELF AND OTHERS

ELECTRIC SHOCK can kill.
- Do not permit electrically live parts or electrodes to contact skin . . . or your clothing or gloves if they are wet.
- Insulate yourself from work and ground.

FUMES AND GASES can be dangerous to your health.
- Keep fumes and gases from your breathing zone and general area.
- Keep your head out of fumes.
- Use enough ventilation or exhaust at the arc or both.

ARC RAYS can injure eyes and burn skin.
- Wear correct eye, ear, and body protection.

ENGINE EXHAUST GASES can kill.
- Use in open, well ventilated areas or vent the engine exhaust to the outside.

HIGH VOLTAGE can kill
- Do not touch electrically live parts.
- Stop engine before servicing.

MOVING PARTS can cause serious injury.
- Keep clear of moving parts.
- Do not operate with protective covers, panels, or guards removed.

- Only qualified personnel should install, use, or service this equipment.
- Read operating manual for details.

READ AND UNDERSTAND THE MANUFACTURER'S INSTRUCTIONS AND YOUR EMPLOYER'S SAFETY PRACTICES.

See American National Standard Z49.1, "Safety in Welding and Cutting", published by the American Welding Society, 550 Le Jeune Rd., Miami, Florida 33126; OSHA Safety and Health Standards, 29 CFR 1910, available from U.S. Government Printing Office, Washington, D.C. 20402.

DO NOT REMOVE THIS WARNING E202

Figure 2-14 Note the warning information for electrical shock and high voltage contained on this typical label, which is attached to welding equipment by the manufacturer. (Courtesy of Lincoln Electric Company)

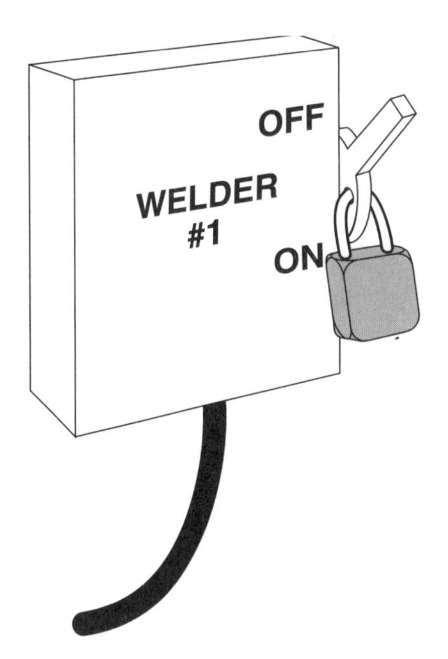

Figure 2-15 Lock off the power when working on a welder.

Figure 2-16 The top button of the shirt worn by the welder should always be buttoned in order to avoid severe burns to that person's neck.

- Pants must have legs long enough to cover the tops of the boots and must be without cuffs that would catch sparks.
- Boots must have high tops to keep out sparks, steel toes to prevent crushed toes (Figure 2-17), and smooth tops to prevent sparks from being trapped in seams.
- Caps should be thick enough to prevent sparks from burning the top of a welder's head.

All clothing must be free of frayed edges or holes. The clothing must be relatively tight fitting to prevent excessive folds or wrinkles that might trap sparks.

CAUTION

There is no safe place to carry butane lighters and matches while welding or cutting. They may catch fire or explode if they are subjected to welding heat or sparks. Butane lighters may explode with the force of a quarter stick of dynamite. Matches can erupt into a ball of fire. Both butane lighters and matches must always be removed from the welder's pockets and placed a safe distance away before any work is started.

SPECIAL PROTECTIVE CLOTHING

In addition to general work clothing, extra protection from hot welding materials is needed for each person who is welding or will come in direct contact with hot materials. Leather is the best material to use for this extra protection because it is lightweight, flexible, flame resistant, and readily available from welding supply stores. Some synthetic insulative materials are available that are often much lighter

Figure 2-17 Safety boots with steel toes are required by many welding shops.

and more flexible than the same item made out of leather. Ready-to-wear leather and synthetic protection materials include items such as capes, jackets, aprons, sleeves, gloves, caps, pants, knee pads, and spats.

Hand Protection

Some type of glove is required for welders' hands. The two preferred types are thin cotton gloves or very soft leather gloves. Both should be gauntlet-type gloves (Figure 2-18). The gauntlet is the material around the top of the glove designed to protect the wrist. All-cotton or leather-palmed cotton gloves work well for most gas tungsten arc welding. They will protect the welder from the heat and the ultraviolet light produced during most light-duty GTAW.

Heavier all-leather welding gloves are needed when making welds in heavier materials at higher welding currents. They may also be needed for production welding, where longer arc-on time is normal.

Gloves need to be thin and flexible to allow the welder the necessary freedom to manipulate the filler metal into the molten weld pool. They also must be free from any type of contamination that would affect the weld. Gloves used for GTAW should be used only for that purpose and not for any other welding or shop operation. The gloves should be put on at the beginning and taken off at the conclusion of welding operations to avoid possibly contaminating the gloves. It is a good idea to have pliers or other gloves available when handling hot metal in preparation for or during postweld cleanup. This will prevent possible damage to the GTAW gloves.

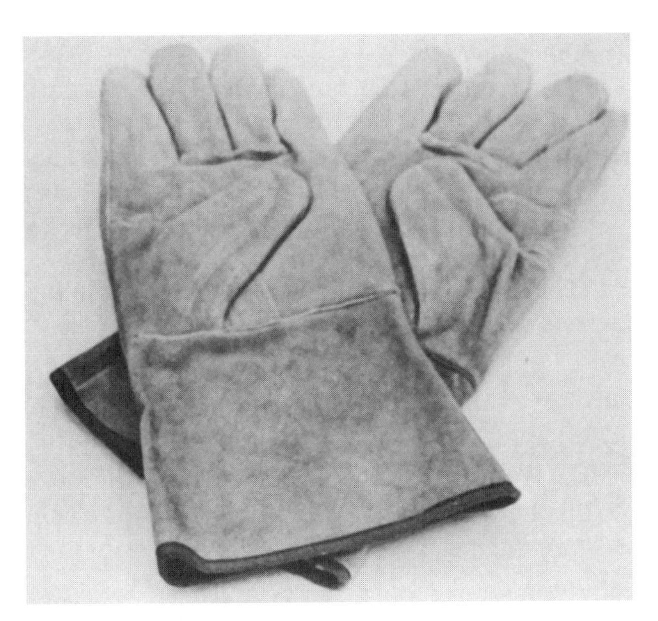

Figure 2-18 All-leather, gauntlet-type, welding gloves.

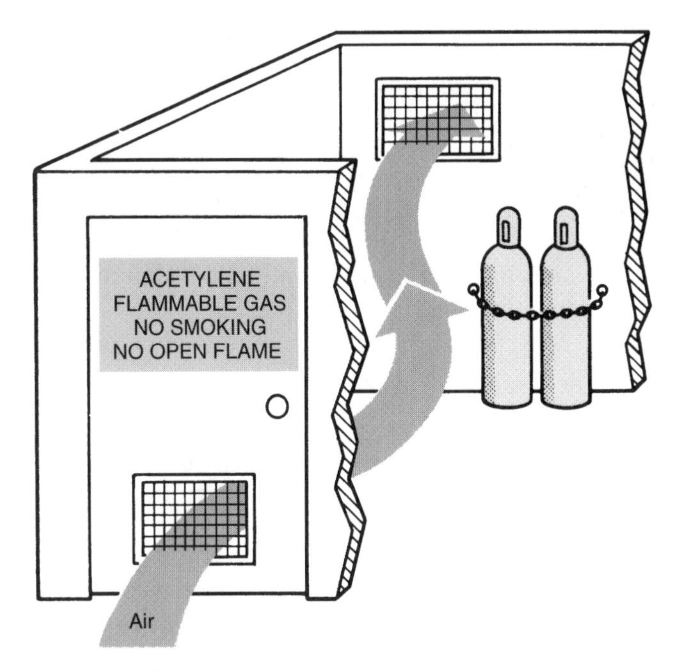

Figure 2-19 A separate room used to store acetylene must have good ventilation and should have a warning sign posted on the door.

HANDLING AND STORING CYLINDERS

Cylinder Storage

Securing gas cylinders

Compressed gas cylinders—whether in use or in storage—must be secured with a chain or other device so that they cannot be accidently knocked over. Cylinders in storage are usually secured to a wall, column, or other stationary object. Cylinders in use may be secured to the welding station or welding machine. Cylinders in use that are attached to a manifold may not need to be secured by a chain if the manifold is located in a separate room specifically designed and used only for manifolding cylinders.

Storage Areas

Cylinder storage areas must be located away from halls, stairwells, and exits so that in an emergency they will not block an escape route. Storage areas must be located away from sources of heat such as radiators and furnaces. They must also be located far enough away from welding areas so that welding sparks cannot accidentally reach the storage area. The location of the storage area should ensure that unauthorized people cannot tamper with the cylinders. A warning sign that reads "Danger! No Smoking, Matches, or Open Lights," for example, must be posted in the storage cylinder area (Figure 2-19).

Cylinder Valve Protection Caps

Cylinder valve protection caps must be in place at all times unless the cylinder is in use. If an unsecured full high-pressure cylinder valve is broken off, the cylinder valve could fly around the shop like a missile.

Never lift a cylinder by the safety cap or cylinder valve. The safety cap could accidentally come off, allowing the cylinder to fall. If the cylinder is lifted by the valve, it can easily be broken off or damaged.

When cylinders are being moved where they are full or empty, the valve protection cap must be place. This is also true for cylinders that are mounted on trucks, trailers, or other vehicles for out-of-shop work.

High pressure gas cylinders must never be dropped or handled roughly.

FIRE PROTECTION

Fire is a constant danger to the welder. The potential for fires cannot always be eliminated, but it should be minimized. Highly combustible materials should be 35′ (10.7 meters) or more away from any welding. When it is necessary to weld combustible materials within this distance, when sparks can reach materials farther than 35′ (10.7 meters) away, or when anything more than a minor fire might start, a fire watch is needed.

Fire Watch

A fire watch can be provided by any one who knows how to sound the alarm and use a fire extinguisher. The fire extinguisher must be able to put out a fire of the type caused by the combustible materials near the welding. Combustible materials that cannot be removed from the welding area should be soaked with water or covered with sand or noncombustible insulating blankets.

Fire Extinguisher

The four types of fire extinguishers are type A, type B, type C, and type D. Each type is designed to put out fires caused by certain types of materials. Some extinguishers can be used on more than one type of fire. However, using the wrong type of fire extinguisher can cause the fire to spread or cause electrical shock or an explosion.

Type A Extinguisher

The symbol for a type A extinguisher is a green triangle with the letter "A" in the center (Figure 2-20).

Type B Extinguisher

The symbol for a type B extinguisher is a red square with the letter "B" in the center (Figure 2-21).

Figure 2-20 Type A fire extinguisher symbol.

Figure 2-21 Type B fire extinguisher symbol.

Type C Extinguisher

These are used on fires involving motors, fuse boxes, and welding machines. The symbol for a type C extinguisher is a blue circle with the letter "C" in the center (Figure 2-22).

Type D Extinguisher

The symbol for a type D extinguisher is a yellow star with the letter "D" in the center (Figure 2-23).

Location of Fire Extinguisher

The fire extinguisher should be appropriate for the types of combustible materials located nearby (Figure 2-24). The extinguisher should be

Figure 2-22 Type C fire extinguisher symbol.

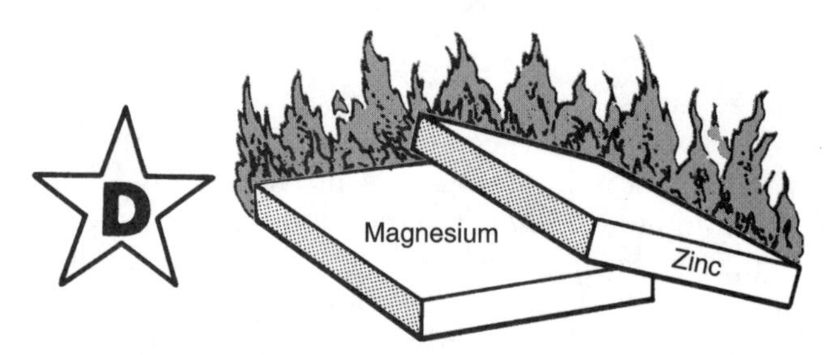

Figure 2-23 Type D fire extinguisher symbol.

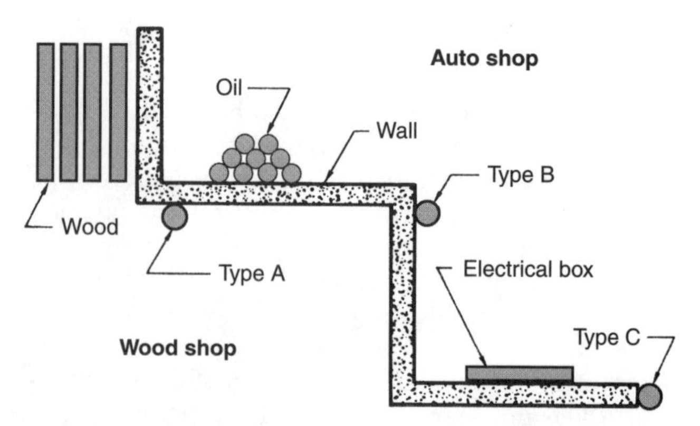

Figure 2-24 The type of fire extinguisher provided should be appropriate for the materials being used in the surrounding area.

located so that it can be easily removed without reaching over combustible material. It should also be placed at a level low enough to be easily lifted off the mounting (Figure 2-25). The location of a fire extinguisher should be marked with red paint and signs high enough so that its location can be seen from a distance over people and equipment. The extinguisher should also be marked near the floor so that it can be found even if a room is full of smoke (Figure 2-26).

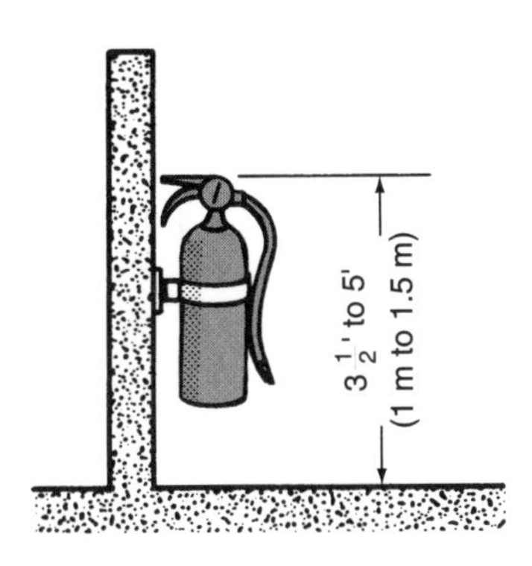

Figure 2-25 Mount the fire extinguisher so that it can be lifted easily in an emergency.

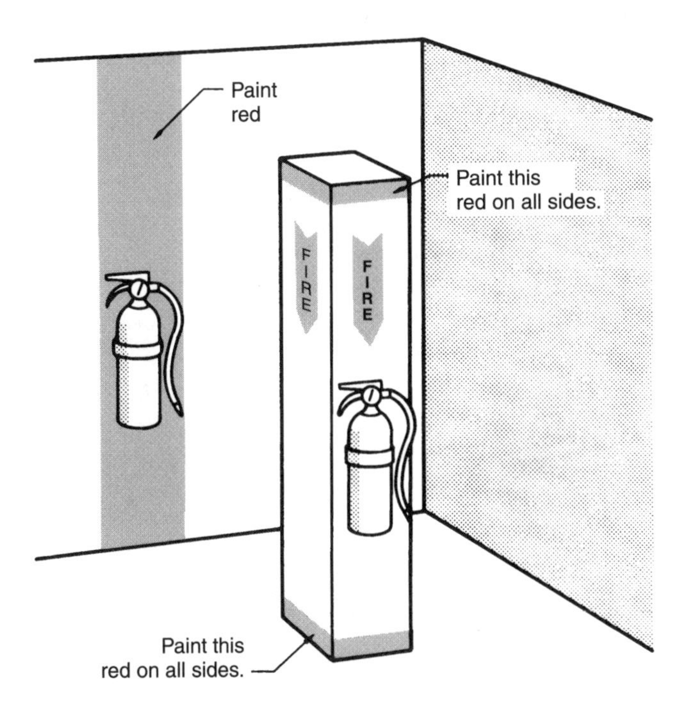

Figure 2-26 The locations of the fire extinguishers should be marked so they can be located easily in an emergency.

Use

A fire extinguisher works by breaking the fire triangle of heat, fuel, and oxygen. Most extinguishers both cool the fire and remove the oxygen. Fire extinguishers use a variety of materials to extinguish the fire. The major ones found in welding shops use foam, carbon dioxide, a soda-acid gas cartridge, a pump tank, or dry chemicals.

When using a foam extinguisher, do not spray the stream directly into the burning liquid. Allow the foam to fall lightly onto the fire.

When using a carbon dioxide extinguisher, direct the discharge as close to the fire as possible—first at the edge of the flames and gradually to the center.

When using a soda-acid gas cartridge extinguisher, place your foot on the footrest and direct the stream at the base of the flames.

When using a dry chemical extinguisher, direct it at the base of the flames. In the case of type A fires, follow up by directing the dry chemicals at the remaining material that is burning.

SUMMARY

The safety of the welder is of utmost importance. A great deal of time, money, and effort is spent each year in providing welders with the safest possible working environment. Usually manufacturers have a safety department with an individual in charge of plant safety. That person's job is to ensure that all welders comply with safety rules and regulations during production. The proper clothing, shoes, and eye protection are emphasized to the welders in these plants. Any welder who does not follow established safety rules is subject to dismissal.

If an accident does occur, proper and immediate first aid must be taken. All welding shops should establish plans of action in the event of accidents. Take time to learn the proper procedure for accident response and reporting before an accident occurs and your prompt action is needed. After the situation has been properly taken care of, you must fill out an accident report. Written accident reports can help improve working conditions so that future similar accidents do not occur.

Equipment must be checked periodically to be sure that it is safe and in proper working condition. Maintenance workers are employed to see that equipment is in proper working condition at all times.

Additional safety information is available in *Safety for Welders,* by Larry F. Jeffus (published by Delmar Publishers Inc.), from the American Welding Society, and from the U.S. Department of Labor (OSHA) Regulations.

POWER TOOLS

All power tools must be properly grounded to prevent accidental electrical shock. If even a slight tingle is felt while using a power tool, stop and have the tool checked by a technician. Power tools should never be used with force or allowed to overheat from excessive or incorrect use. Extension cords should have a large enough current rating to carry the load (Table 2-2). An extension cord that is too small will cause the tool to overheat.

Safety glasses must be worn at all times when using any power tools.

Grinders

Grinding using a pedestal grinder or a portable grinder is required for many welding jobs. Often it is necessary to grind a groove, remove rust, or smooth a weld. Grinding stones have the maximum revolutions per minute (r/min) listed on the paper blotter (Figure 2-27). They must never be used on a machine with a higher rated number of revolutions per minute. If grinding stones are turned too fast, they can explode.

Grinding Stone

Before a grinding stone is put on the machine, it should be tested for cracks. This is done by tapping the stone in four places and listening for a sharp ring, which indicates that it is good (Figure 2-28). A dull sound indicates that the grinding stone is cracked and should not be used. Once a stone has been installed and used, it may need to be trued and balanced by using a special tool designed for that purpose (Figure 2-29). Truing keeps the stone face flat and sharp for better results.

Types of Grinding Stones

Each grinding stone is made for grinding specific types of metal. Most stones are for ferrous metals, such as iron, cast iron, steel, and stainless steel. Some stones are made for nonferrous metals, such as aluminum, copper, and brass. If a ferrous stone is used to grind nonferrous metal, the stone will become glazed (that is, the surface clogs with metal) and may explode due to frictional heat building up on the surface. If a nonferrous stone is used to grind ferrous metal, the stone will be quickly worn away.

When the stone wears down, keep the tool rest adjusted to within 1/16″ (2 mm) (Figure 2-30), so that the metal being ground cannot be pulled between the tool rest and the stone surface. Stones should not be used when they are worn down to the size of the paper blotter. If small parts become hot from grinding, use pliers to hold them. Gloves should never be worn when grinding. If a glove gets caught in a stone, the whole hand may be drawn in.

The sparks from grinding should be directed down and away from other welders or equipment.

TABLE 2-2

RECOMMENDED EXTENSION CORD SIZES FOR USE WITH PORTABLE ELECTRIC TOOLS

NAMEPLATE	CORD LENGTH IN FEET							
AMPERES	25	50	75	100	125	150	175	200
1	16	16	16	16	16	16	16	16
2	16	16	16	16	16	16	16	16
3	16	16	16	16	16	16	14	14
4	16	16	16	16	16	14	14	12
5	16	16	16	16	14	14	12	12
6	16	16	16	14	14	12	12	12
7	16	16	14	14	12	12	12	10
8	14	14	14	14	12	12	10	10
9	14	14	14	12	12	10	10	10
10	14	14	14	12	12	10	10	10
11	12	12	12	12	10	10	10	8
12	12	12	12	12	10	10	8	8

Note: Wire sizes shown are American Wire Gauge (AWG).

Figure 2-27 Always check to be sure that the grinding stone and the grinder are compatible before installing a stone.

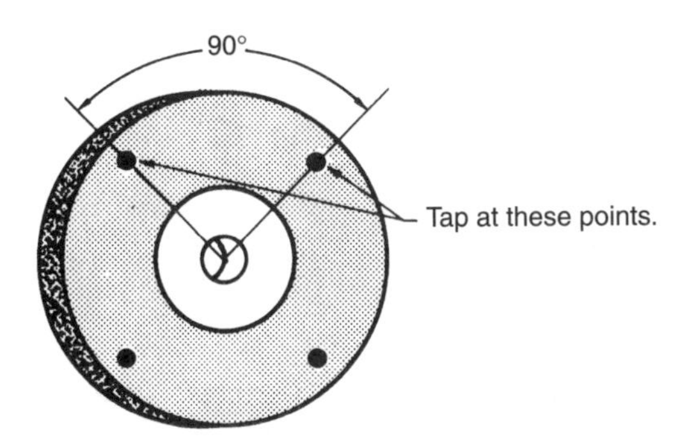

Figure 2-28 Grinding stones should be checked for cracks before they are installed.

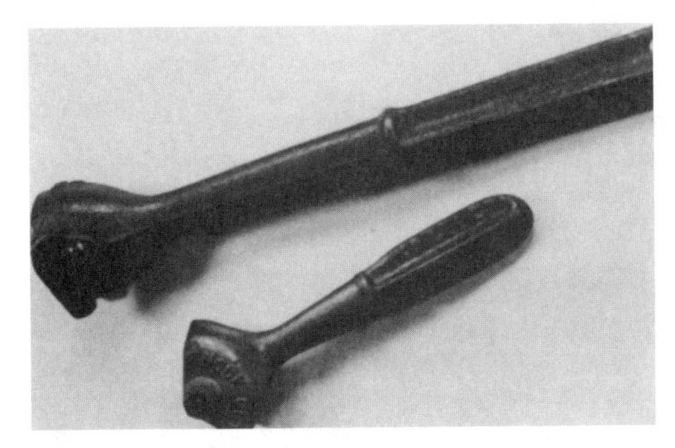

Figure 2-29 Use a grinding stone redressing tool as needed to keep the stone in balance.

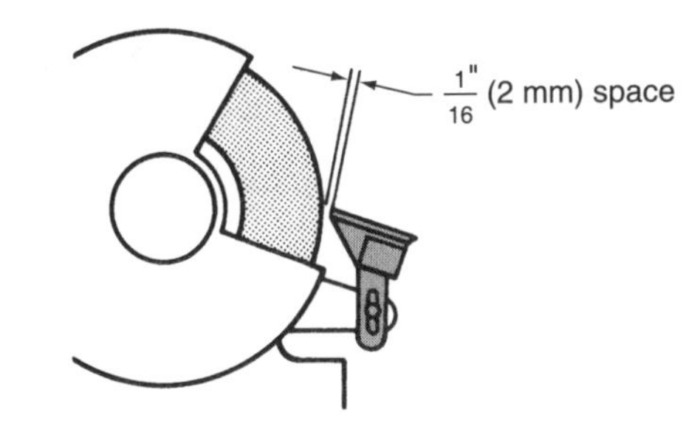

Figure 2-30 Keep the tool rest adjusted.

REVIEW QUESTIONS

1. Describe first-, second-, and third-degree burns.
2. Describe some of the first-aid practices for burns.
3. What types of light can cause burns that are not visible to the human eye?
4. Which type of gas tungsten arc-welding light will burn exposed skin?
5. What type of light can cause first- and second-degree burns to a welder's eyes or exposed skin?
6. What is the name of the light that may cause temporary night blindness or eye strain?
7. Why is it important to wear proper eye protection at all times?
8. Why should you seek medical attention any time you injure the eye?
9. Why must welders check their helmets daily?
10. Why should welders wear ear protection at all times?
11. What does the acronym MSDS mean and what is it used for?
12. List two ways to clean a surface that has been painted or has any grease.
13. When is forced ventilation always required?
14. Welding cables must never be spliced within _____ of the electrode holder.
15. What is the best type of clothing to wear when welding?
16. What is the best type of material to use for extra protection?
17. What is the preferred type of gloves?
18. What are the proper procedures for storing cylinders?
19. What is the purpose of cylinder valve protection caps?
20. What is an A type extinguisher used for?
21. What is a B type extinguisher used for?
22. What is a C type extinguisher used for?
23. What is a D type extinguisher used for?

CHAPTER 3

POWER SUPPLIES AND WELDING CURRENTS

INTRODUCTION

The welding power supplies and currents used for GTAWelding are similar to those used for most shielded metal arc (stick) welding processes. In most cases a basic SMA welding machine can be easily adapted for GTA welding. The conversion usually requires only that the stick welding electrode be replaced with a GTA welding torch and a source for the shielding gas. Obviously more sophisticated GTAWelding machines have components and options not found on the basic SMA welding machine. The addition of these devices on GMA machines provides the welder with the ability to control the weld better.

TYPES OF WELDING CURRENTS

All three types of welding current, often referred to as polarities, can be used for GTA welding. Each type of welding current has unique features that make it more desirable for particular welding applications. Some individual characteristics of the welding currents include joint penetration, heat distribution, and weld metal cleaning action.

The major differences among the current's heat distributions are represented in Figure 3-1. The relative differences in weld bead shape of the three welding currents illustrated are all at the same amperage setting.

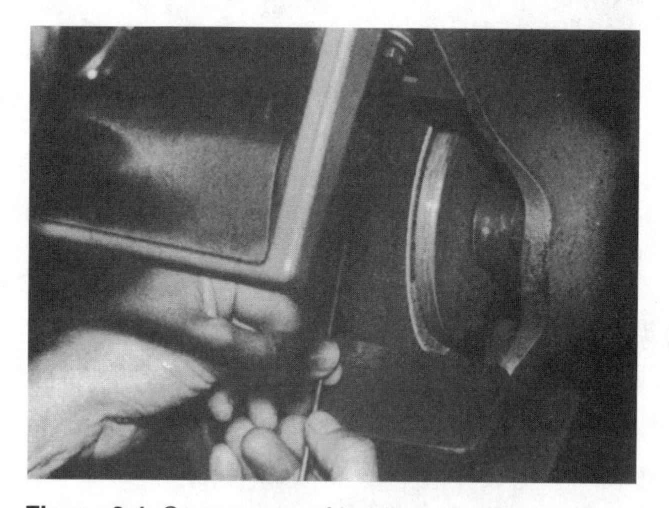

Figure 3-1 Correct way of holding a tungsten when grinding.

Direct Current (DC)

Direct current refers to electricity that flows in one direction through a conductor across an arc. A common example of DC is the power provided by batteries or from the power adapter in most automobiles. DC is the oldest welding power. Examples of using this current date back from the early 1900s when batteries were used with their metal electrodes to produce welds in steel plates.

Direct welding current is identified by the polarity of the electrode. If the electrode is negative, the current is referred to as being *direct-current electrode negative* (DCEN). If the electrode is positive, the welding current is referred to as being *direct-current electrode positive* (DCEP). Changing the electrode from positive to negative or negative to positive has significant effects on the weld produced.

Direct-current electrode negative DCEN, which used to be called *direct-current straight polarity* (DCSP), concentrates about two-thirds of its welding heat on the work and the remaining one-third on the tungsten. The high heat input to the weld results in deeper fusion or penetration into the base metal. The lower level of heat on the tungsten means that a smaller-sized tungsten can be used without overheating. The smaller-sized tungsten electrode may save time, money, and tungsten.

Direct-current electrode positive (DCEP), which used to be called *direct-current reverse polarity* (DCRP), concentrates only about one-third of the arc's heat on the weld metal and two-thirds on the tungsten electrode. DCEP produces welds that are much wider and much shallower than DCEN currents. DCEN produces a strong cleaning action. Because of the high heat input to the tungsten electrode, larger-sized tungsten is required. The lower heat input to these metals and strong cleaning action make this a good choice of currents for thin, heavily oxidized metals. Because the metal being welded does not emit electrons as freely as tungsten, the arc may wander or be somewhat erratic.

There are many theories as to why DCEP arcs produce a cleaning action. One of the more common explanations is that the electrons accumulate under surface oxides and lift them off as they jump free of the surface underway to the tungsten electrode. Additionally, positive ions accelerated to the metal's surface provide additional energy in knocking the oxide free. So in combination, the electrons and ions cause the surface activity needed to produce the cleaning. Although many theories exist as to exactly how this occurs, it is important to note that DCEP provides such metal oxide cleaning and can be used to your advantage (Figure 3-2).

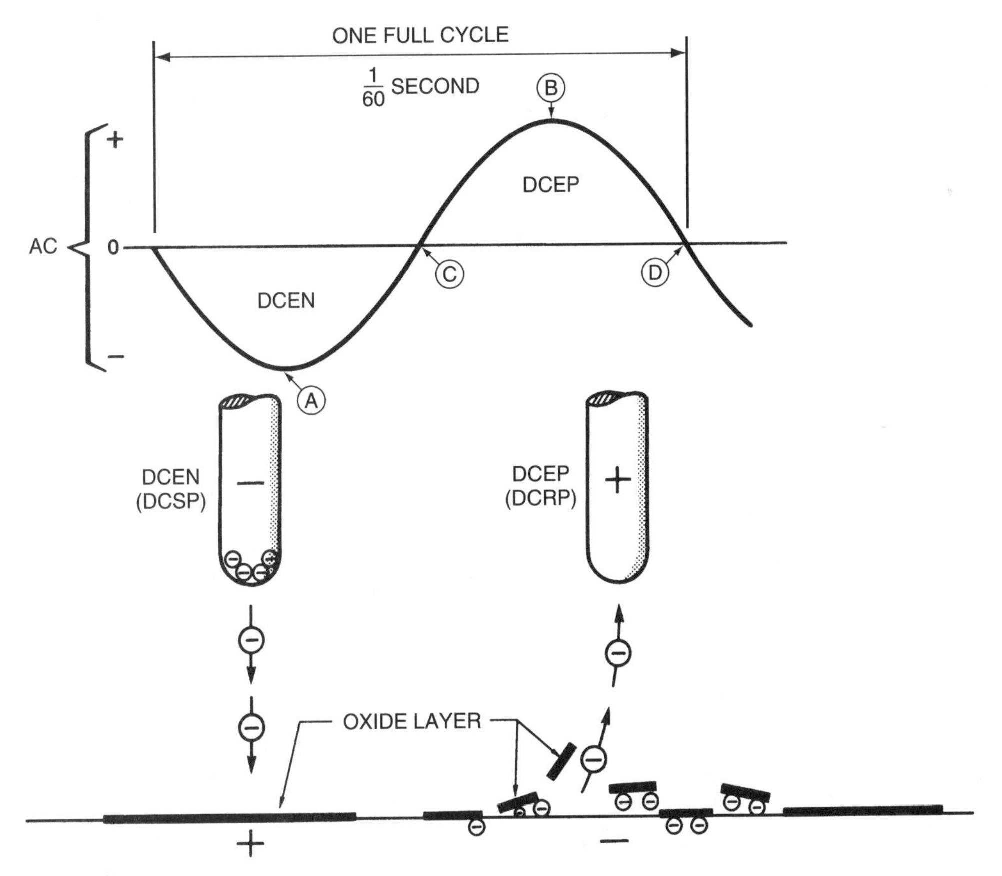

Figure 3-2 Electrons collect under the oxide layer during the DCEP portion of the cycle and lift the oxides from the surface.

Alternating Current (AC)

Alternating current (AC) gets its name from the fact that electrons travel in one direction for a short period of time and then reverse and travel in the other direction; they alternate back and forth. In other words, AC is made up of electricity flowing as DCEN half of the time and electricity flowing as DCEP the other half of the time. The rate at which this change in current direction occurs is referred to as its frequency, or cycle. In the United States, current that has a frequency of 60 cycles, or 60 hertz (60 Hz), is supplied to homes and businesses. Figure 3-2 illustrates current that is at its peak at points A and B. The current decreases from its peak until all flow has stopped at points C and D. The arc at this point stops and is reestablished as the current begins to build again toward the maximum.

During SMA welding, the ionized cloud of shielding gases produced by burning the electrode flux makes reestablishing the arc easy (Figure 3-3). However, it is quite difficult to reestablish the gas tungsten arc without this hot gaseous cloud. Thus a voltage assist from another source is needed. A very high voltage but low current is generated using a spark gap oscillator or other such device, which provides the necessary path for the relatively low-voltage, high-amperage welding current to reestablish its arc (Figure 3-4). The high frequency ensures that a voltage peak will occur during the time that the welding current reverses its direction (Figure 3-5). Without this high frequency assist, GTA welding using alternating current cannot be performed.

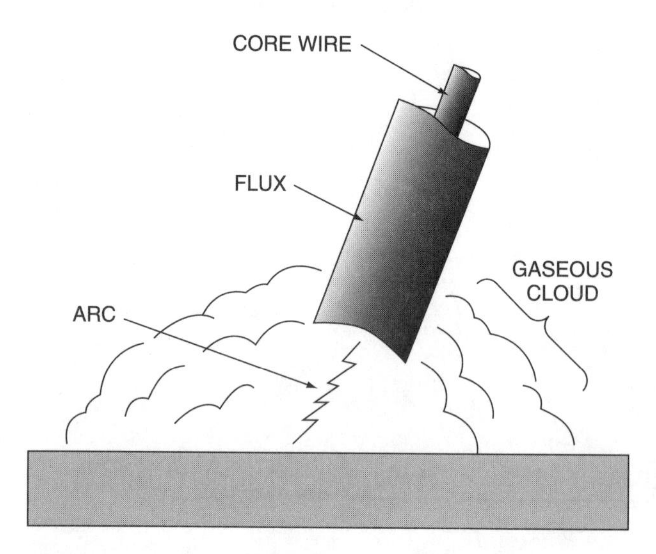

Figure 3-3 SMA welding electrode flux burns to form a gaseous cloud that helps maintain the AC arc.

High-Frequency

The high-frequency current is established in the welder by using a capacitor discharge across a gap set on points inside the machine (Figure 3-6). Changing the gap between the points will change the frequency of the current it generates. The closer the points are, the higher the frequency generated; the wider the gap between the points, the lower the frequency generated. The voltage for this spark gap generator is stepped up with a special transformer located inside the machine. The high-frequency current is induced into the primary welding circuit by a coil.

The high frequency may be set so that it automatically cuts off after the arc is established, usually used with DC. It is set on the continuous mode for AC welding. When used in this manner, it is referred to as *alternating current, high-frequency* stabilized, or ACHF.

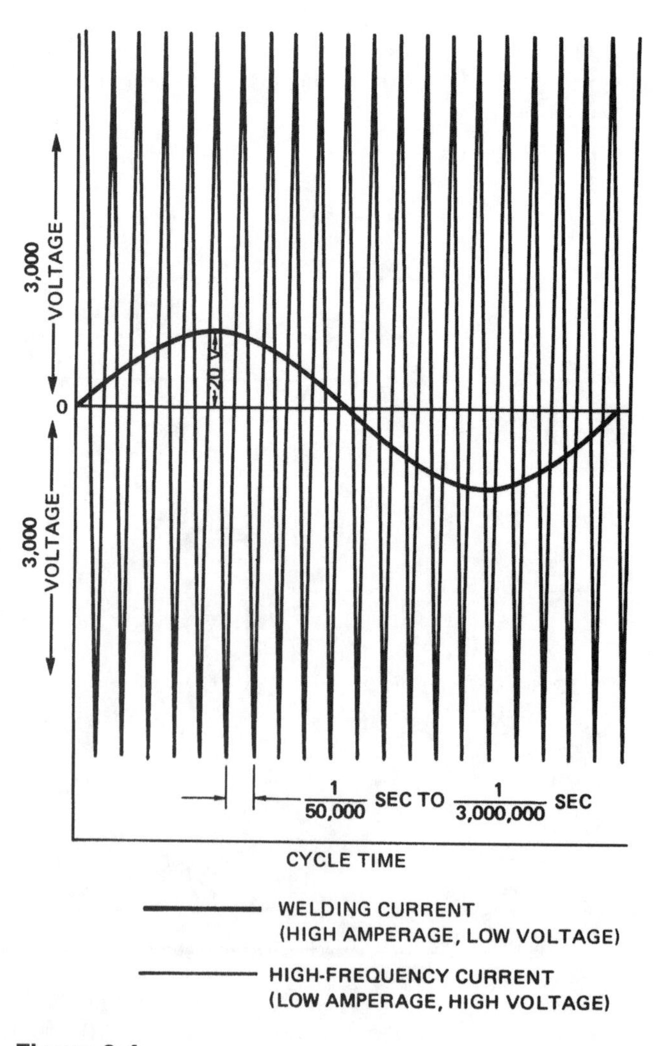

Figure 3-4

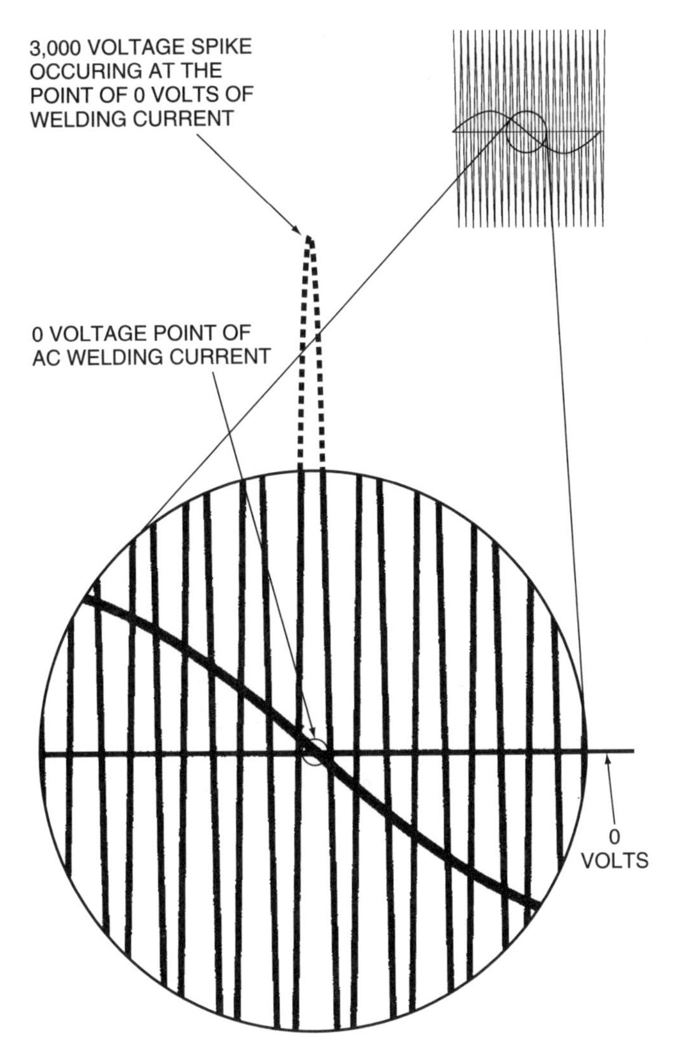

Figure 3-5 High-frequency voltage spike ensures that AC welding current continues.

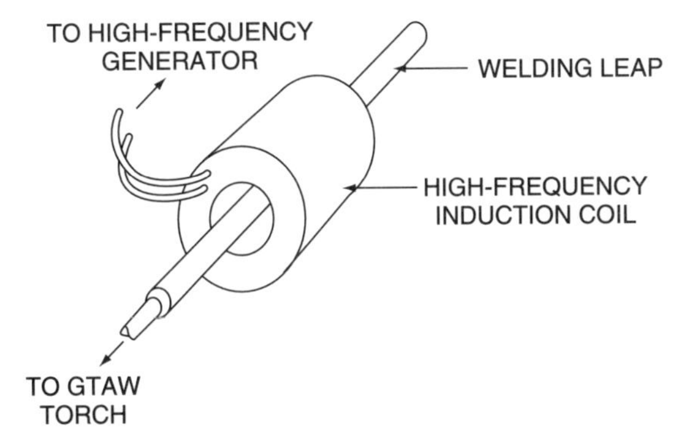

Figure 3-6 A coil is used to induce high frequency onto welding current.

CURRENT SETTING

The amperage set on a machine and the actual welding current are often not the same. The amperage indicated on the machine's control is the same as that at the arc only when the following conditions are met:

- The power to the machine is exactly correct.
- The lead length is very short.
- All cable connections are perfect with zero resistance.
- The arc length is exactly the right length.

If any one of these factors changes, the actual welding amperage will change.

In addition to the difference between indicated and actual welding amperage, there is a more significant difference between amperage and welding power. The welding power, in watts, is based on the formula W = E x I, or volts (E) multiplied by amperes (I) equals watts (W). Thus, the indicated power to a weld from two different types of welding machines set at 100 amperes will vary with the voltage of the machine.

The welding machine setting will vary within a range from low to high (cool to hot). The range for one machine may be different from that of another machine. The setting will also be different for various types and sizes of tungstens, polarities, types and thicknesses of metal, joint position or design, and shielding gas used.

A chart, such as the one in Figure 3-7, and a series of tests can be used to set the lower and upper limits for the amperage settings. As students' welding skills improve with practice, they will become familiar with the machine settings so that a table for these settings is no longer needed. In the welding industry, some welders will mark a line on the dial of the machine to help in resetting the machine. If a welder is required to make a number of different machine setups, a list or chart can be made and taped to the machine. This practice is more professional than marking the machine dials.

In general, the current-carrying capacity at DCEN is about ten times greater than that at DCEP.

The preferred electrode tip shape affects the temperature and erosion of the tungsten. With DCEN, a pointed tip concentrates the arc as much as possible and improves arc starting with either a short, high-voltage electrical discharge or a touch start. Because DCEN does not put much heat on the tip, it is relatively cool, the point is stable, and it can survive extensive use without damage.

CURRENT AND TUNGSTEN ELECTODE SIZE	AMERAGE/MACHINE SETTING				
	TOO LOW	LOW	GOOD	HIGH	TOO HIGH

Figure 3-7 Sample chart used to record GTAW machine settings.

With AC, the tip is subjected to more heat than with DCEN. To allow a larger mass at the tip to withstand the higher heat, the tip is rounded. The melted end must be small to ensure the best arc stability.

DCEP has the highest heat input to the electrode tip. For this reason a slight ball of molten tungsten is suspended at the end of a tapered electrode tip. The larger mass above the molten ball holds it in place like a drop of water on your fingertip (Figure 3-8).

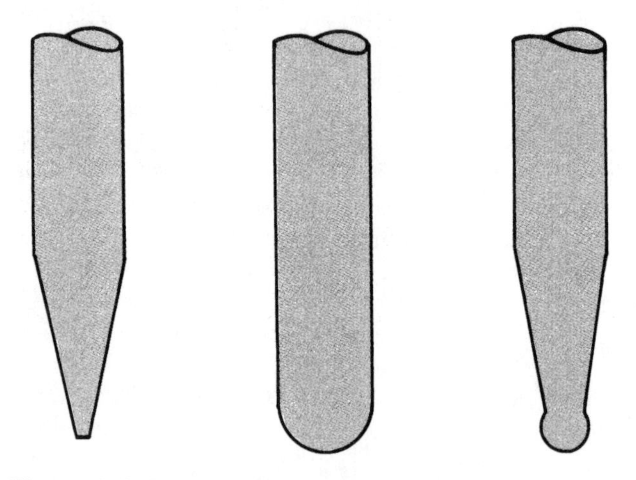

Figure 3-8 Basic tungsten electrode and shapes: (A) pointed, (B) rounded, and (C) tapered with a balled end.

REVIEW QUESTIONS

1. What are characteristics of welding currents?
2. Give an example of direct current.
3. How is the direct welding current identified?
4. What is the heat distribution with DCEN welding current?
5. What does lower level of heat signify in DCEN?
6. What is the heat distribution with DCEP welding current?
7. What type of welds does DCEP produce?
8. What is one of the more common theories as to why DCEP arcs produce a cleaning action?
9. Where does alternating current get its name?
10. What frequency of current is supplied to homes in the United States?
11. Why must AC welding power use high frequency in order to work?
12. What happens to the frequency when the points are closer?
13. When is the amperage on the machine's control the same as that at the arc?
14. What is the formula for the welding power?

CHAPTER 4

GTA WELDING TORCHES, HOSES, AND ACCESSORIES

INTRODUCTION

Gas tungsten arc welding torches are available in a variety of designs and amperage sizes. The most common torch has a head at a 70° angle; some torches are available with 90° heads (right angled) or 180° (straight) (Figure 4-1). Hoses, nozzles, and other accessories must all be designed to withstand the high-temperature environment of the welding and resist deterioration as a result of the intense UV light. Most GTA torches are black so that they do not reflect the intense light of the welding arc.

Because of the high temperatures the torch is subjected to during welding, some method of cooling it must be provided. Torches can generally be divided into two major categories according to the method by which they are cooled. The two methods of cooling a GTA torch are water-cooling and air-cooling (Figure 4-2). Torches specifically designed for one method of cooling are not interchangeable with torches designed for the other method of cooling. Water-cooled torches have internal passageways for water to be circulated, and air-cooled torches have additional surface area to provide air cooling. However, many of the other components such as caps, nozzles, and collets may be interchanged within a single manufacturer's torch model and amperage range. Parts from one manufacturer's torch may not be interchangeable with another manufacturer's torch because there are no standards for these parts.

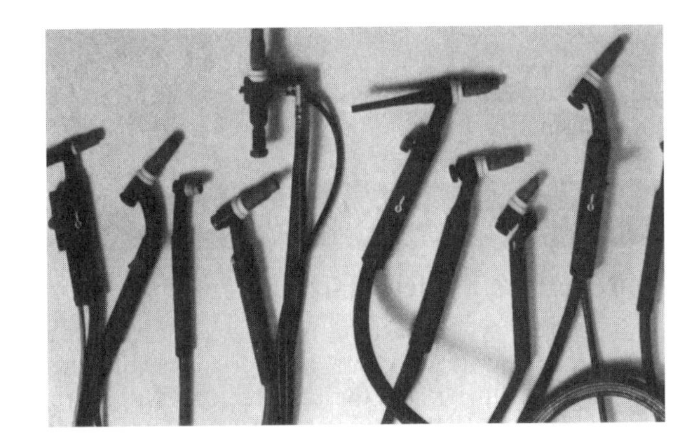

Figure 4-1 GTAW torches. (Courtesy of Miller Electric Manufacturing Co.)

GTA WELDING TORCHES

Most GTAW torches are made up of the following components:

- Torch body
- Back cap
- Collet and collet body
- Nozzle
- Handle

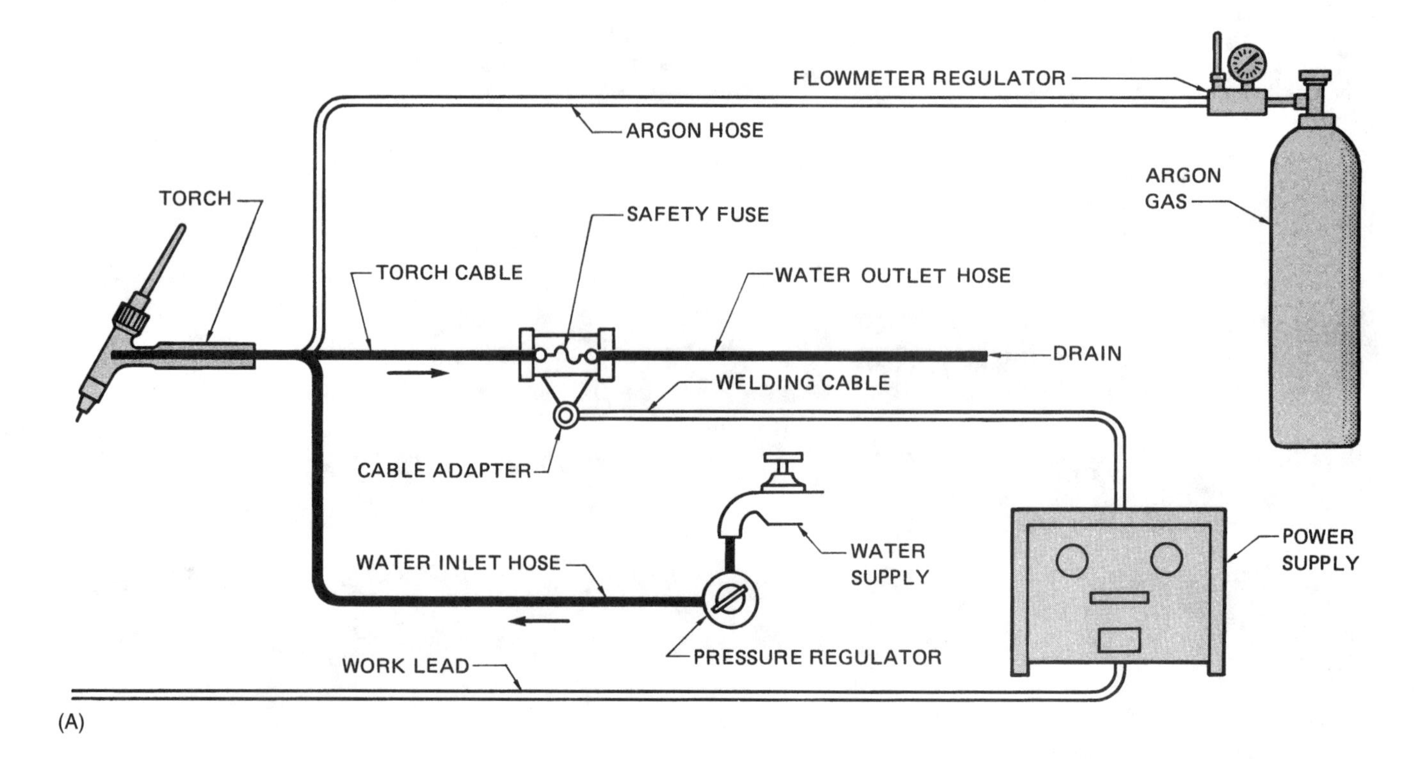

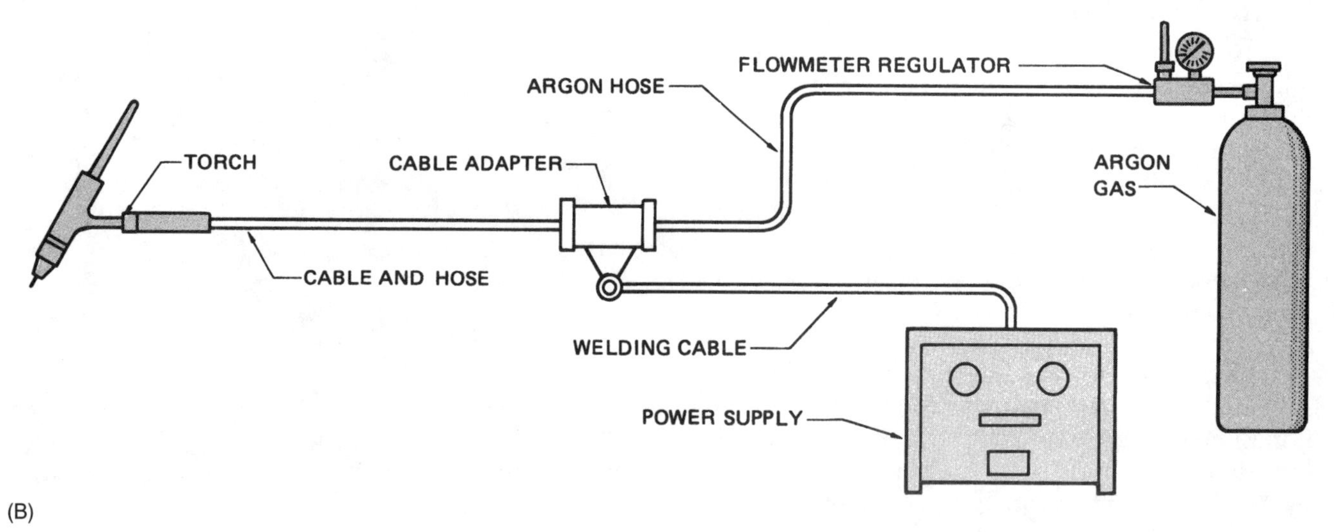

Figure 4-2 Schematic of GTAW setup with a (A) water-cooled torch and an (B) air-cooled torch.

- Assorted insulators and gaskets
- Gas lens (optional)

See Figure 4-3.

Torch Body

A torch body, or welding head, consists of an inner metal part surrounded by an insulative high-temperature, heat-resistant plastic casing (Figure 4-4). The internal metal part is usually manufactured from a copper or brass alloy. Both these metal alloys provide excellent electrical and thermal conductivity. Good electrical conductivity is required in order to carry the high welding currents, and thermal conductivity is required to enable it to dissipate excessive welding heat.

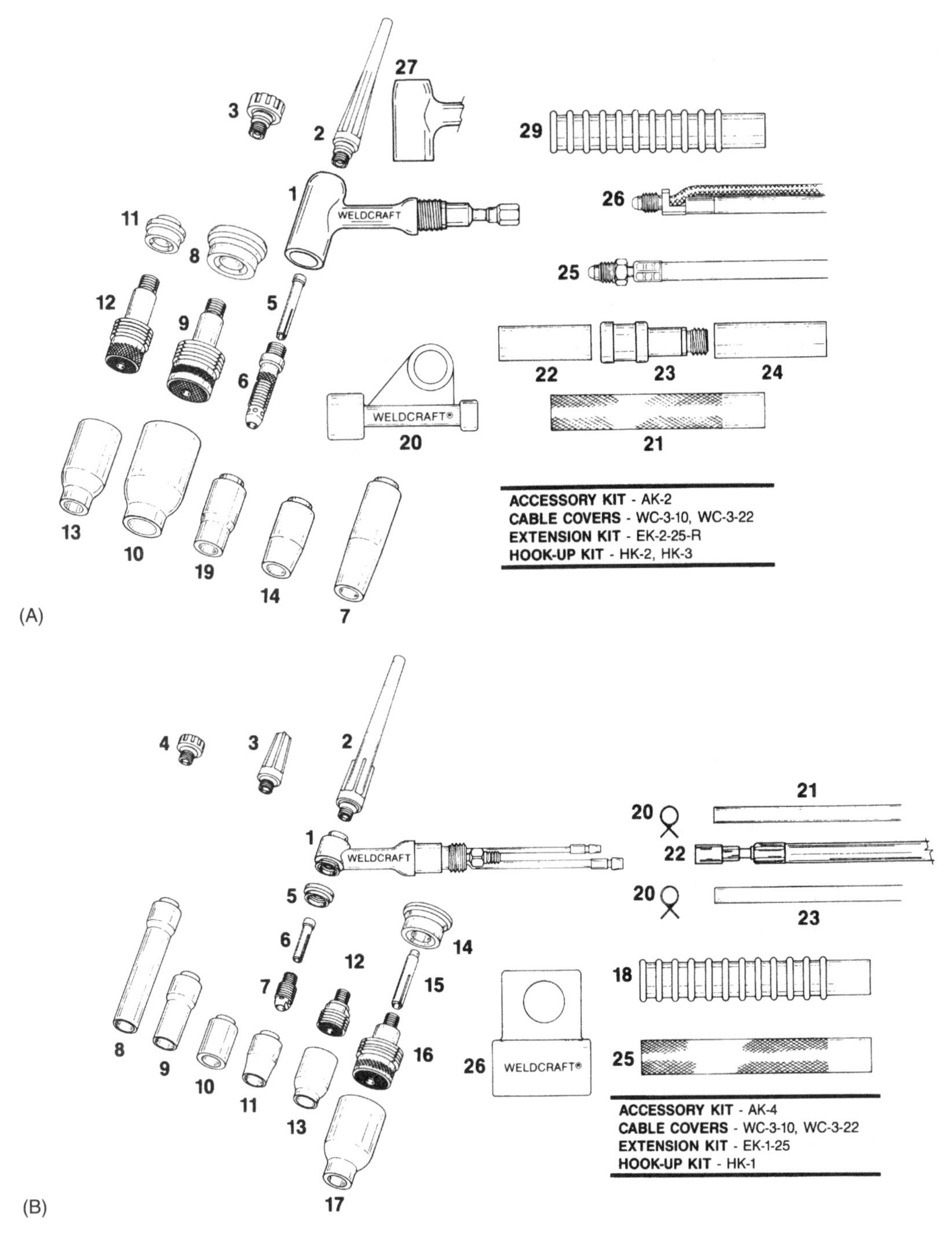

Figure 4-3 (A) Air-cooled torch: (1) torch body, 70° (2) long back cap, includes 98W18 O-ring (3) short back cap, includes 98W18 O-ring (5) collet (6) collet body (7) long lava nozzle (8) large-diameter gas lens insulator (9) large-diameter gas lens (10) large-diameter gas lens nozzle (11) gas lens insulator (12) gas lens (13) alumina gas lens nozzle (14) lava nozzle (19) alumina nozzle (20) power cable adapter (21) handle (smooth) (22) front valve (23) valve body (24) rear handle, valve assembly; includes items 22, 23, 24 (25) power cable (26) gas hose (27) torch body with 90° head (29) handle (ribbed). (B) Water-cooled torch: (1) torch body, includes cup gasket, 70° (2) long back cap, includes 98W77 O-ring (3) medium back cap, includes 98W77 O-ring (4) short back cap, includes 98W77 O-ring (5) cup gasket (6) collet (7) collet body (8) extra long lava nozzle (9) long lava nozzle (10) lava nozzle (11) alumina nozzle (12) gas lens (13) alumina gas lens nozzle (14) large-diameter gas lens insulator (15) large-diameter gas lens collet (16) large-diameter gas lens (17) large-diameter gas lens nozzle (18) handle (ribbed, optional) (20) hose clamp (21) gas hose (22) power cable (23) water hose (25) handle (smooth, standard) (26) power cable adapter (Courtesy of Weldcraft®)

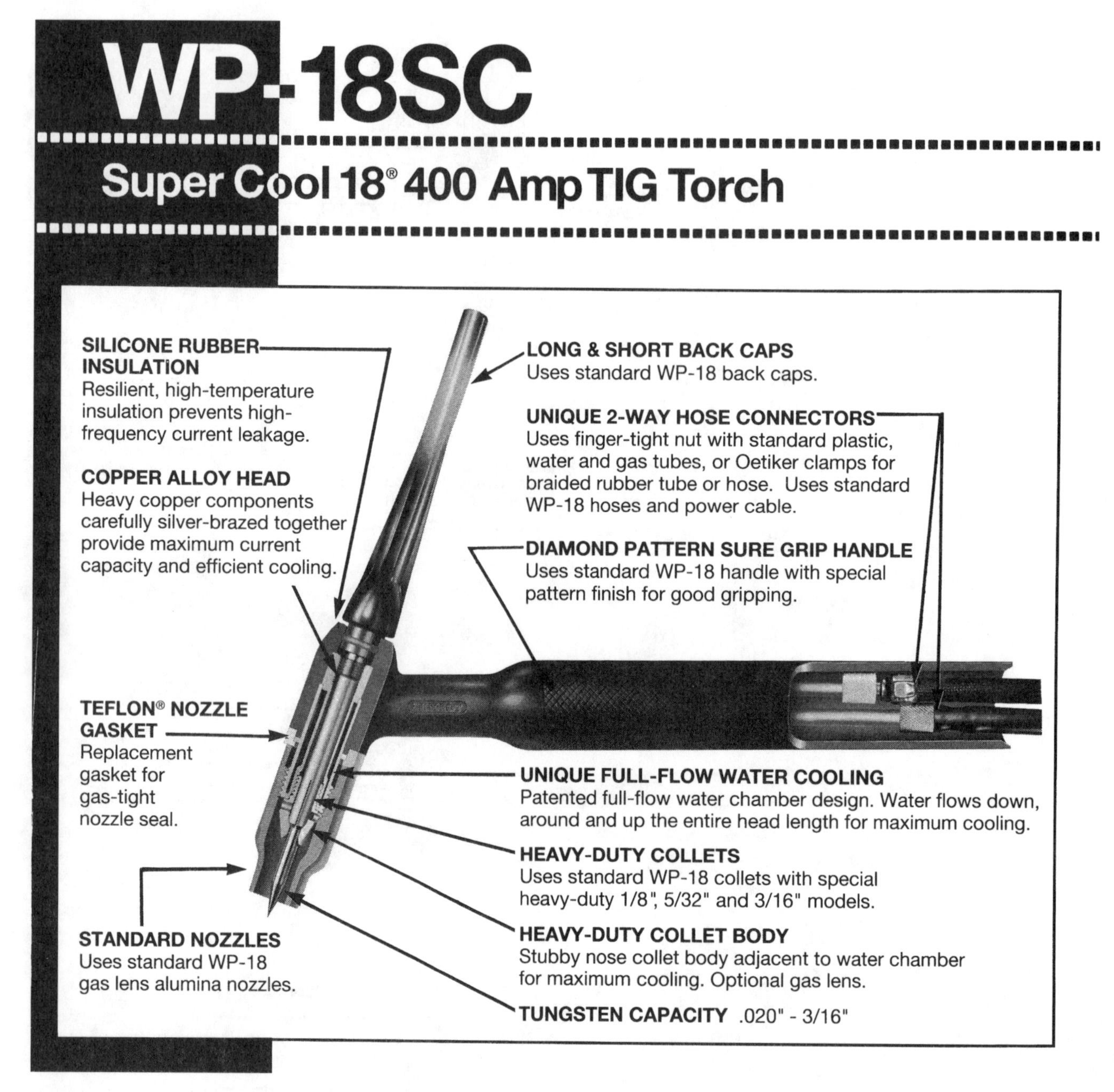

Figure 4-4 Weldcraft's Super Cool 18 400-amp TIG Torch. (Courtesy of Weldcraft®)

The most common welding head angle is 70°; however, torches are available that have 90°, 180°, or flexible angles.

Connections for the hoses and power cables can be either threaded or barbed (Figure 4-5). Barbed connections allow for a smaller, tighter fit of the hose connections, reducing both the size and weight of the head.

The welding head is the major component of the GTAW torch. All of the other parts and pieces are connected to it. Its size, design, and amperage capacity determine whether the torch is air- or water-cooled and its welding capacity.

Back Cap

The back cap serves two major purposes. First, it can be loosened or tightened to secure or release the tungsten electrode. Second, it also covers the back end

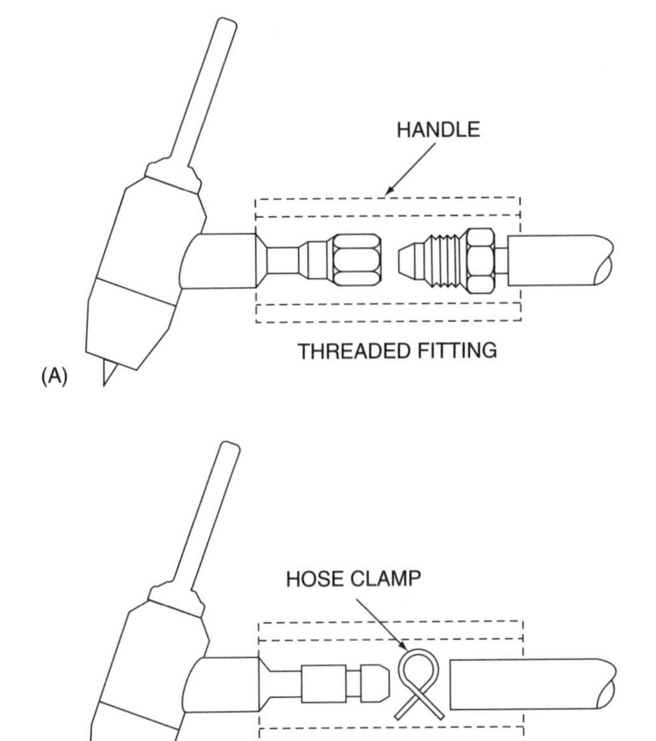

Figure 4-5 (A) 70° torch and (B) flexible torch.

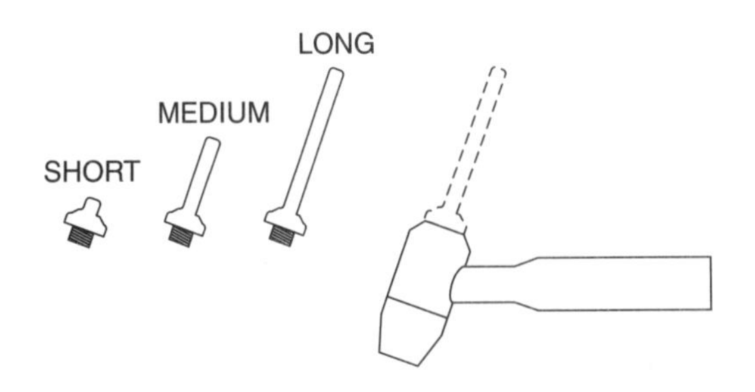

Figure 4-6 Back caps are available in several different lengths.

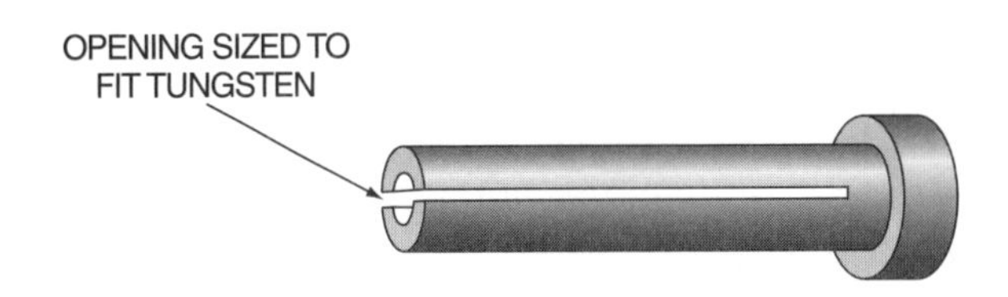

Figure 4-7 Collet.

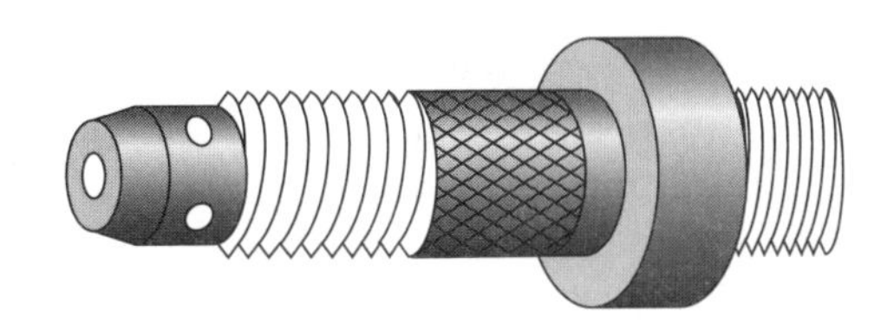

Figure 4-8 Collet body.

of the tungsten electrode, protecting it and sealing off the top of the welding head to prevent shielding gas leaking. Back caps are available in a variety of lengths from less than 1″ to around 8″ (Figure 4-6). The exact length of back caps varies from manufacturer to manufacturer and from torch capacity to torch capacity. Generally they are available in lengths of approximately 1″, 2″, 3″, and 7″ lengths. There may or may not be an O-ring provided with the back cap to provide a gas-tight seal.

Collet and Collet Body

Both the collet and collet body can be made of copper or brass. In some cases these components may be plated to increase electrical and thermal conductivity. Both the collet and collet body must be sized to fit the tungsten diameter being used.

The collet appears to be a long, thin-walled section of tubing with relief slits in its sides. These slits allow the collet to be compressed tightly around the tungsten as the back cap is tightened (Figure 4-7).

The collet body serves both as a gas diffuser and attachment for the nozzle. Several holes are drilled into the collet body to provide even and uniform shielding gas flow. External threads allow the nozzle to be screwed onto the collet body (Figure 4-8).

Nozzles

Welding nozzles or cups come in a variety of sizes and materials. The size of the opening—called the orifice—and the end of the nozzle affects or determines the welding current and accessibility in confined spaces. Beginning welders will often find it easier to use a smaller nozzle for better visibility. However, larger nozzle diameters may be required for adequate shielding gas coverage. Although smaller diameter nozzles may provide greater weld visibility, they may become fused or melted during welding due to the high arc heat. If this occurs, it can adversely affect the shielding gas coverage by disrupting the normal aerodynamic design of the nozzle and disrupting proper gas flow.

Most nozzles are made of ceramic, but for some special applications such as automatic welding, a metal nozzle may be required to withstand the welding heat. Fused quartz nozzles are transparent and are used sometimes when greater visibility is required. Fused quartz nozzles are relatively expensive as compared to ceramic nozzles. Nozzles sizes, gas flow, and current ranges are shown in Figure 4-9.

The choice of nozzle size—both length and diameter—is usually made based on the welder's personal preferences. Welding specifications may indicate the nozzle size to use. Factors that may affect cup size selection are as follows:

- Joint Design: Some joints with restricted access, such as T, corner, and lap joints, may require a smaller diameter and longer nozzle length for both accessibility and visibility.
- Large diameter cups may provide better gas coverage even in drafty places.
- Gas flow: Small diameter cups allow for a lower gas flow rate for better economy while still providing adequate coverage of the weld.

Ceramic nozzles are heat resistant and will provide a relatively long life when used correctly. The useful life of a ceramic nozzle is affected by the current level and its closeness to the work.

Silicon nitride nozzles will withstand much higher temperatures than ceramic nozzles and therefore have a longer useful life. Silicon nitride nozzles are the best choice when the cup-walking technique is used.

The fused quartz (glass) nozzle is used when welding visibility is a problem. Fused quartz nozzles are very durable and not any more easily broken than ceramic nozzles. They are however more expensive than either of the other two types of nozzles.

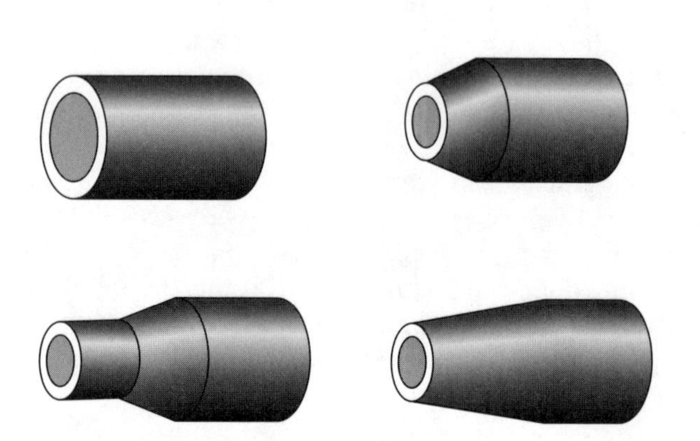

Figure 4-9 Nozzles.

The longer a nozzle, the longer the tungsten must be extended from the collet. This longer tungsten extension can result in higher tungsten temperatures, which in turn can result in more tungsten melting off and dropping into the weld. When longer nozzle sizes and lengths are needed, it is better to use low amperages or larger diameter tungsten sizes.

Handle

The handle on the torch is made of a high temperature plastic and is either slipped or threaded onto the welding head. The handle is approximately 11/2″ (35 mm) in diameter. Handles are usually either smooth or ridged for better gripping. Often they have a flat side that allows the welder to index, or hold, the torch in the desired position. This flat spot enables welders to feel the position of the welding torch in their hand without having to raise their helmet to check its position (Figure 4-10). The choice of handle design is strictly a personal preference.

Assorted Insulators and Gaskets

An assortment of nylon and neoprim insulators and gaskets is provided by the manufacturers of the various welding torches. These parts are not as resistant to the high temperatures of the welding environment and may become damaged and need to be replaced periodically. Their primary purpose is to provide a soft seal between more rigid components, such as between the nozzle and torch head

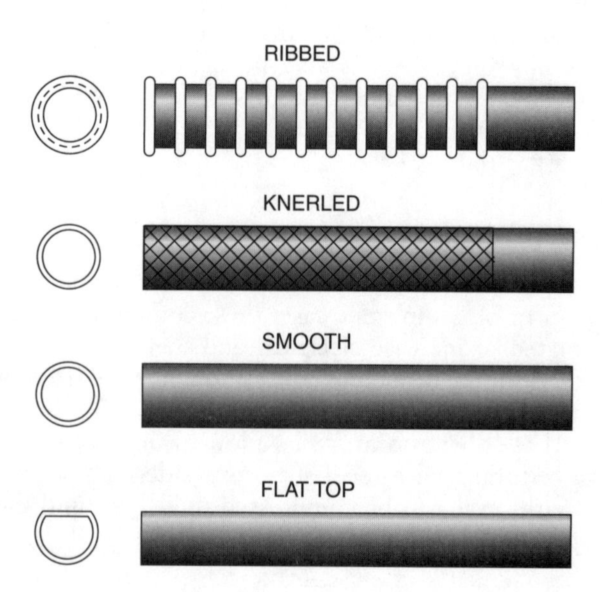

Figure 4-10 GTAW torch handle types.

(Figure 4-11). For safety reasons, do not operate the torch without the insulators and gaskets in place, and do not attempt to tape over damaged or missing insulators or gaskets.

Gas Lens

The gas lens is a device that is screwed onto the torch head and replaces the collet body on most torches. It produces a very smooth, even flow of shielding gas due to a large number of small openings on the gas lens (Figure 4-12). The smooth shielding gas flow is free of turbulence, thus providing better coverage of the weld pool. Spatter or other debris that becomes lodged on the gas lens can disrupt this smooth gas flow. Such material must be removed or the gas lens replaced in order to reestablish the desired effect of the lens.

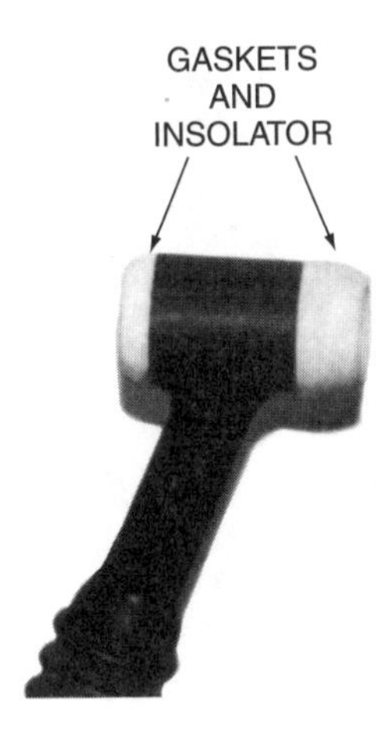

Figure 4-11 Gasket insulators on a GTAW torch head.

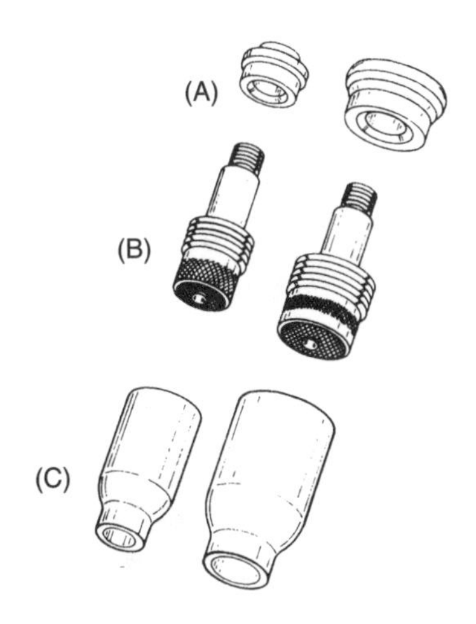

Figure 4-12 (A) Special torch gasket/insulator, (B) gas lens, (C) Special nozzle for gas lens (Courtesy of Weldcraft®)

Gas lenses are considered to be an optional component for most GTA welding. They are most often required when welding in drafty areas where the shielding gas coverage is subject to being blown away from the weld or disrupted. They are not normally used for most GTA welding training due to their expense and nominal benefit for welds produced in controlled environments.

HOSES AND CABLES

The specific number of hoses and cables connected to the GTAW torch varies depending on the torch type. Air-cooled torches have the fewest (with a minimum of one combination power shielding gas cable), and water-cooled torches have the greatest number.

Hoses

Air-cooled torches sometimes have a single hose that connects to the torch and surrounds the welding power cable. Other air-cooled torches have a separate shielding gas supply hose and power cable (Figure 4-13). The single-hose system has the advantage of easier torch manipulation because the single gas/power cable is much less likely to become snagged or caught. The advantage of the dual system is that if either the hose or the cable becomes damaged, only that damaged cable or hose needs to be repaired.

A water-cooled torch has three wires connecting it to the welding machine. One of the hoses carries shielding gas; another provides cooling water to the torch; and the third hose is a combination of return water and power cable.

The shielding gas hose must be made from a heat-resistant plastic that must both resist damage from the hot welding environment and not contaminate the shielding gas. Rubber hoses contain oils and other solvents that can be picked up by the gas, resulting in weld contamination.

Hoses that have been used for water or other fluids may not be used as shielding gas hoses because they may also contain residue that will contaminate the weld.

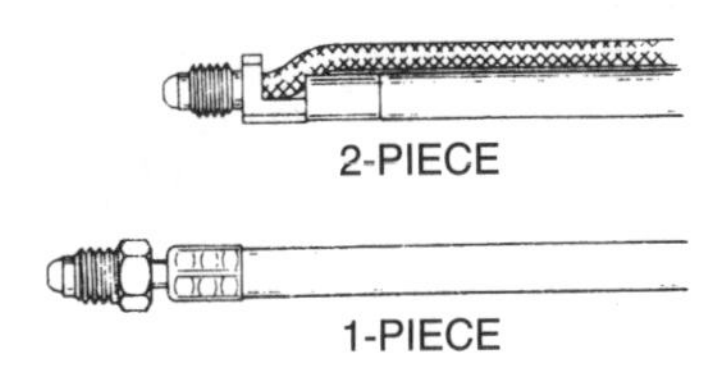

Figure 4-13 One- or two-piece power cables. (Courtesy of Weldcraft®)

The supply for water-in hose may be made of any sturdy material. Water hose fittings have left-hand threads, and gas fittings have right-hand threads (Figure 4-14). This prevents the water and gas hose from accidentally being reversed when attaching them to the welder.

The return or water-out hose surrounds the power cable. This permits a much smaller-sized power cable to be used because of the water cooling. A problem can occur with the power cable if cooling water from the torch becomes stopped or restricted. If this occurs, the smaller-diameter power cable inside the return water hose will rapidly overheat and possibly melt the surrounding hose. When this occurs, the power cable will become unsafe because of potential water leaks and must be replaced. Some welding machines are equipped with water flow sensors to shut off the welding power if a loss in coolant water flow is detected. Water supplied to a water-cooled torch must be first supplied to the torch head and then return around the cable. This allows the head to receive the coolest water before the power cable warms it. Running the water through the torch first is advantageous because, when the water solenoid is closed, there is no water pressure in the hoses. This prevents water from leaking when you may be away from the equipment.

Cables

Some torch systems that have handheld remote controls have an additional pair of wires running to the torch. The small-diameter wires are used to stop, start, and possibly control the welding power setting. The maximum allowable voltage in these circuits is 80 volts.

Figure 4-14 Left-hand threaded fittings are identified with a notch.

Protective Coverings

A protective covering can prevent the hoses and cables from becoming damaged by hot metal (Figure 4-15). These protective coverings are made of a heat-resistant material that can be zipped, snapped, or otherwise fastened around the hoses and cables. Even with this protection, the hoses should be supported (Figure 4-16), so that they are not underfoot or on the floor. Any support provided to the cables should

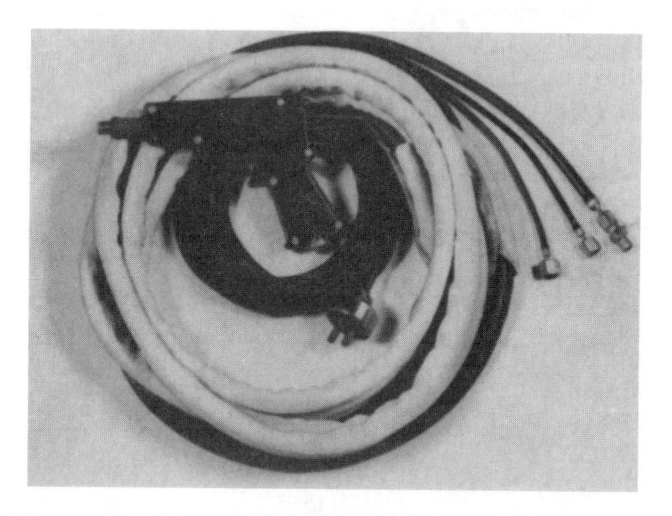

Figure 4-15 Zip-on protective covering also helps keep the hoses neat. (Courtesy of Hobart Brothers Company)

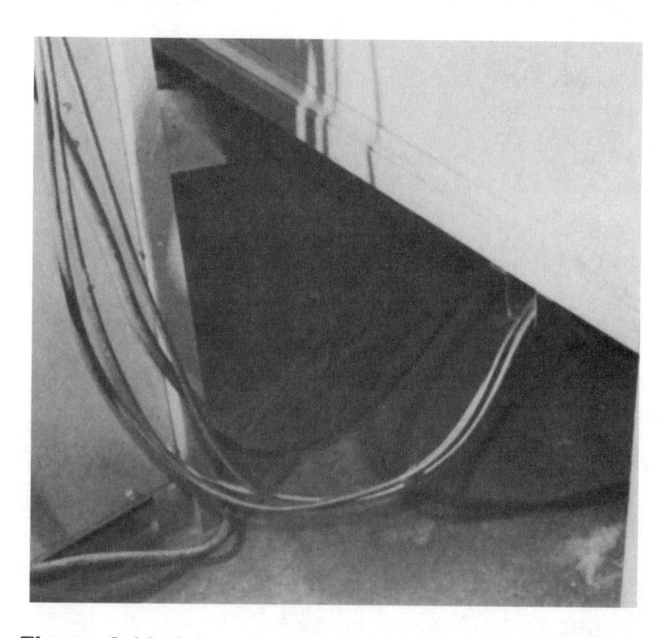

Figure 4-16 A bracket holds the leads off the floor.

be near the floor and not overhead. Supporting the hoses near the floor both prevents their possible damage and reduces potential radio or electronic interference from the welding current.

Flexible Heads

Most handles and weld heads are rigidly attached. However, special welding torches have flexible heads that allow the welder to change the position of the welding head. This allows for easier access and greater control of some welds (Figure 4-17). The flexible welding head makes joints such as the horizontal fillet weld (2F) much easier. Welders can hold their wrist at a more comfortable angle while directing the arc at the most ideal angle into the weld pool.

The welding head—which actually consists of the main body and the attached cap tungsten, tungsten collet, and nozzle—is made of copper and encased in a high-temperature plastic.

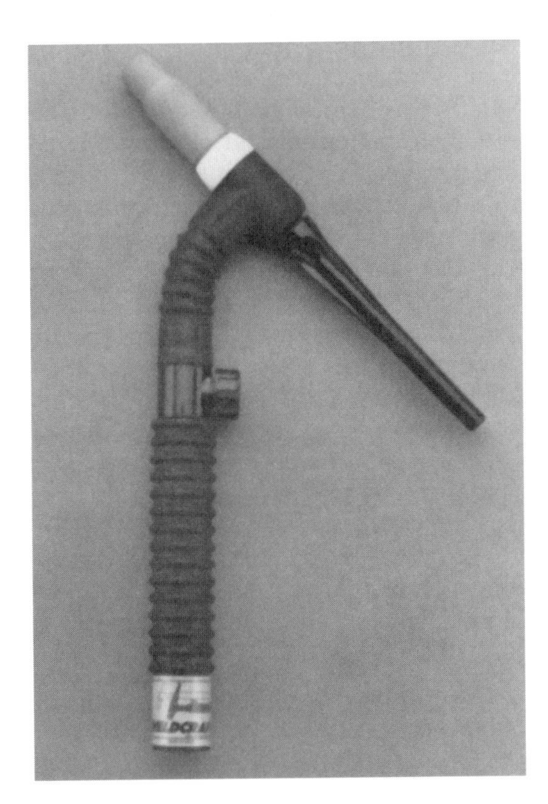

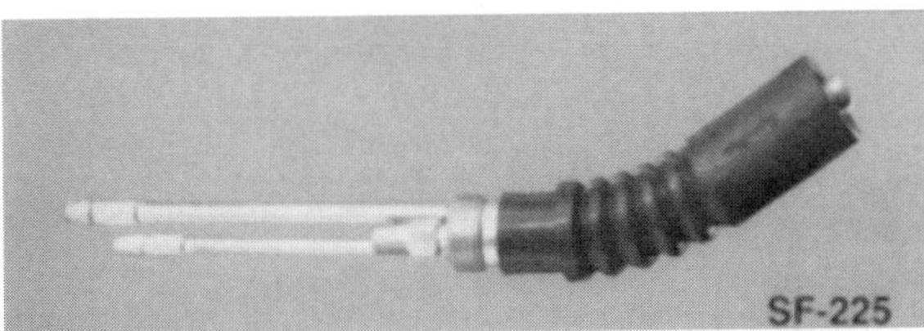

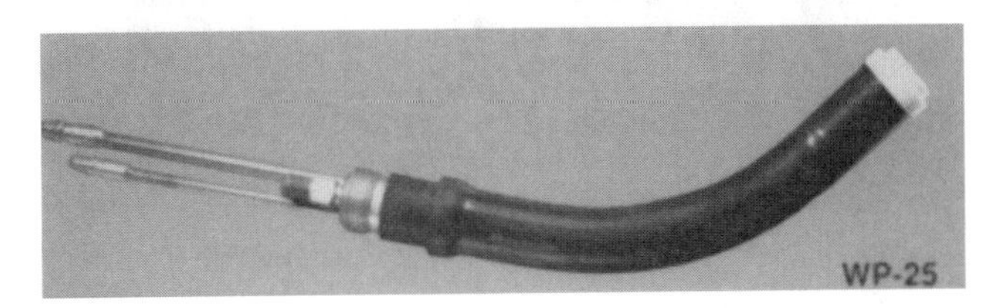

Figure 4-17 Flexible heads. (Courtesy of Weldcraft®)

Welding Cables

Two welding cables are required in this process: One is the power cable that goes from the welding machine to the torch, and the other is the work cable, or ground cable, which goes from the workpiece back to the welding power supply. The welding power cables are made of very thin strands of copper wire so that they are more flexible. Their insulation must resist oil, heat, and flames to help protect them from the welding environment. Water-cooled cables have special fittings on the end that allow for the connection of power and water to the cable in a special leakproof configuration.

Welding power cables are available in lengths from 10 to 25 feet. However, because of the high frequency that is sometimes used in GTAWelding the Federal Communications Commission (FCC) restricts the lengths of these cables to 25 feet or less. Cables longer than 25 feet may act as an antenna and transmit static to radios, television, and other electronic devices. The ground cable length is not restricted by the FCC but should not be any longer than is necessary to conveniently reach from the work to the welder. Ground cables are larger in diameter and should not be excessively long to reduce voltage drop. A voltage drop can occur when smaller or longer cables are used and will affect the welding arc adversely.

CAUTION

Water-cooled welding power cables that become damaged must be repaired or replaced immediately. Water is extremely dangerous around electricity.

CAUTION

There can be no splice or repair on any welding leave within 10 feet of the welding end.

GTA WELDING EQUIPMENT

Figure 4-18 shows two industrial applications of gas tungsten arc welding.

Torches

GTA welding torches are available water cooled or air cooled. The heat transfer efficiency for GTA welding may be as low as 20%. This means that 80% of the heat generated does not enter the weld. Much of this heat stays in the torch and must be removed by some type of cooling method.

The water-cooled GTA welding torch is more efficient than an air-cooled torch at removing waste heat. Compared to the air-cooled torch, the water-cooled torch operates at a lower temperature, resulting in a lower tungsten temperature and less erosion.

The air-cooled torch is more portable because it has fewer hoses and may be easier to manipulate than the water-cooled torch. The water-cooled torch requires a water reservoir or other system to give the needed cooling. The cooling water system should contain some type of safety device to shut off the power if the water flow is interrupted. The power cable is surrounded by the return water to keep it cool, so a smaller size cable can be used. Without the cooling water, the cable quickly overheats and melts through the hose.

The water can become stopped or restricted for a number of reasons, such as a kink in the hose, a heavy object sitting on the hose, or failure to turn on the system. Water pressures higher than 35 psi (241 kg/mm^2) may cause the water hoses to burst. When an open system is used, a pressure regulator must be installed to prevent pressures that are too high from damaging the hoses.

GTA welding torch heads are available in a variety of amperage ranges and designs (Figure 4-19). The amperage listed on a torch is the maximum rating and cannot be exceeded without possible damage to the torch. The various head angles allow better access in tight places. Some of the heads can be swiveled easily to new angles. The back cap that both protects and tightens the tungsten can be long or short (Figure 4-20).

A remote control can be used to start the weld, increase or decrease the current, and stop the weld. The remote control can be either a foot-operated or hand-operated device. The foot control works adequately if the welder can be seated. Welds that must be performed away from a welding station must have a hand or thumb control.

Most remote controls have an on-off switch that is activated at the first or last part of the control movement. A variable resistor increases the current as the control is pressed more. A variable resistor

(A)

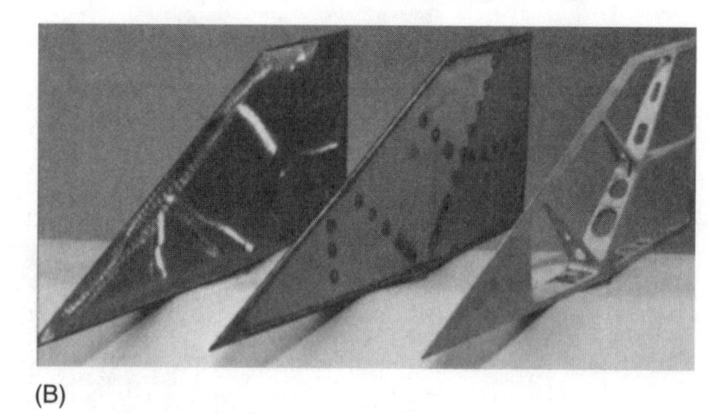

(B)

Figure 4-18 (A) An operator GTA welds a cap ring on a pneumatic tank. (Courtesy of Miller Electric Manufacturing Company) (B) Patriot missile air vanes GTA plug-welded skin to subframe work.

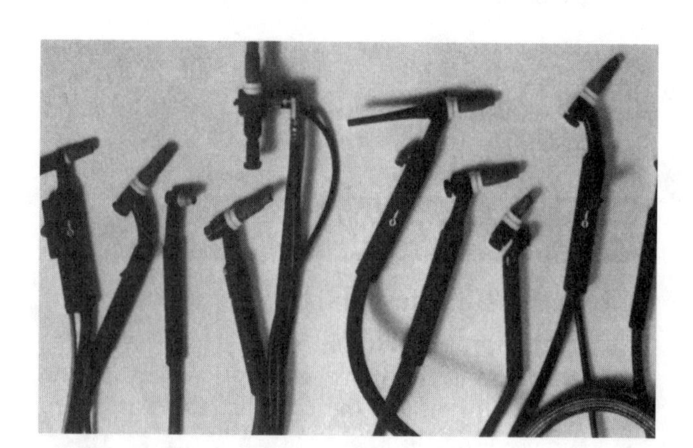

Figure 4-19 GTAW torches. (Courtesy of Miller Electric Manufacturing Co.)

works in a manner similar to the accelerator pedal on a car to increase the power (current) (Figure 4-21). The operating amperage range is determined by the value that has been set on the main controls of the machine.

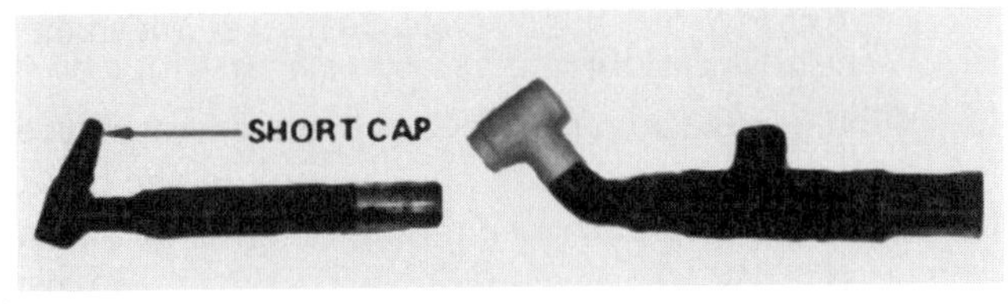

(A)

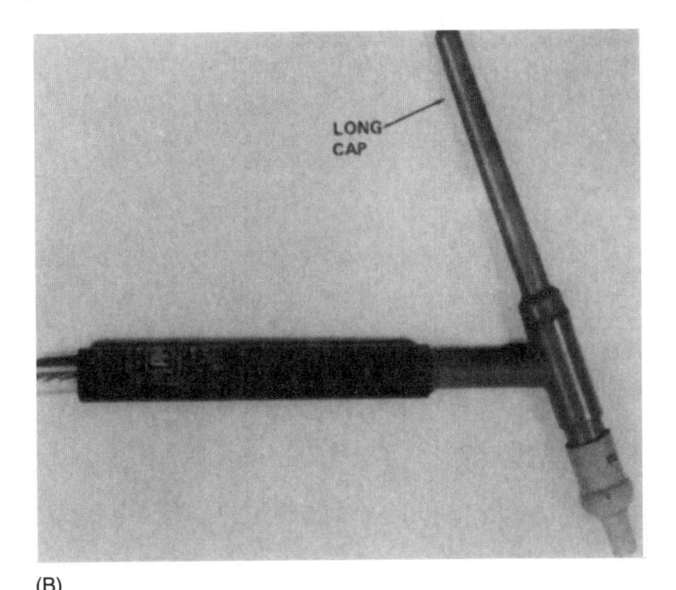

(B)

Figure 4-20 (A) Short back caps are available for torches when space is a problem. (B) Long back caps allow use of tungstens that are a full 7″ (177 mm) long. (Courtesy of Hobart Brothers Company)

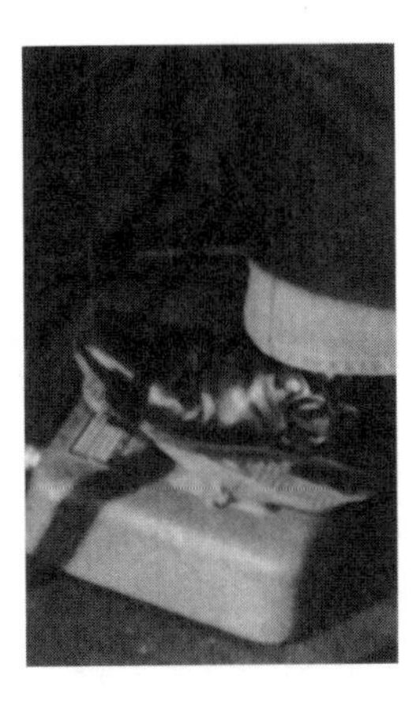

Figure 4-21 A foot-operated device can be used to increase the current.

Water and Gas Solenoids

Some welding machines have built-in water and gas solenoid valves that stop and start the flow of cooling water and shielding gas at the same time that the welding power is started and stopped. Solenoid valves are electrical mechanical devices that, when energized, electrically will open and allow flow through them. These valves are usually wired in connection with the welding power supply so that they will automatically open, allowing for flow when the welder is energized for welding.

Being able to start and stop the water flow automatically has its advantages. If the cooling water continues to circulate when the torch is not being used, condensation may form inside the torch. This condensation will seriously affect the weld's quality. Starting and stopping the shielding gas automatically will dramatically reduce its consumption, thus saving money.

Add-on or external solenoid valves are available for welders that did not come equipped with them from the manufacturer (Figure 4-22). These solenoid valves can be added as part of an external high-frequency generator. Such devices allow for the conversion of many basic welding machines into GTA welders.

Quick Disconnects

Quick disconnects are available for GTAWelding hoses and cables. These allow for the removal of the GTAW torch when it is not being used. Such disconnects are a good idea when a single welding power supply is to be used when both GTA and SMA welding. Removing the GTA torch and hose will prevent it from accidentally becoming damaged by the SMA welding sparks.

Figure 4-22 GTAW unit that can be added to a standard power supply so that it can be used for GTAW. (Courtesy of Lincoln Electric Company)

REVIEW QUESTIONS

1. At what angle are GTAW torch heads available?
2. What color are most GTA torches and why?
3. What are the two methods of cooling a GTA torch?
4. List seven components found as part of most GTAW torches.
5. What is the most common GTAW torch head angle?
6. What two functions does the GTAW torch back cap serve?
7. What factors may affect selection of a GTA welding torch nozzle size?
8. Why is silicon nitride used to make some GTA welding nozzles?
9. Why is a fused quartz (glass) nozzle used for some GTAW?
10. How is a gas lens attached to a GTAW torch?
11. Why are gas lenses used for GTAW?
12. What is the advantage of using a dual hose system?
13. What can prevent hoses and cables from being damaged by hot metal?
14. What does supporting a hose near the floor do?
15. What is the advantage of the flexible welding head?
16. What are some of the common lengths of available GTAW power cables?
17. Which GTAW torch can withstand the most heat for the longest time?
18. What are solenoid valves used for?

CHAPTER 5

TUNGSTEN

INTRODUCTION

Tungsten is a metal whose properties make it an ideal choice for use as a nonconsumable electrode. Its major properties for this purpose are that it is a good electrical conductor, has a high melting temperature, and readily emits an arc. These properties can be further enhanced by properly selecting its size and its end shape and by adding other specific elements.

Tungsten resists oxidation under normal atmospheric conditions, although at elevated temperatures it does oxidize rapidly. Tungsten is used in its pure form primarily as an element in television or other electronic tubes. It found wide use early on as filament material for light bulbs and is used extensively as electrodes for gas tungsten arc welding. As an alloy, its most frequent form is tungsten carbide. Tungsten carbide provides a very hard tip for machine tools, or as a grinding compound for such things as sandpaper and grinding stones.

TUNGSTEN

Some of its characteristics that make it a particularly good electrode are as follows:

- Resists oxidation: Tungsten does not oxidize under normal atmospheric conditions. This ability to resist oxidation means that tungsten will stay clean and bright for a long period of time while it is in storage. Remaining oxide free helps reduce tungsten contamination of the weld. Very hot tungsten will rapidly oxidize if exposed to the air, so during welding it must be protected by inert gas atmosphere to prevent oxidation.
- Has good current-carrying characteristics: Tungsten has a low resistance to electric current flow, which allows the welding current to pass through it without its becoming overheated due to electrical resistance. The tip of the tungsten does become very hot during the welding process as a result of its contact with the arc. The main body of the tungsten remains well below 1,000° F (540° C) and does not glow from the heat.
- Is a good electron emitter: White-hot tungsten has an ability to emit electrons freely. The electrons are literally sprayed from the tip of the hot tungsten to the workpiece. This free emission of electrons helps establish a smooth arc as the electrons that compose the arc travel across the gap to the work (Figure 5-1).

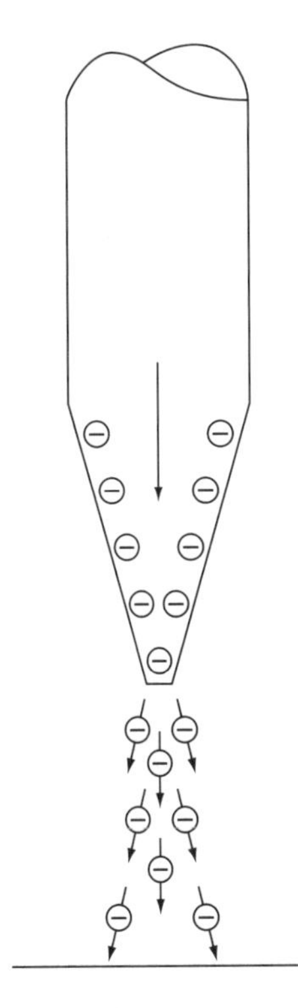

Figure 5-1 Electrons (–) will freely leave the tip of a hot tungsten to form an electric arc.

- Has a high melting temperature: Tungsten has the highest melting temperature of any commercial metal. Its melting temperature is 6,170° F (3,410° C). In addition to its high melting temperature, it has an even higher boiling temperature: 10,700° F (5,630° C). Tungsten emits a brilliant white light well before it reaches its melting temperature. This brilliant white light can be as bright as the arc itself. Tungsten's high boiling point is only slightly lower than the temperature of the arc, which means that, during the welding process, only small amounts of tungsten on the surface of the electrode may actually boil off and transfer across the arc. The transferring of the small part is called *tungsten erosion.*
- Has a desirable hardness: Tungsten has a hardness of 421 on the standard Brinell hardness test as compared with 111 for mild steel.

Tungsten is produced mainly by reduction of its oxide ore with hydrogen. Because tungsten is very hard and has a very high melting temperature, it is extremely difficult to weld it into usable shapes using common manufacturing techniques. It is necessary, therefore, to take powdered tungsten that has a purity of at least 99.95% and compress this powder into the desired shape. The densely packed powder is then heated in an inert-gas atmosphere at a temperature below its melting point but at a temperature at which its grained structure will grow together forming a usable ingot. This process is called *centering.* It is very similar to what happens to crushed ice if it is tightly compacted and sticks together without ever having melted. The tungsten ingot is then heated to increase its ductility and then is swaged and drawn through dies to produce electrodes. Graphite is used as a lubricant to facilitate the drawing process (Figure 5-2). After drawing, the tungsten wire is cut into standard lengths of 3, 6, 7, 12, 18, and 24 inches with blunt ends. It is available in a variety of diameters ranging from 0.010 to 0.250 inches (0.30 to 8.00 mm).

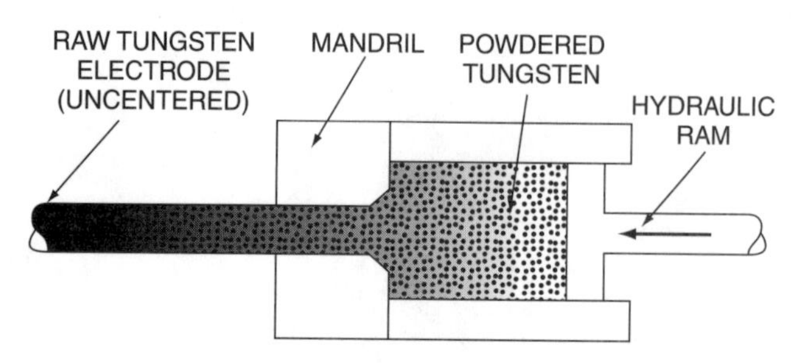

Figure 5-2 Extruding tungsten for electrodes.

The high melting temperature and good electrical conductivity make tungsten the best choice for a nonconsumable electrode. The arc temperature, around 11,000° F (6,000° C), is much higher than the melting temperature of tungsten but not much higher than its boiling temperature of 10,220° F (5,660° C).

As the tungsten electrode becomes hot, the arc between the electrode and the work will stabilize. Because electrons are more freely emitted from a hot tungsten, the highest possible temperature at the tungsten electrode tip is desired. Maintaining a balance between the heat required for a stable arc and that high enough to melt the tungsten requires an understanding of the GTA torch and electrode.

The thermal conductivity of tungsten and the heat input are prime factors in the use of tungsten as an electrode. In general, tungsten is a good conductor of heat. This conductive property allows the tungsten electrode to withstand the arc temperature well above its melting temperature. The heat of the arc is conducted away from the electrode's end so fast that it does not reach its melting temperature. For example, a wooden match burns at approximately 3,000° F (1,647° C). Because aluminum melts at 1,220° F (971° C), a match should easily melt an aluminum wire. However, a match will not even melt a 1/16″ (2-mm) aluminum wire. The aluminum, like a tungsten electrode, conducts the heat away so quickly that it will not melt.

Because of the intense heat of the arc, some erosion or loss of the tungsten electrode will occur. This eroded metal is transferred across the arc (Figure 5-3). Slow erosion of the electrode results in a limited amount of tungsten inclusions in the weld, which are acceptable. Standards and codes give the size and amount of tungsten inclusions that are allowable in various types of welds. The tungsten inclusions are hard spots that cause stresses to concentrate, possibly resulting in weld failure. Although tungsten erosion cannot be completely eliminated, it can be controlled. A few ways of limiting erosion include the following:

- Having good mechanical and electrical contact between the electrode and the collet
- Using as low a current as possible
- Using a water-cooled torch
- Using as large a size of tungsten as possible
- Using DCEN current

- Using as short an electrode extension from the collet as possible
- Using the proper electrode end shape
- Using an alloyed tungsten electrode

The torch end of the electrode is tightly clamped in a collet, which is the cone-shaped sleeve that holds the electrode in the torch. The collet inside the torch is cooled by air or water. Heat from both the arc and the tungsten electrode's resistance to the flow of current must be absorbed by the collet and torch. To ensure that the electrode is being cooled properly, be sure the collet connection is clean and tight. For water-cooled torches, make sure water flow is adequate.

Collet-tungsten connection efficiency is shown in Figure 5-4 and Figure 5-5.

Large-diameter electrodes conduct more current because the resistance heating effects are reduced. However, excessively large sizes may result in too low a temperature for a stable arc.

During the centering process, the hot tungsten is protected from oxidation by an inert-gas atmosphere. However, some contamination occurs and must be removed with acid. The surface that remains is slightly rough. Tungsten at this point is often sold and is called *cleaned tungsten.*

Some tungsten receives an additional finishing process called *centerless grind.* The centerless grinding process places the tungsten between two rotating

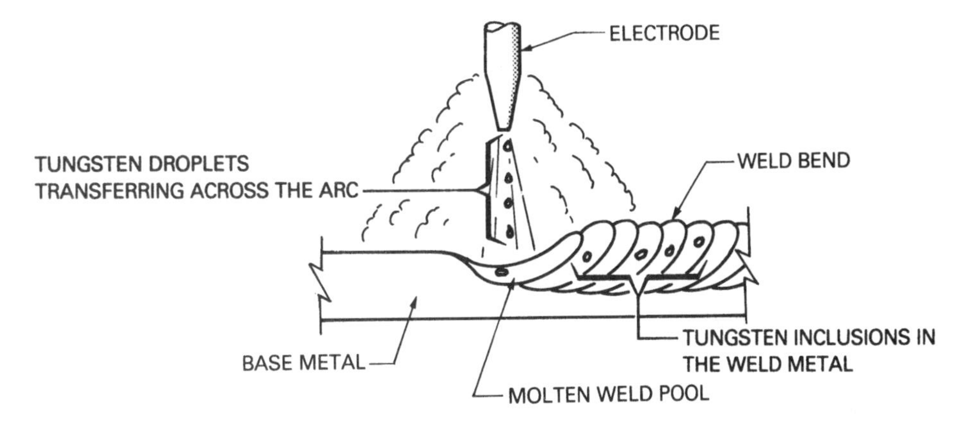

Figure 5-3 Some tungsten will erode from the electrode, be transferred across the arc, and become trapped in the weld deposit.

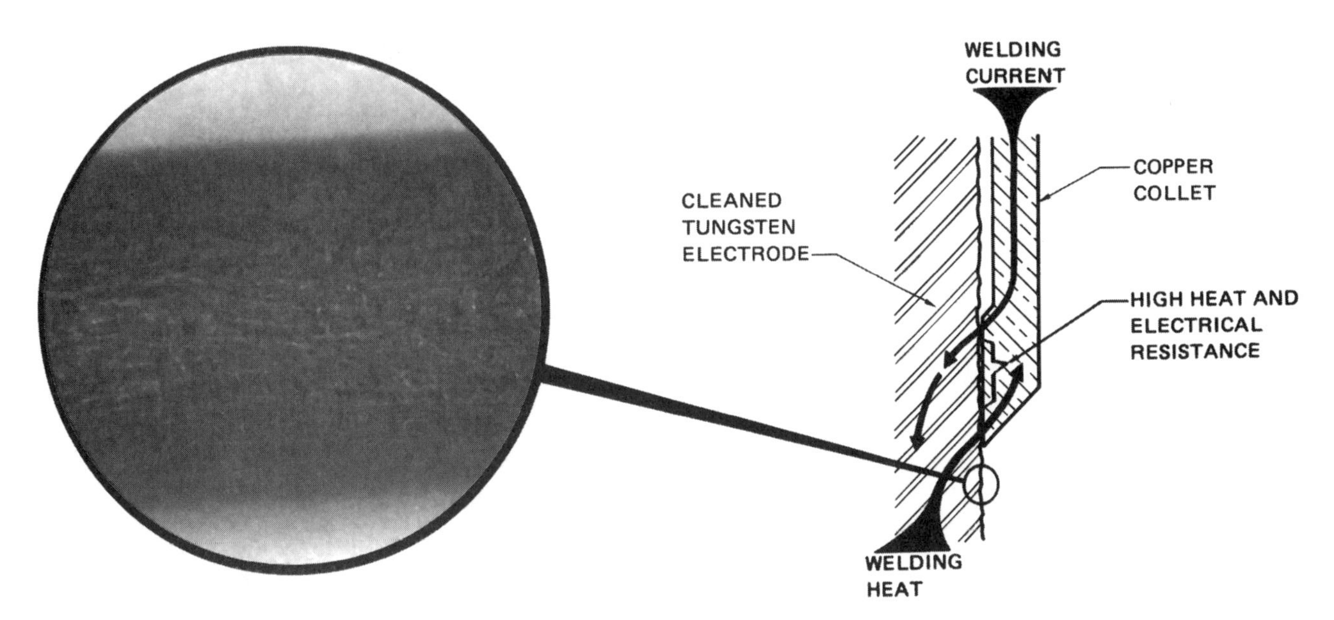

Figures 5-4 Irregular surface of a cleaned tungsten electrode (poor heat transfer to collet).

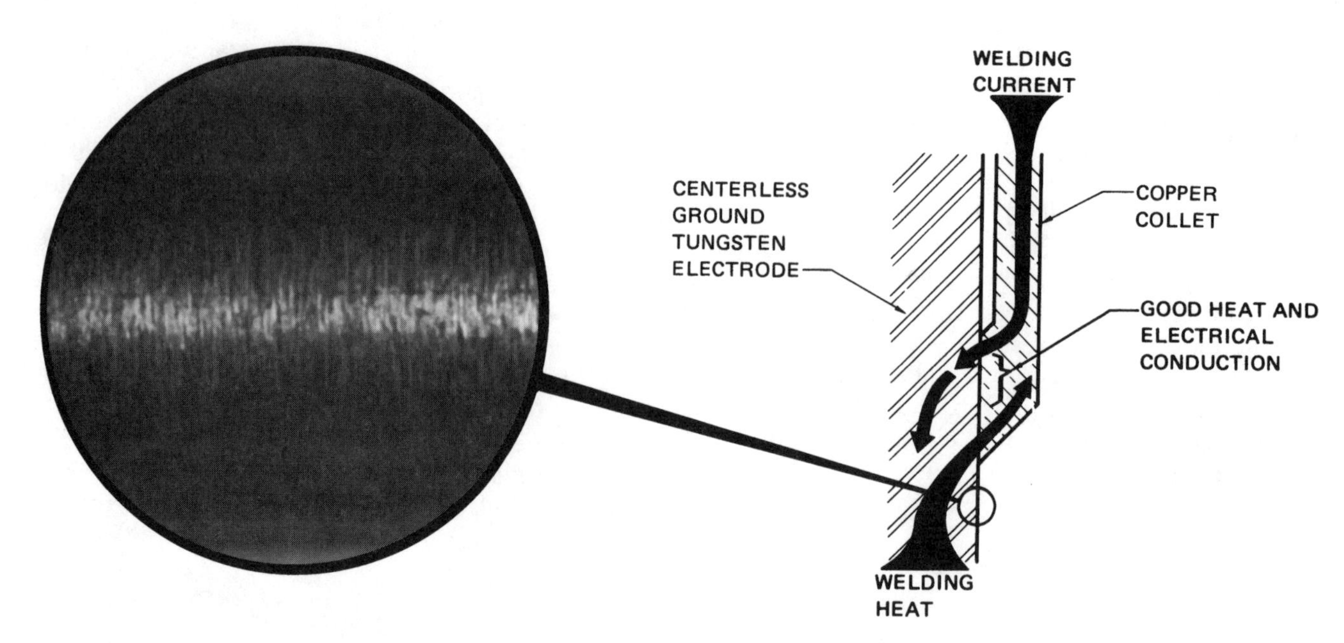

Figure 5-5 Smooth surface of a centerless ground tungsten electrode (good heat transfer to collet).

grinding stones. Tungsten produced in this way has a very smooth polished surface.

The tungsten in both the cleaned and centerless ground is exactly the same. The difference between cleaned and centerless ground tungsten is its electrical and thermal conductivity between the collet and tungsten surface during welding. Centerless ground tungsten has a much greater contact surface and therefore better conductivity (Figure 5-6). This improved conductivity both prevents heat buildup due to the high welding current and helps to conduct arc-generated heat away from the tungsten.

For the beginning welder, clean tungsten will work most adequately. It is slightly less expensive than centerless ground, and most beginning welders have much shorter arc-on times than production welders, thus eliminating excessively high tungsten temperatures.

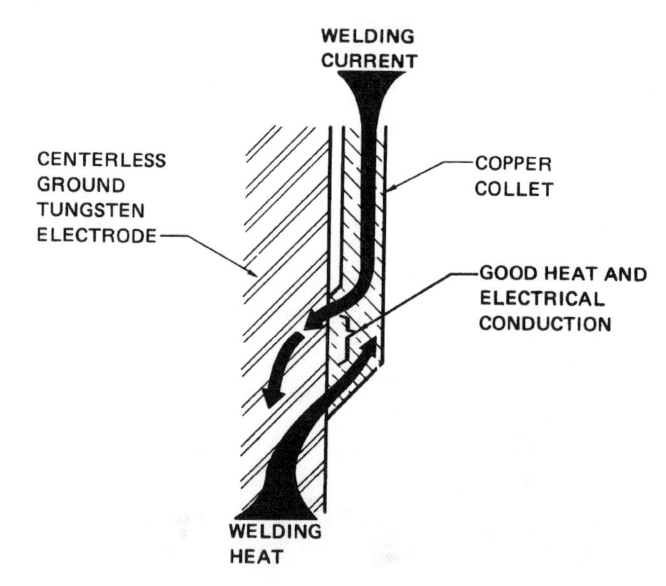

Figure 5-6 Electrical and thermal conductivity are both improved with a smooth, tight fit.

TYPES OF TUNGSTEN

The desirable properties of pure tungsten can be improved to make it an even better nonconsumable electrode. These properties can be improved by adding cerium (Ce), lanthanum (La), thorium (Th), or zirconium (Zr) to the tungsten.

For GTAW, tungsten electrodes are available in the following American Welding Society (AWS) classifications:

- Pure tungsten (EWP)
- 1% thorium tungsten (EWTh-1)
- 2% thorium tungsten (EWTh-2)
- 1/4% to 1/2% zirconium tungsten (EWZr)
- 2% cerium tungsten (EWCe-2)
- 1% lanthanum tungsten (EWLa-1)
- Alloys not specified or designated (EWG)

The type of finish on the tungsten is specified as cleaned or ground. For more information on composition and other manufacturing requirements and specifications for tungsten welding electrodes, refer to AWS publication A5.12, *Specifications for Tung-*

AWS CLASSIFICATION	TUNGSTEN COMPOSITION	TIP COLOR
EWP	Pure	Green
EWTh-1	1% thorium added	Yellow
EWTh-2	2%thorium added	Red
EWZr	1/4% to 1/2% zirconium added	Brown
EWCe-2	2% cerium added	Orange
EWLa-1	1% lanthanum added	Black
EWG	Alloy not specified	

Figure 5-7 Tungsten electrode types and identification.

sten and Tungsten Alloy Electrodes for Arc Welding and Cutting.

Tungsten electrodes are identified by a standardized color code (Figure 5-7). One end of each tungsten electrode is marked with the appropriate color. The color marking is made in such a manner that it does not increase the tungsten diameter, which might interfere with the mechanical contacts between the tungsten and collet during use.

The AWS has established a standard identification system for tungsten electrodes. The letter *E* in the classification stands for electrode. The *W* is the atomic symbol for tungsten, also called Wolfram. The last letter or letters refer to the alloy or oxide that has been added to the tungsten. The letter *P* denotes pure tungsten without alloying elements. The other designations that may be used include Th for thorium, Zr for zirconium, Ce for cerium, La for lithium, and the letter *G* for alloys that have not been standardized. A dash followed by the number 1 or 2 is used for some alloys to indicate the approximate percentages of the alloying element or oxide.

ELECTRODE SIZE

As they heat up, the tungsten electrodes used in gas tungsten arc welding improve its arc characteristics. Cold tungsten will carry a current; however, the arc between the tungsten and the work is not very stable. As the tungsten tip heats up and begins to glow red, the arc jumping from the tungsten's tip to the workpiece will begin to stabilize. Initially there may be an erratic sound as the arc heats the tungsten tip to a bright red temperature. Once the arc has stabilized, any erratic sound should have disappeared, and there may be a smooth, steady hum from the torch.

Maintaining a balance between having the tungsten provide a smooth arc yet not allowing excessive melting of the end of the tip will come with experience. Selecting an appropriately sized tungsten for the work will help in reaching this balanced temperature.

Excessively large electrodes will not have sufficient heat to stabilize the electron emission characteristic of the tungsten discussed earlier. Excessively small electrodes may result in extremely high surface temperatures at the point, and tungsten migration can be excessive as the erosion of the end of the electrode occurs. This erosion, or moving of tungsten across the arc destabilizes the arc in addition to showing up as impurities in the weld metal. The correct tungsten should allow for rapid arc stability on initial start-up yet not overheat to the extent that it erodes the tungsten tip.

Large tungsten will slowly stabilize the arc as additional power is applied, but the excessive current may cause weld burn-through or other root face defects. Small tungsten diameters may overheat before a large enough weld pool is established, resulting in inadequate penetration or even incomplete fusion.

A properly sized tungsten will allow the tip to become very hot while within a very short distance, the cooler center picks up the higher surface temperature and conducts the heat into the torch body (Figure 5-8). By having this type of temperature gradient from the very hot tip of the tungsten to the torch body, tungsten erosion or migration will be minimized. Tungsten erosion or migration occurs when small particles of tungsten that melt from the surface of the tungsten are transferred across the arc into the molten weld pool. Some tungsten migration will occur even under the most ideal conditions. The parts that transfer may be so small that they will be difficult to see on an X ray. However, excessive heat will cause these particles of tungsten to become much larger, and they will therefore show up as small white spots in an X ray (Figure 5-9).

Often beginning welders assume that such tungsten inclusions can be melted and will become dissolved into the base weld metal. This will not happen because, once the tungsten has been transferred into the molten weld pool, it immediately solidifies. It will not remelt even with the arc directed straight at its location. Remelting cannot occur because tungsten's melting temperature is several times greater than that of most base metals such as mild steel, stainless steel, or aluminum. The small tungsten inclusions are much denser than the base metal and will quickly sink to the bottom of the weld pool. Also, because of

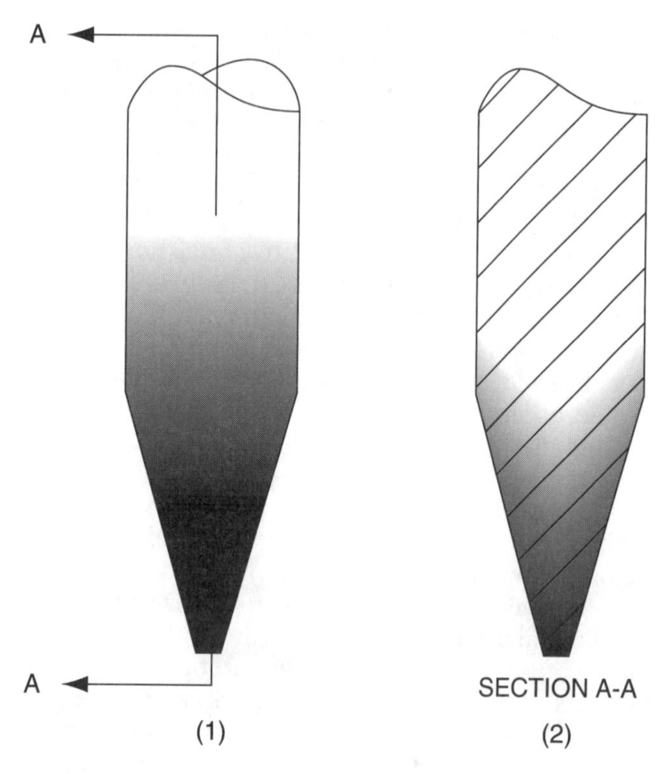

Figure 5-8 (1) External temperature, and (2) internal temperature.

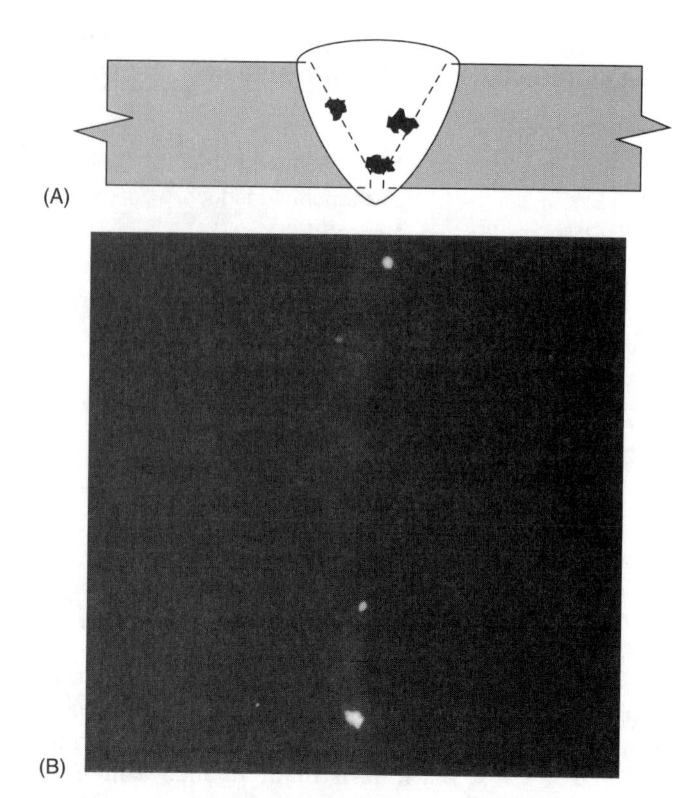

Figure 5-9 Welding defect: tungsten inclusions. Random bits of tungsten fused, but not melted, into weld metal. Radiographic image: irregularly shaped lower density spots randomly located in the weld image.

tungsten's higher density, fewer X rays pass through it, and it therefore appears as white spots on the X-ray negative.

TUNGSTEN END SHAPES

The blunt end of a new tungsten electrode must be shaped for welding. The three shapes correspond with specific welding currents. These shapes and currents are tapered point for DCEN, rounded end for AC, and tapered with a small rounded end for DCEP.

Tapered Tungsten

Tapered tungsten tip or pointed-end shape is used with DC electrode negative (DCEN), sometimes referred to as straight polarity (Figure 5-10). The taper may end in a point or have a small flat spot at the end, and its length is usually three times the tungsten's diameter (Figure 5-11). The angle of the point will effect the weld bead's penetration. A longer taper will produce narrower weld beads with deeper penetration. Blunter angled tapers will produce wider weld beads with shallower penetration (Figure 5-12).

The point of the tungsten will become very hot, whereas only a short distance back from the tip, the tungsten is much cooler. The two major advantages of this tungsten shape are that the end of the tungsten will become hot enough to freely emit electrons, whereas the remainder of the tungsten is cool enough to slow tungsten erosion.

Rounded End

The rounded-end tungsten is used with AC (Figure 5-13). The rounded end of the tungsten should be only slightly larger than the diameter of the tungsten (Figure 5-14).

The rounded end of the tungsten allows it to withstand the higher heat generated during the DCEP phase of the AC. The larger area of the bald end also provides for improved arc stability for arc reignition during the AC cycles. Surface tension holds the molten tungsten in place even though there is a substantial quantity of molten tungsten.

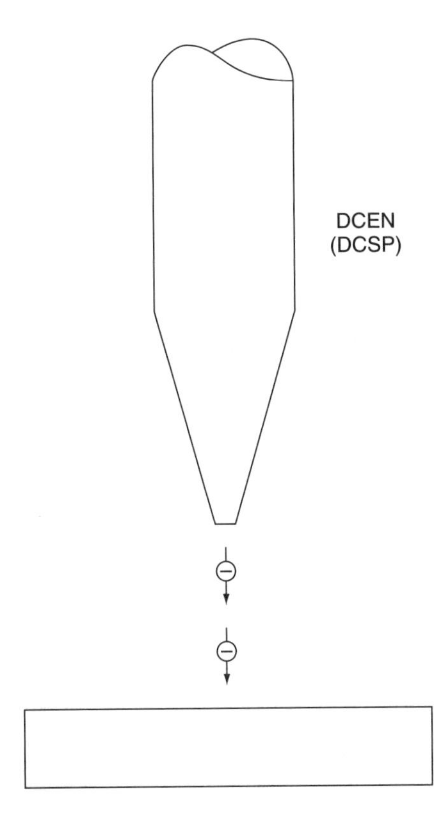

Figure 5-10 Electron flow for DCEN (DCSP).

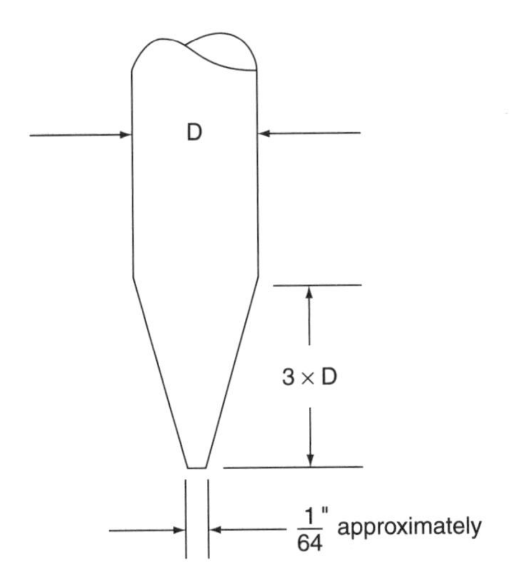

Figure 5-11 DCEN tungsten tip shape.

Tapered With a Rounded End

This somewhat unusual end shape is used for direct current electrode positive (DCEP), sometimes referred to as reverse polarity (Figure 5-15). DCEP welding current has approximately 2/3 of the heat concentrated on the tungsten. This end shape allows the tungsten to withstand that level of heat without having excessive tungsten erosion. As the welding current increases, the size of the rounded tip will increase. Surface tension draws it back up toward the larger mass of tungsten. The welder can use this as a visual que, indicating when the tungsten tip is becoming excessively hot. If the tungsten does become overheated, a larger-sized tungsten can be used if DCEP current is required for the welding process.

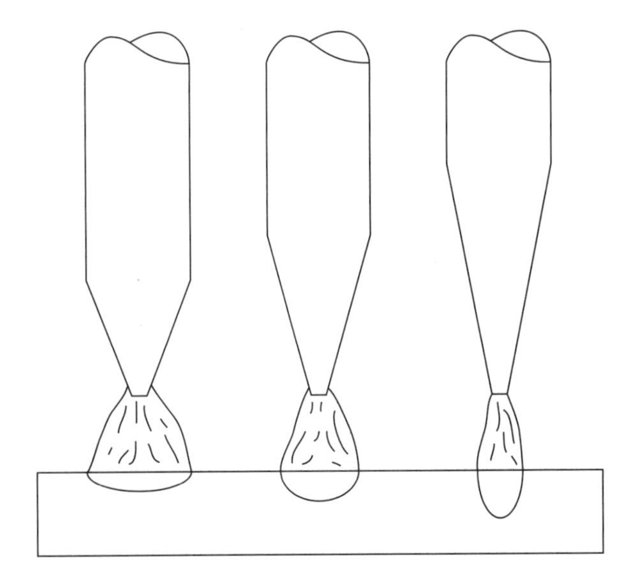

Figure 5-12 The same amount of heat is produced with each of the tungsten shapes, but the longer the taper, the more concentrated the arc's heat becomes.

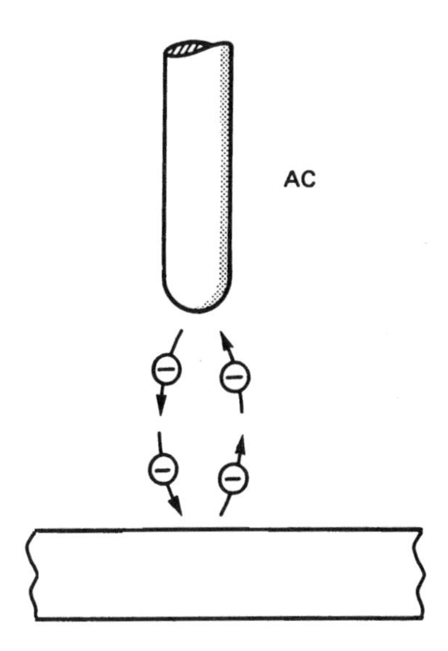

Figure 5-13 Electron flow for AC.

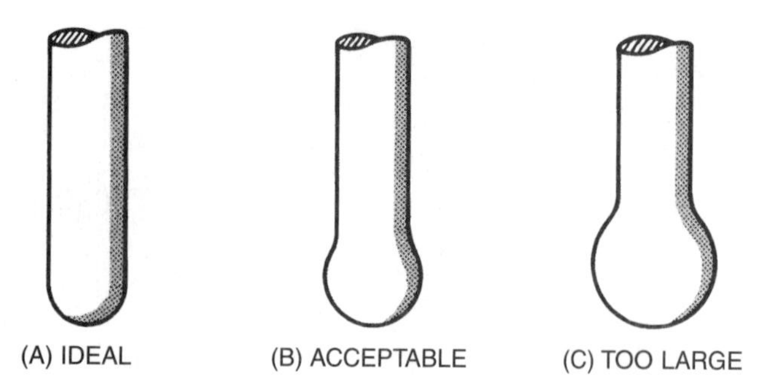

Figure 5-14 Melting the tungsten end shape.

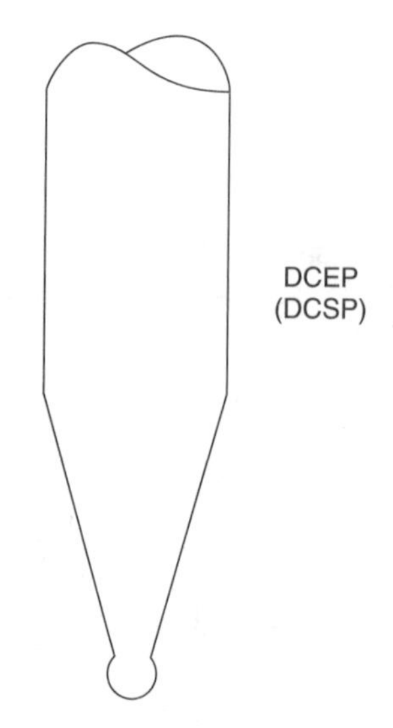

Figure 5-15 DCEP tungsten tip shape.

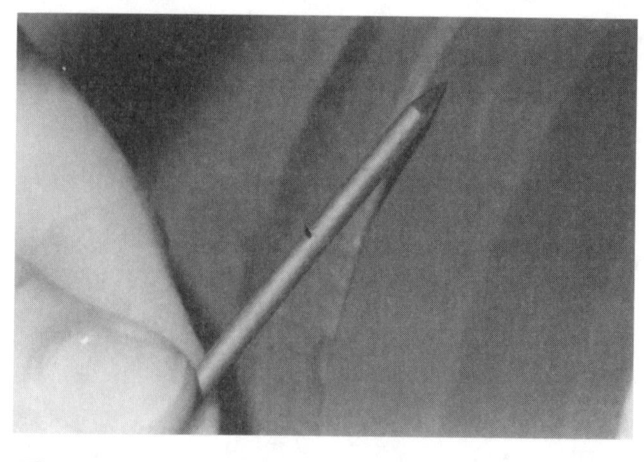

Figure 5-16 Correct method of grinding a tungsten electrode.

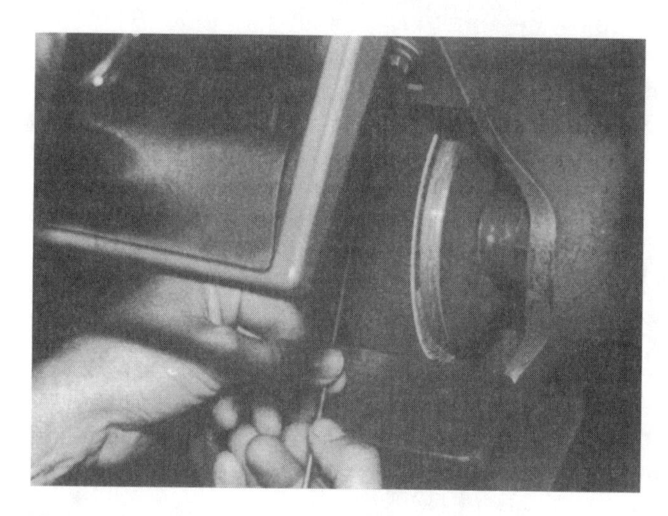

Figure 5-17 Correct way of holding a tungsten when grinding.

SHAPING THE TUNGSTEN

The desired end shape of a tungsten can be obtained by grinding, breaking, and remelting the end or by adding chemical compounds. Tungsten is brittle and easily broken. Welders must be sure to make a smooth, square break where they want to locate it.

Grinding

A grinder is often used to clean a contaminated tungsten or to point the end of a tungsten (Figure 5-16 and Figure 5-17). The grinding stone used to sharpen tungsten should have a fine, hard stone and it should be used for grinding tungsten only. Because of the hardness of the tungsten and its brittleness, the grinding stone chips off small particles of the electrode. A coarse grinding stone will result in more tungsten breakage and a poorer finish. If the grinder is used for metals other than tungsten, particles of these metals may become trapped on the tungsten as it is ground. The metal particles will quickly break free when the arc is started, resulting in contamination. Never grind the tungsten tip so that the grind marks would go around the tip (Figure 5-18).

Breaking and Remelting

Tungsten is hard but brittle, resulting in a low impact strength. If tungsten is struck sharply, it will break without bending. When it is held against a sharp corner and hit, a fairly square break will result. Figure 5-19, Figure 5-20 and Figure 5-21 show how to break the tungsten correctly on a sharp corner using two pliers or using wire cutters.

Once the tungsten has been broken squarely, the end must be melted back so that it becomes somewhat rounded. This is accomplished by switching the welding current to DCEP and striking an arc under argon

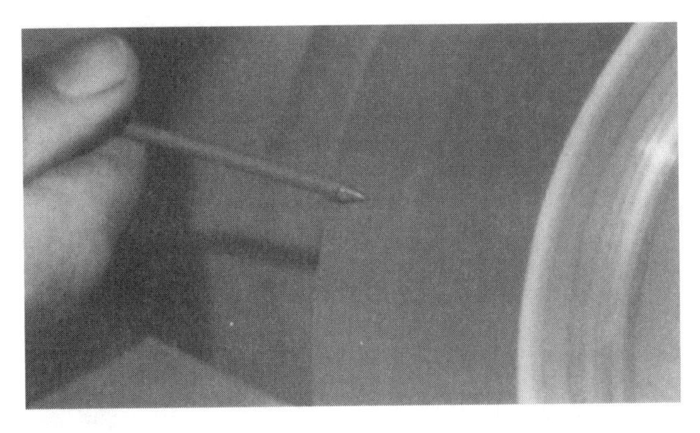

Figure 5-18 Incorrect method of grinding a tungsten electrode.

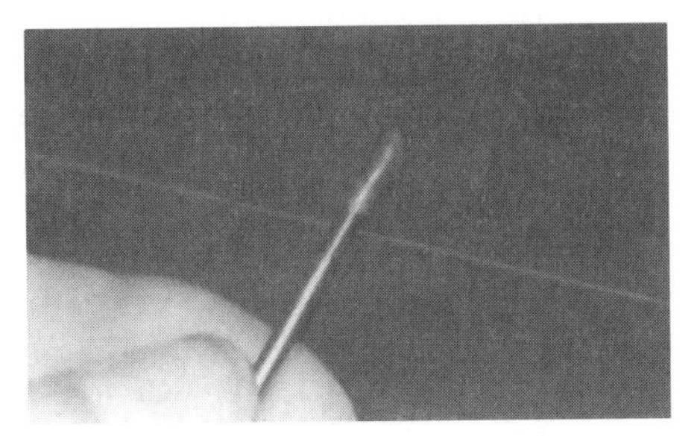

STEP 1

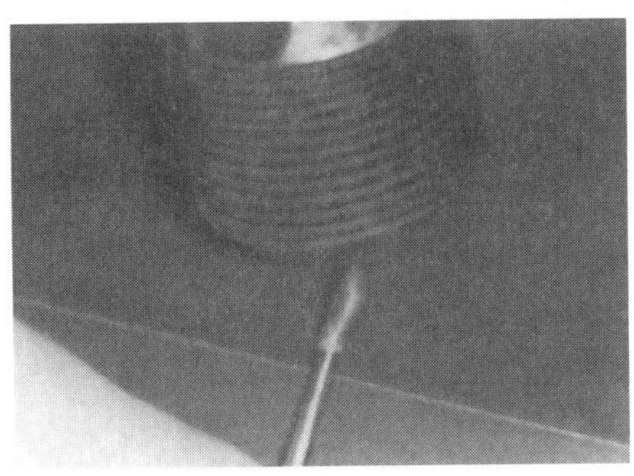

STEP 2

STEP 3

Figure 5-19 Breaking the contaminated end from a tungsten by striking it with a hammer.

Figure 5-20 Correctly breaking the tungsten using two pairs of pliers.

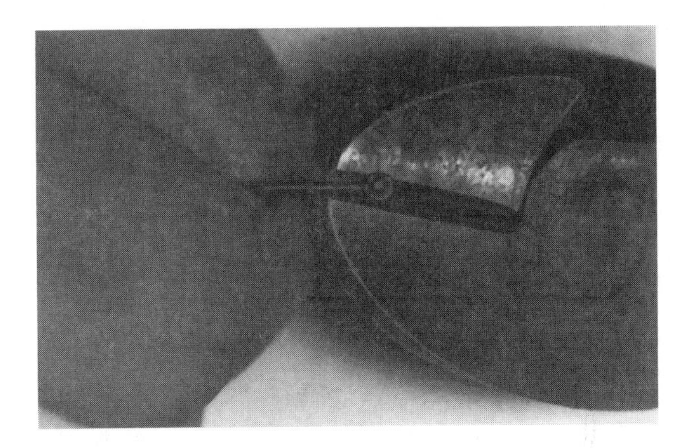

Figure 5-21 Using wire cutters to correctly break the tungsten.

shielding on a piece of copper. If copper is not available, another piece of clean metal can be used. Do not use carbon as it will contaminate the tungsten.

Chemical Cleaning and Pointing

The tungsten can be cleaned and pointed using one of several compounds. The tungsten is heated by shorting it against the work. The tungsten is then dipped in the compound, a strong alkaline, which rapidly dissolves the hot tungsten. The chemical reaction is so fast that enough additional heat is produced to keep the tungsten hot (Figure 5-22). When the tungsten is removed, cooled, and cleaned, the end will be tapered to a fine point. If the electrode is contaminated, the chemical compound will dissolve the tungsten, allowing the contamination to fall free.

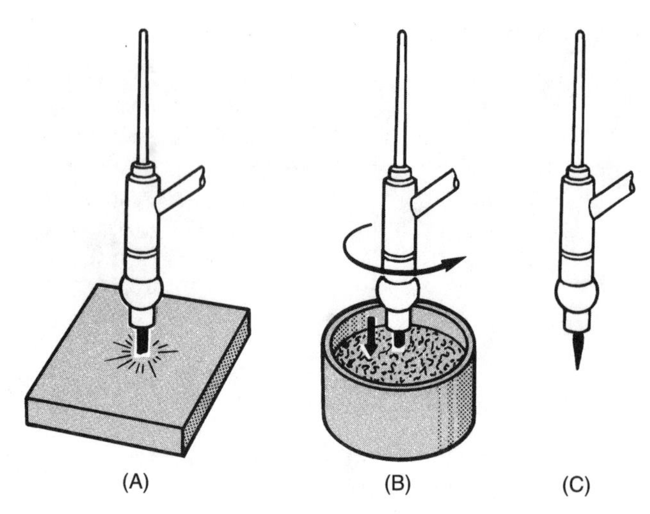

Figure 5-22 Chemically cleaning and pointing tungsten: (A) shorting the tungsten against the work to heat it to red hot, (B) inserting the tungsten into the compound and moving it around, and (C) cleaned and pointed tungsten ready for use.

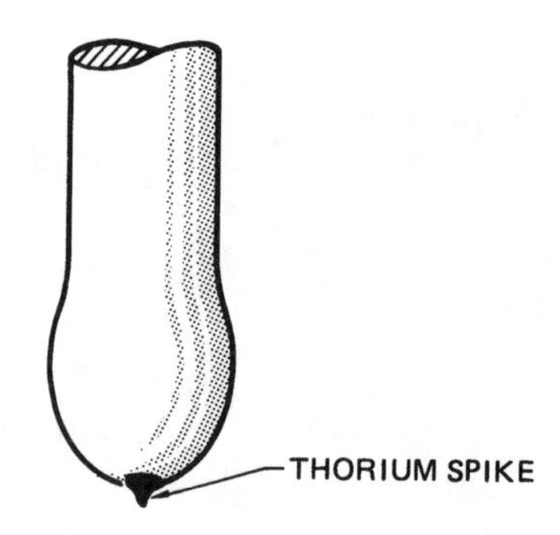

Figure 5-23 Thorium spike on a balled tungsten electrode.

Pointing and Remelting

The tapered tungsten with a balled end, a shape sometimes used for DCEP welding, is made by first grinding or chemically pointing the electrode. Using DCEP, as in the procedure for the remelted broken end, strike an arc on some copper under argon shielding and slowly increase the current until a ball starts to form on the tungsten. The ball should be made large enough so that the color of the end stays between dull red and bright red. If the color turns white, the ball is too small and should be made larger. To increase the size of the ball, simply apply more current until the end begins to melt. Surface tension will pull the molten tungsten up onto the tapered end. Lower the current and continue welding. If the tip is still too hot, it may be necessary to increase the size of the tungsten. DCEP is seldom used for welding.

Pure Tungsten (EWP)

Pure tungsten has the lowest heat resistance and electron emission characteristic of all the tungsten electrodes used for GTAW. It has a limited use with AC welding currents on metals, such as aluminum and magnesium.

Thoriated Tungsten (EWTh-1 and EWTh-2)

Thorium oxide (ThO_2), when added in percentages up to 0.6% to tungsten, improves its current-carrying capacity. The addition of 1% to 2% of thorium oxide does not further improve current-carrying capacities; it does however help with electron emission. This can be observed by a reduction in the electron force (voltage) required to maintain an arc of a specific length. Thorium also increase the serviceable life of the tungsten. The improved electron emission of the thoriated tungsten allows it to carry approximately 20% more current. This also results in a corresponding reduction in electrode tip temperature, resulting in less tungsten erosion and subsequent weld contamination.

Thoriated tungstens also provide a much easier arc starting characteristic than pure or zirconiated tungsten. Thoriated tungstens work well with DC electrode negative (DCEN) and can maintain a sharpened point well. They are very well suited for making weld on steel, steel alloys (including stainless), nickel alloys, and most other metals other than aluminum or magnesium.

Thoriated tungsten does not work well with AC. It is difficult to maintain a bald end, which is required for AC welding. A thorium spike (Figure 5-23) may also develop on the bald end, disrupting a smooth arc.

Thorium is a very low-level-radioactive oxide. The level of radioactive contamination from thorium electrodes however, has not been found to be a health hazard during welding. It is, however, recommended that grinding dust be contained. Because of concern in other countries regarding radioactive contamination to the welder and welding environment, thoriated tungstens have been replaced with other alloys.

Zirconium Tungsten (EWZr)

Zirconium oxide (ZrO_2) also helps tungsten emit electrons freely. The addition of zirconium to the tungsten has the same effect on the electrode characteristic as thorium but to a lesser degree. Because zirconium tungstens are more easily melted than thorium tungstens, ZrO_2 electrodes can be used with both AC and DC.

Because of the ease in forming the desired bald end on thorium (versus zirconium tungstens), they are normally the electrode chosen for AC welding of aluminum and magnesium alloys. Zirconiated tungstens are more resistant to weld pool contamination than pure tungsten, thus providing excellent weld qualities with minimal contamination.

Zirconiated tungstens also have the advantage over thoriated tungstens in that they are not radioactive.

Cerium Tungsten (EWCe-2)

Cerium oxide (CeO_2) is added to tungsten to improve the current-carrying capacity in the same manner as thorium does. These electrodes were developed as replacements for thoriated tungstens because they are not made of a radioactive material. Cerium oxide electrodes have a similar current-carrying capacity to that of pure tungsten; however, they have improved arc-starting and arc-stability characteristics similar to those of thoriated tungstens. They can also provide a longer life than most other electrodes, including thorium.

Cerium tungsten electrodes have a slightly higher arc voltage for a given length than does thoriated tungsten. This very slight increase in voltage does not cause problems for manual welding. The higher voltage, however, may require that new weld tests be performed to requalify welding procedures. Cerium tungsten may be used for both AC and DC welding currents. Cerium electrodes contain approximately 2% of cerium oxide.

Lithium Tungsten (EWLa-1)

Lithium oxide (La_2O_3) in about 1% concentration is added to tungsten. Lithium oxide tungstens are not radioactive. They have similar current-carrying characteristics to the thorium tungstens, except that they have a slightly higher arc voltage than do thorium or cerium tungstens. This does not normally pose a problem for manual arc welding; however, it will usually require that new test plates be produced to recertify weld procedures.

Alloy Not Specified Tungsten (EWG)

This classification is for tungstens whose alloys have been modified by manufacturers. Such alloys have been developed and tested by manufacturers to meet specific welding criteria. Specific alloy compositions are not normally available from manufacturers; however, they do provide welding characteristics for these electrodes.

REVIEW QUESTIONS

1. What are the major properties of tungsten that make it a good nonconsumable electrode metal?
2. What is an advantage of tungsten's resistance to oxidation?
3. What is the approximate melting temperature of tungsten?
4. What is the boiling temperature of tungsten?
5. What allows the tungsten electrode to withstand the arc temperature well above its melting temperature?
6. How can tungsten erosion be controlled?
7. What is cleaned tungsten?
8. What is centerless grind?
9. What is the advantage of using centerless ground tungsten as opposed to using cleaned tungsten?
10. List six different types of tungstens used for GTAW.
11. What does the letter *E* in the tungsten classification system stand for?
12. What is the disadvantage of having an excessively large electrode?
13. What is a properly sized tungsten?
14. What is tungsten erosion or migration?
15. Why must a grinding stone used to sharpon tungsten not be used with other metals?
16. Which one of the types of tungsten has the lowest heat resistence?
17. Because they are more easily melted, which tungsten can be used with both AC and DC?
18. What are the advantages of zirconiated tungstens over thoriated tungstens?
19. Which has a higher arc voltage: cerium or thoriated tungsten?
20. When would the tungsten designation EWG be used?

CHAPTER 6

SHIELDING GAS AND FLOW METERS

INTRODUCTION

An inert, or nonreactive, gas must be provided through the torch to surround the hot tungsten and weld pool to prevent atmospheric contamination. Both the type of gas and the flow rate affect the quality of the weld produced.

FLOWMETER

Shielding gas flow is controlled by a flowmeter. Flowmeters may be simply a flowmeter or a combination flowmeter/pressure regulator (Figures 6-1 and Figure 6-2). Combination flowmeter/regulators can be connected directly to high-pressure gas cylinders. Flowmeters, however, must be connected to a manifold system that has a controlled working pressure within the designed operating pressure of the flowmeter. The working pressure for a flowmeter is marked on the instrument (Figure 6-3).

Flowmeters can be divided into two general types based on their methods of displaying the flow rate. One method uses a dial, from wich the flow rate is read directly. The other type uses a small ball that floats on top of a column of flowing gas. The flow is metered or controlled on both types of flowmeters by opening a small valve on the meter's base. The flow rate is read in units of cubic feet per hour (cfh) or liters per minute (L/min).

The floating-ball-type metering device is read by comparing the small floating ball in the glass tube to a fixed scale. Meters from various manufacturers may

Figure 6-1 Flowmeter. (Courtesy of Concoa Controls Corporation of America)

be read differently. For example, they may read from the top, center, or bottom of the ball (Figure 6-4). The ball floats on top of a stream of gas inside a tube. The inner tube's inside diameter is smaller at the bottom and larger at the top. The increased size allows for more room for the gas to flow past the ball. It therefore takes a larger volume of gas flowing to force the ball to rise higher in the tube (Figure 6-5). If the tube

Figure 6-2 Flowmeter regulator. (Courtesy of Concoa Controls Corporation of America)

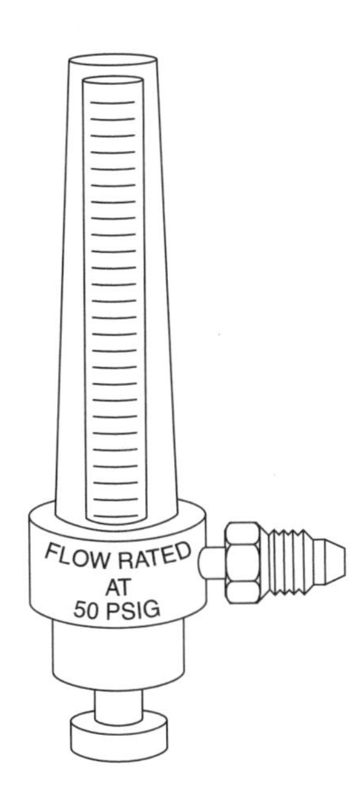

Figure 6-3 Flowmeters must be supplied with a specific pressure to operate correctly.

is not vertical, the reading is not accurate, but the flow rate will remain unchanged. Also, when using a flowmeter connected to a manifold, it is important to have the correct pressure because changes in pressure will affect the accuracy of the flow reading. To get accurate readings, be sure the gas being used is read on the proper flow meter scale (Figure 6-6). Less dense gases, such as helium and hydrogen, will not support the ball on as high a column with the same

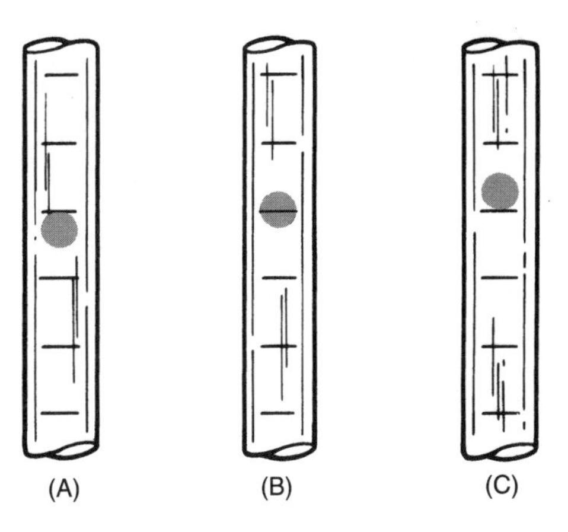

Figure 6-4 Three methods of reading a flowmeter: (A) top of ball, (B) center of ball, and (C) bottom of ball.

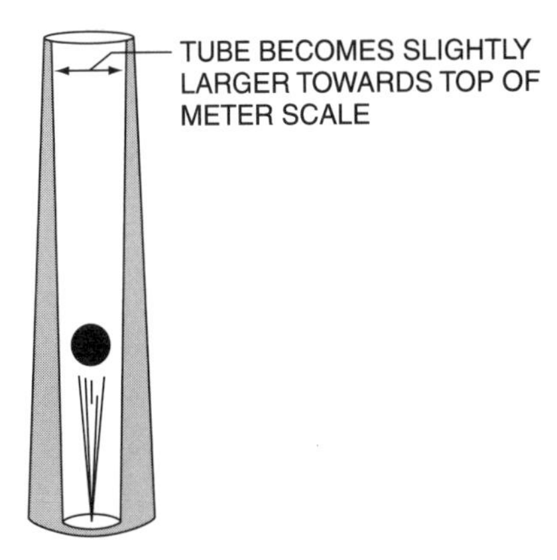

Figure 6-5 The Flowmeter ball floats on top of a stream of shielding gas.

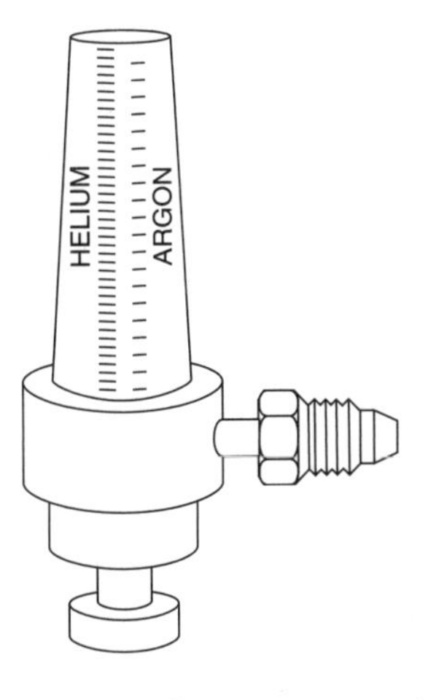

Figure 6-6 Dual-scale flowmeter regulator.

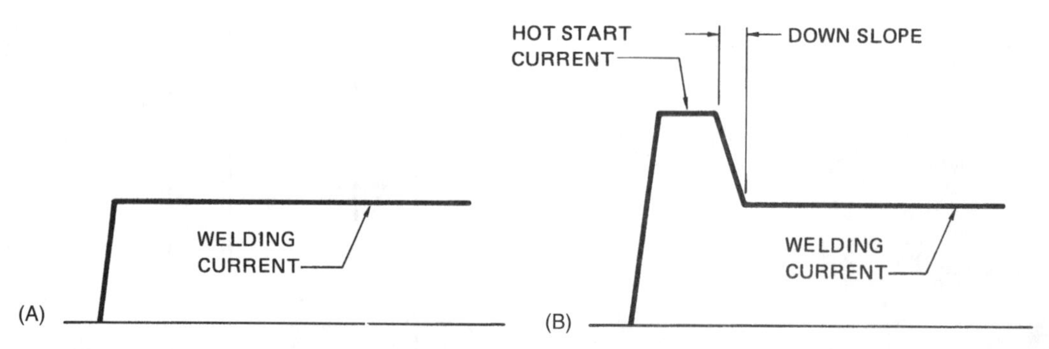

Figure 6-7 (A) Standard method of starting welding current; (B) hot start method of starting welding current.

flow rate as a more dense gas, such as argon. Flowmeters designed to be used with multiple gases will have multiple scales, one for each gas.

Hot Start

The hot start allows a controlled surge of welding current, as the arc is started, to establish a molten weld pool quickly. Establishing a molten weld pool rapidly on metals with a high thermal conductivity is often hard without this higher-than-normal current. Adjustments can be made in the length of time and the percentage above the normal current (Figure 6-7).

SHIELDING GASES

The shielding gases used for the GTA welding process are argon (Ar), helium (He), hydrogen (H), nitrogen (N), or a mixture of two or more of these gases. In addition to protecting the molten weld pool from harmful effects of the atmosphere, the shielding gas also affects the amount of heat produced by the arc and the resulting weld bead appearance.

Argon and helium are noble, or inert, gases. This means that they will not combine or react chemically with any other material. Both argon and helium may not be found in any compound but exist only as mixtures. Examples of compounds are water (H_2O), acetylene (C_2H_2), and carbon dioxide (CO_2). Compounds are made up of elements that have a combined fixed ratio to produce a unique material. Mixtures, however, can exist as any ratio and may be easily separated. Therefore, because they are inert, they will not affect the molten weld pool in any way.

Shielding gas purity is critical to weld quality. Argon for example is supplied at 99.999% pure. Because of the very high level of purity required of the shielding gas, never allow noninert gases such as O_2, CO_2, or nitrogen to come in contact with your inert gas system. Very small amounts will contaminate the inert gas in the system, which may result in the weld failing.

Argon

Argon is a by-product of oxygen production in air separation plants. Air is taken into these plants and cooled to a temperature that causes it to condense into a liquid. The liquid is a mixture of all of the gases that make up air, which are oxygen (20.9%), nitrogen (78.0%), with less than 1% of other gases, including argon. This mixture is then fractionally distilled. The process of fractional distillation results in the various products vaporizing at different specific temperatures, allowing them to be captured individually. Although argon represents a very small percentage of the total air, substantial quantities of it are produced because this process is undertaken on large scales. The argon produced in this manner is distributed in cylinders as gas or in containers as a liquid.

Because argon is denser (heavier) than air, it efficiently shields welds in most positions. However, its higher density may be a problem when welding in the overhead position. Slightly higher fluorides may be necessary at this position.

Argon is relatively easy to ionize because a lower arc voltage is required for argon-shielded welds. This characteristic makes starting arcs easier and thus suitable for AC applications. This property also results in a very slight change in arc voltage with the change in arc length. Changes in arc voltage affect the quantity of heat produced by the arc and therefore also affect the weld that is produced. The stable arc voltage lends itself very well to manual GTAW, where arc lengths may vary as filler metals are added (Figure 6-8).

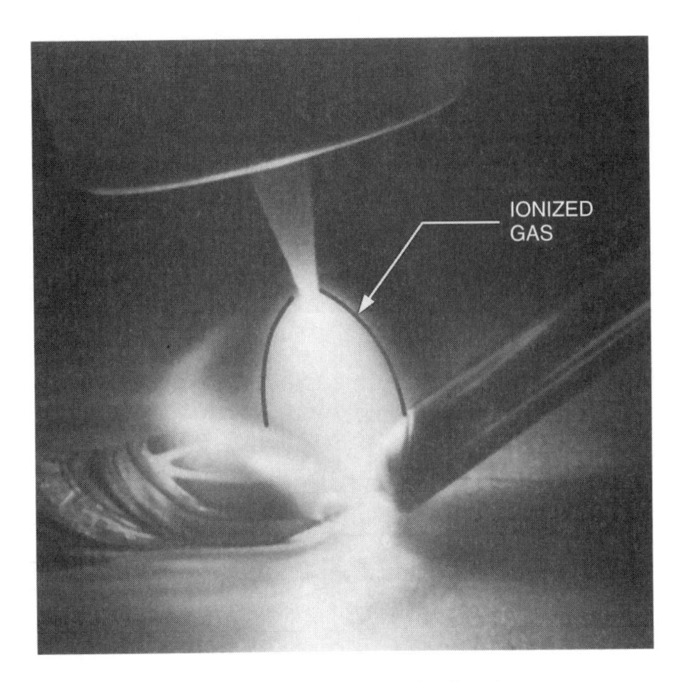

Figure 6-8 Highly concentrated ionized argon gas column.

Helium

Helium is a by-product of the natural gas industry. It is removed from natural gas during the refinement of natural gas. Helium has a very high ionization potential, which means that the arc in helium must have a high voltage. Welds produced with helium have the advantage of much deeper penetration due to the higher heat produced by the higher arc voltage. The arc force with helium is sometimes sufficient to displace the molten weld pool when very short arcs are used (Figure 6-9). In some mechanical applications, therefore, the tip of the tungsten can be positioned below the workpiece surface to obtain very deep and narrow weld beads. This technique is especially effective for welding aged-hardened aluminum alloys prone to excessive hardening as a result of welding. It also is very effective at high welding speeds such as those used for producing tubing.

Because of the significant change in arc voltage relative to arc length, helium is less forgiving for manual welding. Weld bead penetration and weld bead profile are very sensitive to arc length changes. Because it is very difficult to maintain an unchanging arc length when manual welding, filler metal wire is added.

Helium has been mixed with argon to gain the combined benefits of both gases. The most common of these mixtures is 75% helium and 25% argon.

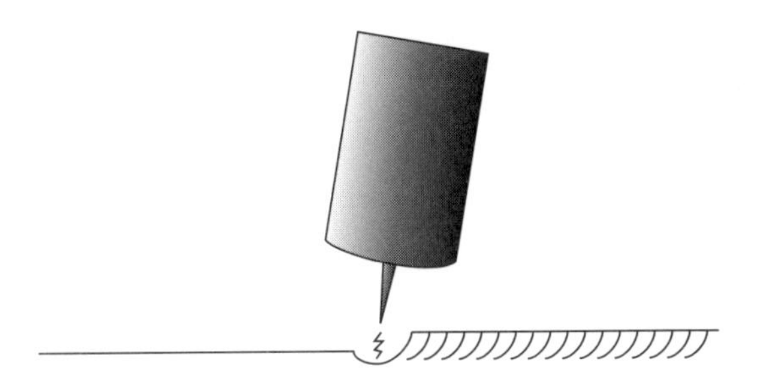

Figure 6-9 Helium's arc force can cause the molten weld pool to be depressed.

Although the GTA process was developed with helium as the shielding gas, argon is now used whenever possible because it is much cheaper. Helium also has the disadvantage of being much lighter than air; therefore higher flow rates may be required to provide good shielding. Its flow rate must often be twice as high as argon's for a similar application. Welding with helium is also prone to disruption due to drafts.

Because the AC voltage cycles back and forth 16 times per second, it produces a very unstable arc with helium shielding. Helium, therefore, is not used with alternating current.

Hydrogen

Hydrogen is not an inert gas and is not used as a primary shielding gas. However, it can be added to argon when deep penetration and high welding speeds are needed. It also improves the welding surface cleanliness and bead profile on some grades of stainless steel that are very sensitive to oxygen. Hydrogen–argon mixtures are restricted to stainless steels welds because hydrogen may be dissolved into molten, aluminum, steel, or other alloys. If this occurs, the hydrogen can cause porosity and, in the case of steels, may result in underbead cracking (Figure 6-10).

Nitrogen

Nitrogen is not an inert gas; however, it may be added to argon. But it cannot be used with some materials, such as ferritic steels because it produces porosity. In other cases, such as with Austenitic stainless steels, nitrogen is helpful as an Austenite stabilizer in these alloys. Nitrogen is used to increase penetration when welding on copper. Unfortunately, because of the potential metallurgical problems, nitrogen has not received wide acceptance as a GTA welding shielding gas.

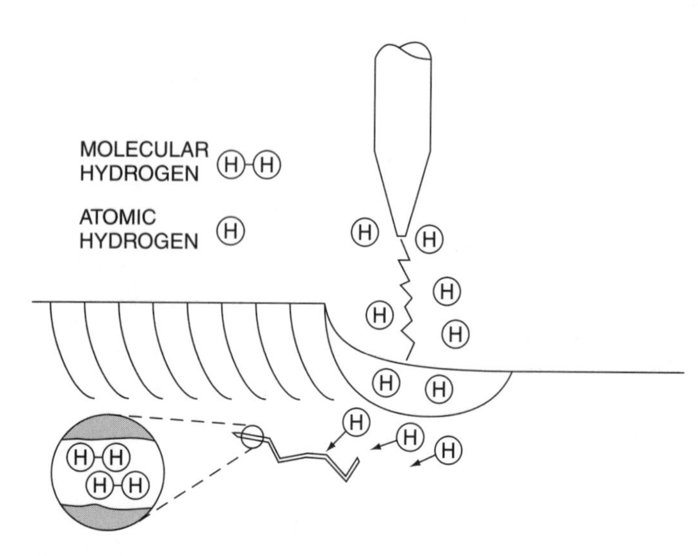

Figure 6-10 Atomic hydrogen can dissolve into molten metal and then collect in small cracks, causing them to grow larger.

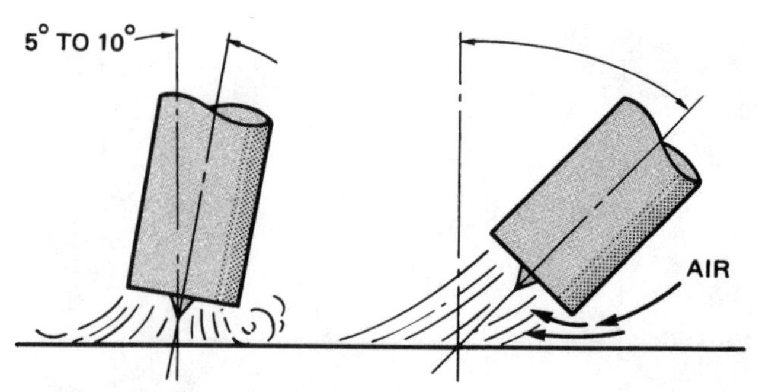

Figure 6-11 Too steep an angle between the torch and work may draw in air.

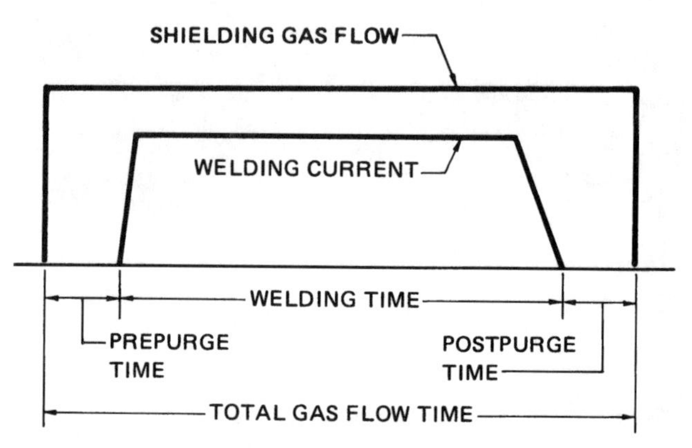

Figure 6-12 Typical welding cycle with pre- and post-gas purging.

GAS FLOW RATES

The weld quality can be adversely affected by improper gas flow settings. The rate of flow should be as low as possible and still give adequate coverage. High gas flow rates waste shielding gas and may lead to contamination, which comes from turbulence produced in the gas at these rates. Air is drawn into the gas envelope by a venturi effect around the edge of the nozzle. Also, the air can be drawn in under the nozzle if the torch is held at too sharp an angle to the metal (Figure 6-11).

Various factors affect the shielding gas flow rate, including the following:

- Nozzle size
- Tungsten size
- Amperage
- Joint design
- Welding position
- Metal type

Additionally, wind speed or drafts will affect gas flow rates. Ideally GTAW should never be performed in windy or drafty areas; however, this is not always possible if wind or draft speeds are excessive. An air curtain should be placed upwind from the weld to reduce or prevent their effects. Under most circumstances, GTA welding should never be performed if the wind speeds are greater than 5 miles per hour.

PREPURGE AND POSTPURGE

Purging is the process of allowing shielding gas to flow before or after the actual weld. Prepurging is the process of allowing the gas to flow before welding begins. This allows the shielding gas to remove any unwanted atmosphere around the weld before the arc begins.

Some machines have built-in timers that allow a specific amount of time for purging to occur before welding current is supplied to the torch. Depending on the machine, the prepurge dial may be marked in seconds or merely divided into ten or more uniform spaces (Figure 6-12). The specific length of prepurging time is usually established by the code or standard. If no prepurged time has been established, 2–5 seconds is generally adequate; however, this will be affected by the nozzle diameter, flow rate, and other welding conditions. The shortest possible time is most desirable because it will waste less shielding gas. This time can be established for your specific welding application through trial-and-error testing on scrap plates.

Some machines do not have prepurge, and many welders find it hard to hold a position while waiting for the current to start. One solution to this problem is to use the postpurge for prepurging. Switch on the current to engage the postpurge. Now, with the current off, the gas is flowing, and the GTA torch can be lowered to the welding position. The welder's helmet should be lowered and the current restarted before the postpurge stops. This allows the welder to prepurge and to start the arc when the welder is ready.

The postpurge is also established by a timer that allows the gas to continue flowing after the welding current has stopped. This period serves to protect both the hot welding metal and the tungsten electrode until they have cooled to a temperature below which they will not rapidly oxidize. The time of the flow is determined by the welding current and tungsten size.

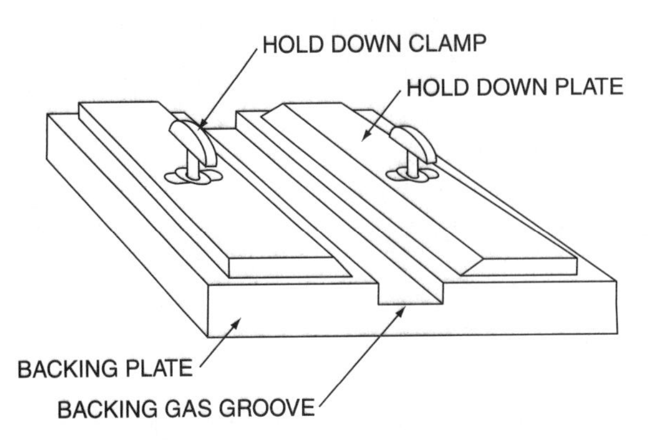

Figure 6-13 Jig for holding plates for welding that has a groove for backing gas coverage.

BACKING GAS

The hot exposed surface of the root of a weld may become contaminated by the surrounding atmosphere. This is not normally a problem when welding on most mild steels; however, severe oxidation may occur on the root of low alloyed steels, stainless steels, aluminum, copper, titanium, and other highly reactive metals. Reactive metals are metals that are easily oxidized at high temperatures. The most common method of protecting the root of a weld from atmospheric contamination is to use a backing gas.

The type of gas used for backing will depend upon the type of material being welded. Nitrogen and CO^2 are often acceptable, depending upon the code or intended use of the weldment. Argon, although expensive, can be used to back any type of material being welded and should be selected if the welder is unsure about a less expensive substitute.

Several methods are available for providing backing gas for plates. The most common is to use a strip of metal, usually copper, that has been grooved in the center to allow for shielding gas (Figure 6-13). This strip is securely clamped to the root side of the weld, and the appropriate gas is allowed to flow continuously during the welding process.

Several methods are in common use for containing the backing gas for pipe welds. On a small diameter of short sections of pipe, the ends of the pipe are capped (Figure 6-14). Gas is allowed to purge the complete pipe section. This method requires too much purging time and gas to be practical on large-diameter pipes. For larger diameters, the pipe is plugged on both sides of the joint to be welded so that a small area can be purged. If the pipe system is complex, consisting of valves and numerous turns, a water-soluble plug or soft plastic gas bag are suggested (Figure 6-15). They can be blown out with air or water when the system is completed.

When a backing gas is used, the gas must have enough time to purge the weld zone completely. When welding on large-diameter pipes or long pipe sections, it is necessary to establish the length of time required to purge the air. To estimate the purged time, you must first establish the interior volume of the pipe or backing strip. Next you must convert the flow rate into cubic inches per minute (cm^3/min.). Then divide the flow rate by the estimated volume to get the time required. Finally, round any fractions of minutes off to the next highest minute.

Standard Units

Volume (in^3) = Length (in inches) × π × $Radius^2$ (in inches)
Flow rate (in^3/min) = cfh × 29
Flow time (minutes) = Volume ÷ Flow rate

SI (Metric Units)

Volume (cubic cm) = Length (in cm) × π × $Radius^2$ (in cm)
Flow rate (cm^3/min.) = L/min × 1,000
Flow time (minutes) = Volume ÷ Flow rate

Standard Unit Example

How long would it take to purge the air out of a 10-foot-long (305 cm) section of 4″ diameter (10 cm) pipe if the flow rate is 20 cfh (9.4 L/min).

Solution

Volume = $120 \times 3.14 \times 2^2$
Volume = 1507 in^3
Flow rate = 20×29
Flow rate = 580 in^3/min
Flow time = 1507 ÷ 580
Flow time = 2.59 min. (approximately 3 min.)

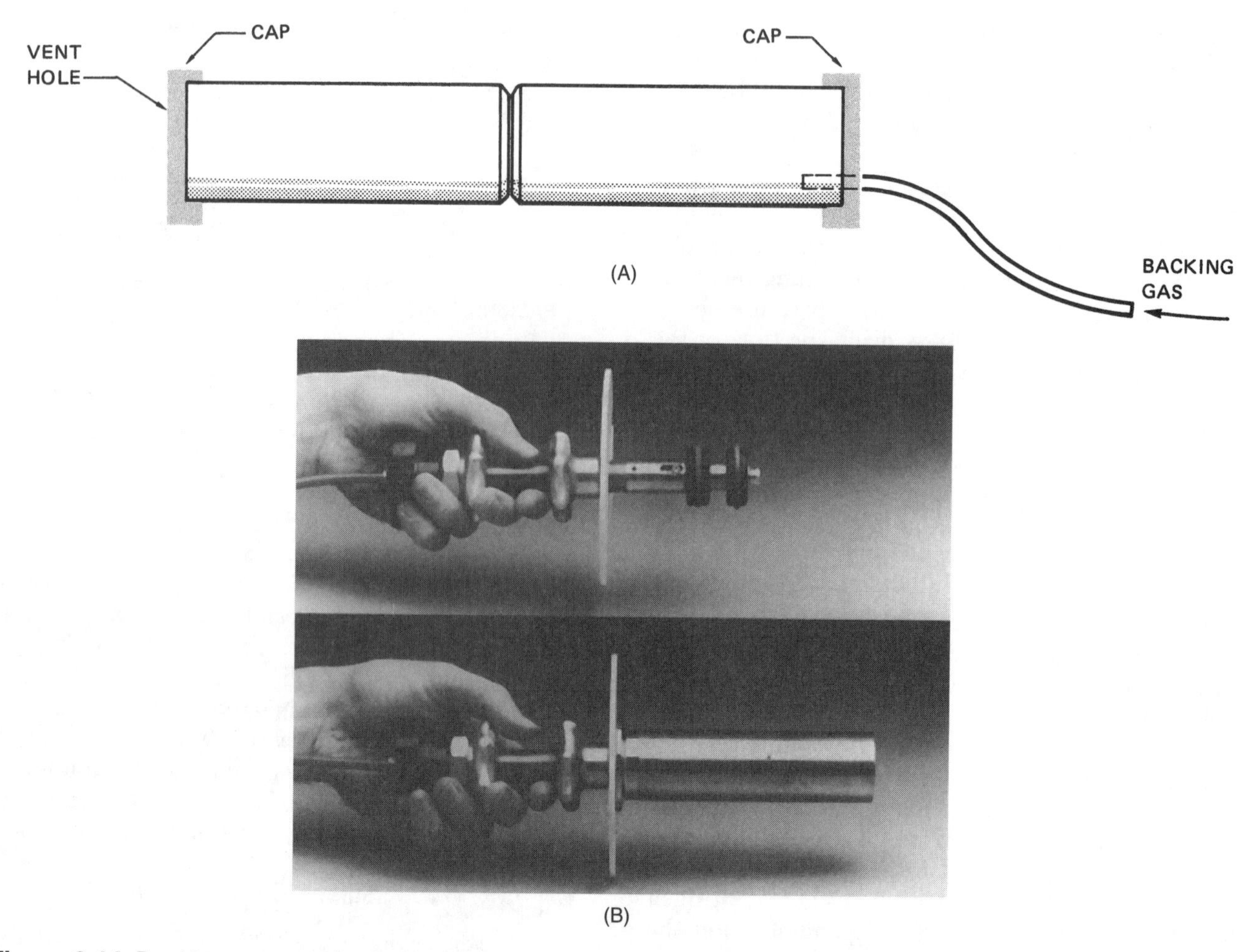

Figure 6-14 Backing gas can be fed into a pipe for welding by (A) capping the ends externally or (B) inserting a purge mandrel inside the pipe (Courtesy of ARC Machines, Inc.)

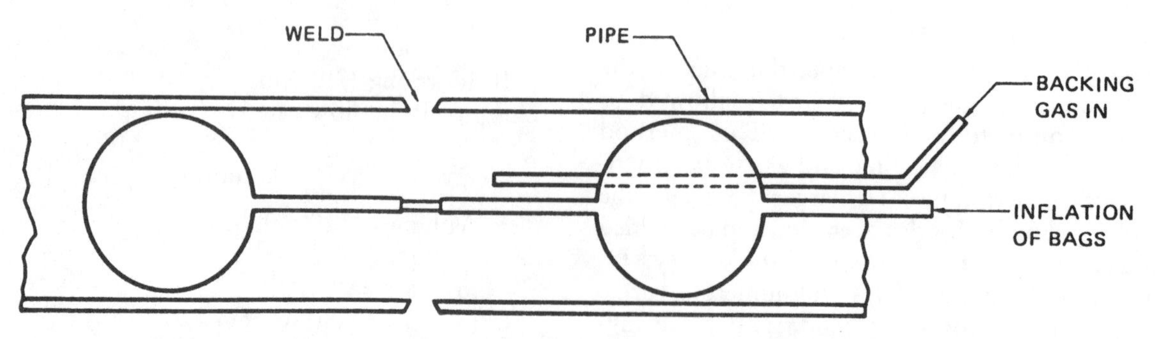

Figure 6-15 Two inflatable bags can be used to seal a pipe so that a backing gas can be used. (Reproduced by permission of Safety Mine Stopper Co., Inc.)

SI (Metric Unit) Example

How long would it take to purge the air out of a 3-meter-long (10 foot) section of 10-centimeter-diameter (4″) pipe if the flow rate is 9 L/hour (20 cfh)?

Solution

Volume = $300 \times 3.14 \times 5^2$
Volume = 23550 cm^3
Flow rate = 9×1000
Flow rate = 9000 cm^3/min
Flow time = $23550 \div 9000$
Flow time = 2.61 min. (approximately 3 min.)
$\pi = 3.14$
Radius = 1/2 Diameter*
Length in inches = Length in feet × 12
Length in cm = Length in meters × 100

*For estimating, either the inside diameter (ID) or outside diameter (OD) of the pipe can be used.

Notice that, in the preceding examples, the lengths, diameters, and flow rates are about the same and so are the minutes required for purging.

Taping over the joint prevents the gas from being blown out too fast. This allows a slower flow rate to be used on the purging gas once the pipe has been purged. The tape is removed just ahead of the weld (Figure 6-16).

In Experiment 6-1, the minimum and maximum gas flow settings for each nozzle size, tungsten size, and amperage setting will be determined. The chart a welder prepares based on experiments is to improve that welder's skills. Charts may differ slightly from one welder to another. As a welder's skills improve, the chart may change. As experience is gained, a welder will learn how to set the gas flow effectively without the need for this chart.

The minimum flow rates and times must be increased to weld in drafty areas or for out-of-position welds. The rates and times can be somewhat lower for tee joints or welds made in tight areas. The maximum flow rates must never be exceeded. Doing so would cause weld contamination and increase the rejection rate.

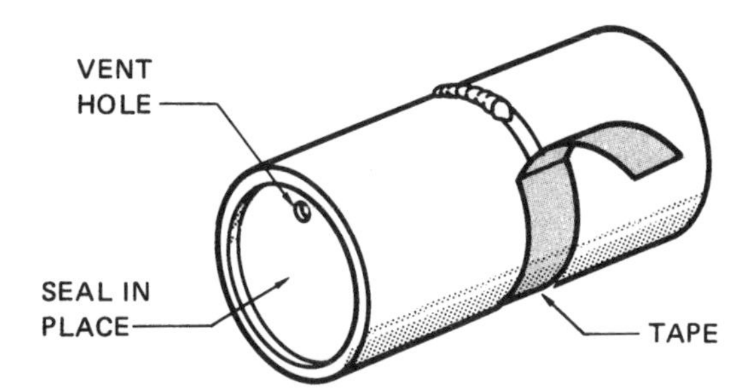

Figure 6-16 Typical setup to allow purging of the inside of pipe for welding.

EXPERIMENT 6-1

Setting Gas Flow

Using a properly prepared GTA welding machine and torch, proper safety protection, one of each available tungsten size, 16-gauge to quarter-inch (6 mm) thick metal, and the welding current chart developed in Experiment 6-1, you will make a chart of the minimum and maximum flow rates and times for each nozzle size, tungsten size, and amperage setting. You will also need an assistant to change and record the flow rate while you work.

Set the machine welding power switch for DCEN (DCSP). Set the amperage to the lowest setting for the size of tungsten used. Set the prepurge time to 0 and postpurge to 20 seconds (Figure 6-17). Turn on the main power. With the torch held so that it cannot short out, depress the remote control to start the shielding gas flow, and set the flow at 20 cfh (9 l/min). Switch the high frequency to start. All other functions, such as pulse, hot start, slope, and so on should be in the off position.

Starting with the smallest nozzle and tungsten size, strike an arc and establish a molten pool on a piece of metal in the flat position. Watch the molten weld pool and tungsten for signs of oxide formation

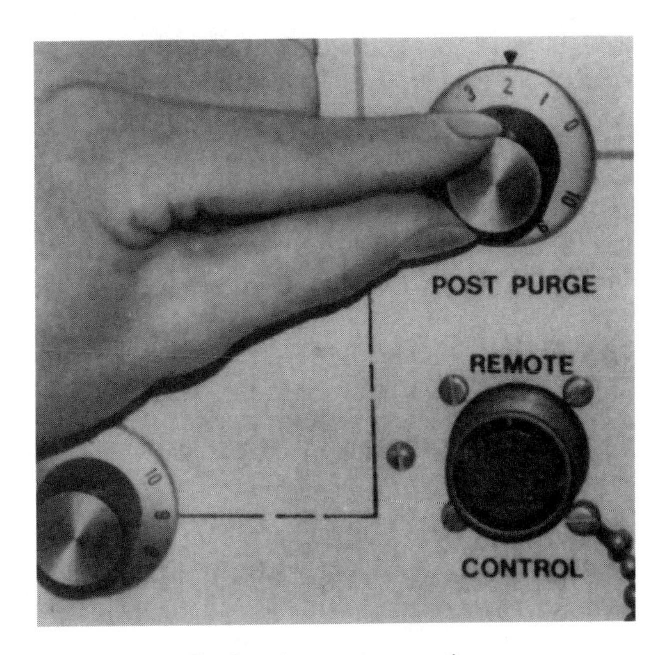

Figure 6-17 Setting the postpurge timer.

as another person slowly lowers the gas flow rate. Have that person note this setting (where oxide formation begins) (Figure 6-18), as the minimum flow rate on the chart next to the nozzle size and current setting. Now slowly increase the flow rate until the molten pool starts to be blown back or oxides start forming. This setting should be noted on the chart as the maximum flow rate for this current and nozzle size (Figure 6-19). Lower the flow to a rate of 2 cfh or 3 cfh (1 l/min or 2 l/min) above the minimum value noted on the chart, and then stop the arc. Record the length of time from the point when the arc stops and the tungsten stops glowing as the postpurge time. Repeat this test at a medium and then high current setting for this nozzle and tungsten size. When using high current settings, it may be necessary to move the torch or use thicker plate to prevent burn-through.

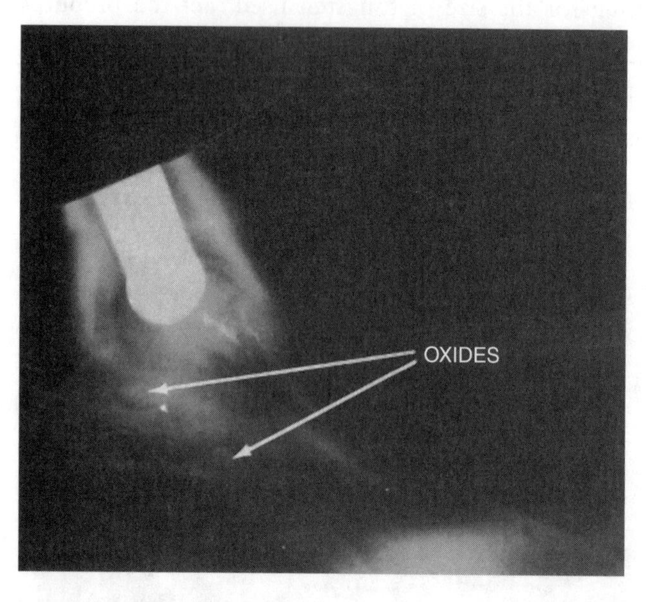

Figure 6-18 Oxides forming due to inadequate gas shielding.

Repeat this test procedure with each available nozzle and tungsten size. Stainless steel or aluminum is preferred for this experiment because the oxides are more noticeable than when mild steel is used. If aluminum is used, the welding current must be AC, and the high-frequency switch should be set on "continuous."

To establish the minimum prepurge time for each nozzle and tungsten size, set the amperage to a medium-high setting. Hold the torch above the metal so that an arc will be instantly started. Set the prepurge timer to 0 and the gas flow to just above the minimum value noted on the chart. Quickly strike an arc on metal thin enough to cause a weld pool to form instantly at that power setting. Stop the arc and examine the weld pool and tungsten for oxides. Repeat this procedure, increasing the prepurge time until no oxides are formed on either the plate or tungsten. Record this time on the chart as the minimum prepurge time. Repeat this test with each available nozzle and tungsten size. Turn off the welding machine, shielding gas, and cooling water, and clean up your work area when you are finished welding.

REVIEW QUESTIONS

1. Why must an inert or nonreactive gas be provided?
2. What are the main types of flowmeters, and how are they read?
3. Why must the tube of a flowmeter be vertical?
4. What is the purpose of a hot start?
5. What are the types of shielding gases used for the GTA welding process?
6. Why are argon and helium called inert gases?
7. Why does argon efficiently shield welds in most positions, and how can this also be a disadvantage?

ELECTODE AND	FLOW RATE					POST FLOW RATE		
NOZZLE SIZE	TOO LOW	LOW	GOOD	HIGH	TOO HIGH	TOO SHORT	OK	TOO LONG

Figure 6-19 Sample chart for setting shielding gas flow rate and time.

8. Why is argon's ease of ionization a benefit?
9. Why do welds with helium have an advantage?
10. Why is helium not used with AC?
11. Why would you add hydrogen to argon for welding?
12. Why are hydrogen—argon mixtures restricted to stainless steel welds?
13. Why is nitrogen not accepted as a GTA welding shielding gas?
14. How can air be drawn into the shielding gas?
15. What allows the removal of any unwanted air?
16. What is the most common method for protecting the rest of a weld from atmospheric contamination?
17. Which gas can be used to back any type of material being welded?
18. What is the most common method for providing gas plates?

CHAPTER 7

FILLER METAL SELECTION

INTRODUCTION

The most widely used filler metal identification system is the one developed by the American Welding Society (AWS). Other identification systems have been developed by other professional organizations such as the American Society of Testing and Materials (ASTM) and the American Iron and Steel Institute (AISI). In addition, most manufacturers have their own trade terms to identify their specific filler metal. Most of these systems use a format very similar to that of the AWS.

The AWS classifications system of filler metals allows manufacturers to produce their filler metals and identify them within a specific classification, yet the exact chemistry each manufacturer selects for a specific filler metal may vary within the AWS classification. Such detailed information about the filler metals' chemistry and characteristics are available from the manufacturers in the form of charts, pamphlets, or pocket filler metal guides. Usually the manufacturers will include the AWS classification for their specific filler metal.

The AWS publishes a variety of books, pamphlets, and charts showing the chemical, mechanical, and weld title specifications for each of the filler metal groups. The AWS also publishes comparison charts and guides that include filler metals that are available from the various manufacturers that fall within the AWS's guidelines.

The AWS classification system is based on both the physical and chemical requirements of the filler metal and weld produced. These specifications are given as ranges that list the minimum and maximum allowable limit for each property or alloying element. Filler metals manufactured within an AWS grouping may have slightly different chemical mechanical properties when compared from manufacturer to manufacturer and still fall within a single AWS classification. Manufacturers add or modify elements within the filler metal to improve or change specific characteristics. Unfortunately, however, as one characteristic is improved, such as strength, another characteristic, such as brittleness, may degrade. Because of this interconnection of properties, some manufacturers produce more than one filler metal within a single classification. This specific information is usually included in the data supplied by the manufacturer. Most major manufacturers include both the AWS identification number and their own identification on the filler metal package. In addition to the identification on the package, each cut length or rod of filler metal used in making critical or certified welds must be identified. This identification may be stamped into the side of the filler metal if the filler metal is large enough, or it may be written on a small flag attached to one end of the filler metal. The ability to identify each rod's specific classification is so critical in some industries, such as the nuclear industry, that during postweld inspections, filler metal stubs are checked for the appropriate classification.

Selecting the best filler metal for a job is seldom delegated to the welder in large shops. The selection of the correct process and filler metal is a complex process. If the choice is given to the welder, it is one of the most important decisions the welder will make.

Covering all of the variables for selecting a filler metal would be well beyond the scope of this text. The following section provides a sample of the types of things that must be considered for the selection of a submerged arc welding (SAW) electrode. To further complicate things, some filler metals have more than one application, and many may be used for the same type of work.

This chapter addresses issues that the welder should consider when choosing a welding filler metal, but they are not in order of importance. They are also not all inclusive.

FILLER METAL INFORMATION

The information provided by manufacturers for their filler metals ranges from general information about the product to very specific chemical, technical, and mechanical properties. In some cases, because of trade secrets, information such as specific alloy elements may not be given out by manufacturers.

Some manufacturers include general information regarding their specific product, such as recommendations for application, welding procedure, and examples of its use.

Understanding the Filler Metal Data

Information provided by manufacturers on their filler metal may include some or all of the following:

- Number of rods per pound
- Available sizes and lengths of filler metal
- Range of welding amperages for each of the filler metal sizes
- Codes and standards under which the filler metal was manufactured
- Weld joint penetration characteristics
- Types of metals that can be welded
- Weld pool characteristics
- Welding procedure information, such as pre- and postheating temperatures
- Weld deposit physical strengths: ultimate tensile strength, yield point, yield strength, elongation, impact strength
- Percentages of alloys such as carbon, sulfur, phosphorus, manganese, silicon, chromium, nickel, copper, and molybdenum

Data Resulting from Mechanical Testing

The mechanical properties of the weld metal that are obtained from standard testing may be included with the manufacturer's literature. Some of the test data obtained from such testing are as follows:

- Tensile strength, psi: the load in pounds that would be required to break a section of sound weld that has a cross-sectional area of one square inch.
- Yield Point, psi: the point in low- and medium-carbon steels at which the metal begins to stretch when force (stress) is applied, after which it will not return to its original length.
- Elongation, percent in 2 inches: the percentage that a 2-inch piece of weld will stretch before it breaks.
- Charpy V notch, ft-lb: the impact load required to break a test piece of weld metal. This test may be performed on metal below room temperature, at which point it is more brittle.

Data Resulting from Chemical Analysis

The chemical information given is for the metal deposited in the weld and not that of the filler metal itself. Some alloying elements in the filler metal may not be transferred into the weld in the exact same percentiles. Additionally there may be some dilution of the weld's metal as a result of fusion and mixing of the base metal (Figure 7-1).

Knowing how a specific percentage of a single alloying element will affect properties of the weld metal becomes very complex because of the interaction of alloying elements with each other and the base metal. A good example of the interaction of alloys can be seen in the carbon equivalency (ce) formula in the following section. Because of this complexity, it is therefore not as important for the welder to know how varying percentages of a specific alloy will impact the weld as it is for them to know what mechanical changes each alloy affects.

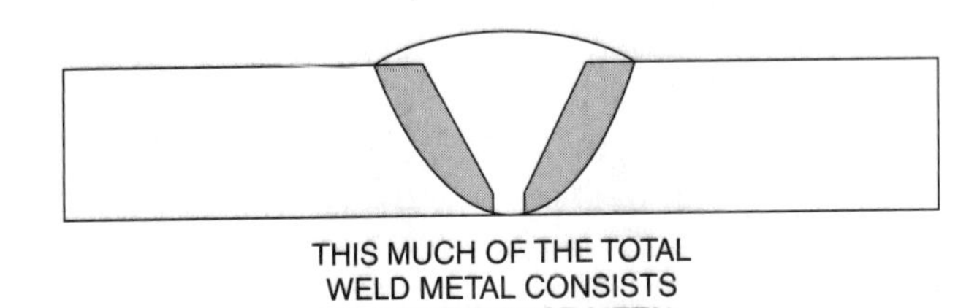

Figure 7-1 Part of the weld metal consists of melted base metal.

The following are the major alloy elements and their effect on the properties of iron:

- Carbon (C) As the percentage of carbon increases, the tensile strength and hardness increase, and ductility is reduced. Carbon also causes Austenite to form.
- Sulfur (S) Suflur is usually a contaminant, and the percentage should be kept as much as possible below 0.04%. As the percentage of carbon increases, sulfur can cause hot shortness and porosity.
- Phosphorus (P) Phosphorusis usually a contaminant, and the percentage should be kept as low as possible. As the percentage of phosphorus increases, weld brittleness, reduced shock resistance, and increased cracking may occur.
- Manganese (Mn) As the percentage of manganese increases, the tensile strength, hardness, resistance to abrasion, and porosity all increase; hot shortness is reduced. Manganese is also a strong Austenite former.
- Silicon (Si) As the percentage of silicon increases, tensile strength increases, and cracking may increase. Silicon is used as a deoxidizer and ferrite former.
- Chromium (Cr) As the percentage of chromium increases, tensile strength, hardness, and corrosion resistance increase with some decrease in ductility. Chromium is also a good ferrite and carbide former.
- Nickel (Ni) As the percentage of nickel increases, tensile strength, toughness, and corrosion resistance increase. Nickel is also an Austenite former.
- Molybdenum (Mo) As the percentage of molybdenum increases, tensile strengthens at elevated temperatures, and creep resistance and corrosion resistance all increase. Molybdenum is also a ferrite and carbide former.
- Copper (Cu) As the percentage of copper increases, the corrosion resistance and cracking tendency increase.
- Columbium (Cb) As the percentage of columbium (niobium) increases, the tendency to form chrome-carbides is reduced in stainless steels. Columbium is also a strong ferrite.
- Aluminum (Al) As the percentage of aluminum increases, the high-temperature scaling resistance improves. Aluminum is also a good oxidizer and ferrite.

Carbon Equivalency (CE)

Because the weldability of iron alloys is affected by the combination of the alloying elements found within a specific metal, it may be necessary to determine the carbon equivalency (CE) of the base metal before welding. A number of different formulas can be used to determine carbon equivalency. Although some formulas are more complex than others, each produces a value for the carbon equivalency. This CE value can then be used to determine the best welding procedure.

Carbon equivalencies are used to determine such things as pre- and or postheat temperatures and times for proper stress relieving of the weldment. The CE of an alloy indicates how the welding will affect the surrounding metal. The area in the base metal surrounding the weld that has been changed during welding is called the heat-affected zone (HAZ) (Figure 7-2). The higher the CE, the more adversely the surrounding base metal will be affected.

By knowing the CE of the base metal, the welder can therefore make the necessary changes in the welding procedure to control potential problems with the heat-affected zone. Some of the most common adjustments to the welding procedure are pre- and/or postheating, filler metal selection, weld size, and welding speed.

$$CE = \%C + \frac{\%MN}{6} + \frac{\%MO}{4} + \frac{\%CR}{5} + \frac{\%NI}{15} + \frac{\%CU}{15} + \frac{\%P}{3}$$

- CE = 0.40% or less — no special welding requirements.
- CE = 0.40% to 0.60% — low-hydrogen welding electrode and the related procedure required
- CE = 0.60% or more — low-hydrogen welding electrode, higher welding heat inputs, preheating, and controlled cooling rates.

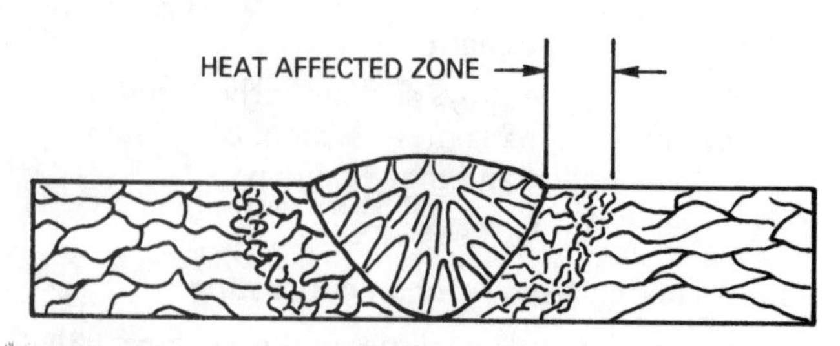

Figure 7-2 Base metal grain structure changes near a weld.

AWS FILLER METAL CLASSIFICATIONS

The AWS classification system uses a series of letters and numbers in a code that gives the important information about the filler metal. The preface letter indicates the filler metal's form or a type of process the filler is to be used with or both. The prefix letters and their meanings are as follows:

- E — Indicates an arc-welding electrode. The filler carries the welding current in the process. We most often think of the E standing for an SMA "stick" welding electrode. It also indicates wire electrodes used in GMAW, FCAW, SAW, ESW, EGW, and so on.
- R — Indicates a rod that is heated by a source other than by electric current flowing directly through it. Welding rods are sometimes referred to as being "cut length" or "welding wire." It is often used with OFW and GTAW.
- ER — Indicates a filler metal that is supplied for use as either a current-carrying electrode or in rod forms. The same alloys are used to produce the electrodes and the rods. This filler metal may be supplied as a wire on a spool for GMAW or as a rod for OFW or GTAW.
- EC — Indicates a composite electrode. These electrodes are used for SAW. Do not confuse an ECu (copper arc-welding electrode) for an $ECNi_2$, which is a composite nickel submerged arc-welding wire.
- B — Indicates a brazing filler metal that is usually supplied as a rod, but it can come in a number of other forms, such as a powder, sheets, washers, and rings.
- RB — Indicates a filler metal that is used as a current-carrying electrode, a rod, or both. The form the filler is supplied in for each of the applications may be different. The composition of the alloy in the filler metal will be the same for all of the forms. This filler can be used for processes such as arc braze welding or oxyfuel brazing.
- RG — Indicates a welding rod used primarily with oxyfuel welding. This filler can be used with all of the oxyfuels, and some of the fillers are used with the GTAW process.
- IN — Indicates a consumable insert. These are most often used for welding on pipe. They are preplaced in the root of the groove to supply both filler metal and support for the root pass. The inserts may provide for some joint alignment and spacing.

WIRE TYPE CARBON STEEL FILLER METALS

The AWS specification for carbon steel filler metals for gas-shielded welding wire is A5.18. Filler metal classified within these specifications can be used for GMAW, GTAW, and PAW processes. Because in GTAW and PAW the wire does not carry the welding current, the letters ER are used as a prefix. The ER is followed by two numbers to indicate the minimum tensile strength of a good weld. The actual strength is obtained by adding three zeros to the right of the number given. For example, ER70S-x is 70,000 psi.

The *S* located to the right of the tensile strength indicates that this is a solid wire. The last number (2, 3, 4, 5, 6, 7) or the letter *G* indicates the filler metal composition and the weld's mechanical properties (Figure 7-3).

ER70S-2
: This is a deoxidized mild steel filler wire. The deoxidizers allow this wire to be used on metal that has light coverings of rust or oxides. A slight reduction in the weld's physical properties may occur if the weld is made on rust or oxides, but the weld will usually still pass the classification test standards. This is a general purpose filler that can be used on killed, semikilled, and rimmed steels.

ER70S-3
: This is a popular filler wire that can be used in single or multiple pass welds in all positions. ER70S-3 does not have the deoxidizers required to weld over rust or oxides or on rimmed steels. It produces high-quality welds on killed and semikilled steels.

ER70S-6
: This is a good general purpose filler wire that has the highest levels of manganese and silicon. The wire can be used to make smooth welds on sheet metal or thicker sections. Welds over rust, oxides, and other surface impurities will lower the mechanical properties but not normally below the specifications of this classification.

Stainless Steel Electrodes

The AWS specifications for stainless steel covered arc electrodes is A5.4 and for stainless steel bare, cored, and stranded electrodes and welding rods is A5.9. Filler metal classified within the A5.4 uses the letter *E* as its prefix, and the filler metal

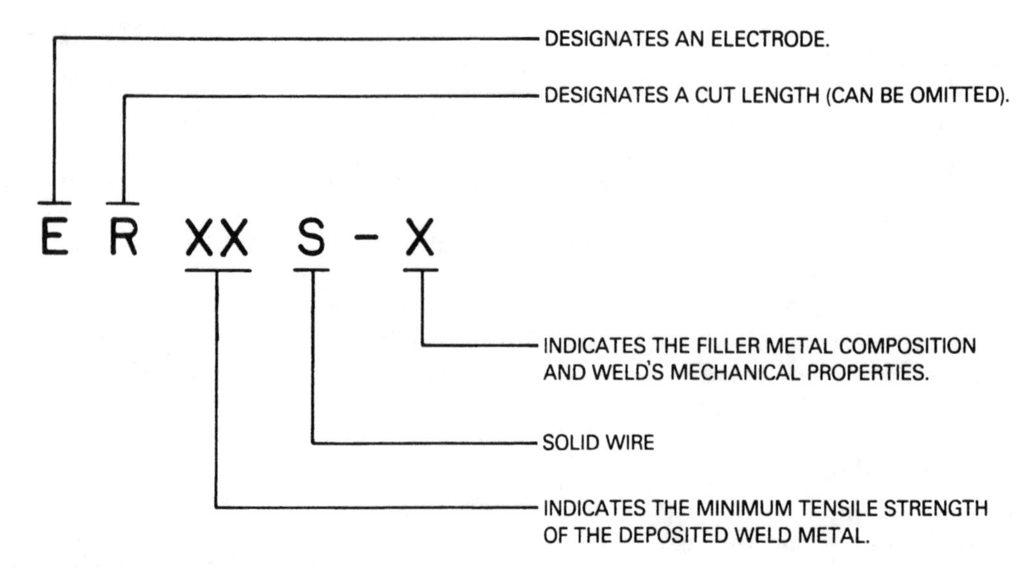

Figure 7-3 AWS numbering system for carbon steel filler metal for GMAW.

within the A5.9 uses the letters ER as its prefix (Figure 7-4).

Following the prefix is the American Iron and Steel Institute's (AISI) three-digit stainless steel number, which indicates the type of stainless steel in the filler metal.

The letter *L* may be added to the right of the AISI number before the dash and suffix number to indicate a low-carbon, stainless welding filler metal. ER308L and ER316L are examples of the use of the letter *L*. (Figure 7-5).

Stainless steel may be stabilized by adding columbium (Cb) as a carbide former. The designation Cb is added after the AISI number for these electrodes, such as E309Cb. Stainless steel filler metals are stabilized to prevent chromium-carbide precipitation.

ER308 and ER308L

All are filler metals for 308 stainless steels. These 308 stainless steels are used for food or chemical equipment, tanks, pumps, hoods, and evaporators. All E308 and ER308 filler metals can be used to weld on all 18-8-type stainless steels such as 301, 302, 302B, 303, 304, 305, 308, 201, and 202.

ER309 and ER309L

All are filler metals for 309 stainless steels, which are used for high-temperatures service, such as furnace parts and mufflers. All E309 filler metals can be used to weld on 309 stainless or to join mild steel to any 18-8-type stainless steel.

ER310

All are filler metals for 310 stainless steels, which are used for high-temperature service where low creep is desired, such as for jet engine parts, valves, and furnace parts. All E310 filler metals can be used to weld 309 stainless or to join mild steel to stainless or to weld most hard-to-weld carbon and alloy steels.

E316-15, E316-16, E3116L-15, E316L-16, ER316, ER316L, and ER316L-Si

All are filler metals for 316 stainless steels, which are used for high-temperature service where high strength with low creep is desired. Molybdenum is added to improve these properties and to resist corrosive pitting. E316 filler metals are used for welding tubing, chemical pumps, filters, tanks, and furnace parts. All E316 filler metals can be used on 316 stainless steels or when weld resistance to pitting is required.

ALUMINUM AND ALUMINUM ALLOYS

The AWS specifications for aluminum and aluminum alloy filler metals is A5.10 for bare welding rods and electrodes. Filler metal classified within the A5.10 with the prefix *ER* uses the Aluminum Association number for the alloy (Figure 7-6).

AISI TYPE NUMBER	442 446	430F 430FSE	430 431	501 502	416 416SE	403 405 410 420 414	321 348 347	317	316L	316	314	310 310S	309 309S	304 L	303 303S E	201 202 301 302 302B 304 305 308	MILD STEEL
201-202-301 302-302B-304 305-308	310 312 309	310 312 309	310 312 309	310 312 309	309 310 312	309 310 312	308	308	308	308	308	308	308	308	308	308	312 310 309
303 303SE	310 309 312	310 309 312	310 309 312	310 309 312	309 310 312	309 310 312	308	308	308	308	308	308	308	308	308 15	308	312 310 309
304-L	310 309 312	310 309 312	310 309 312	310 309 312	309 310 312	309 310 312	308	308	308-L	308	308	308	308	308 -L	308	308	312 310 309
309 309S	310 309 312	310 309 312	310 309 312	310 309 312	309 310 312	309 310 312	308	317 316 309	316	316	308	308	308	308	308	308	309 310 312
310 310S	310 309 312	310 309 312	310 309 312	310 309 312	310 309 312	310 309 312	308	317 316 309	316	316	310	310	308 310	308	308	308	310 309 312
314	310 312 309	310 312 309	310 312 309	310 312 309	310 312 309	310 309 312	309 310 308	309 310	309 310	309 310	310- 15	310	308 310	308 310	309 310	308 310	310 309 312
316	310 309 312	310 309 312	310 309 312	310 309 312	309 310 312	309 310 312	308	316	316	316	309 310 316	310 309 316	309 310 316	308 316	308 316	308 316	309 310 312
316L	310 309 312	310 309 312	310 309 312	310 309 312	309 310 312	309 310 312	308	316 317 306	316-L	316	309 310 316	310 309 316	316 309	308 316	308 316	308 316	309 310 312
317	310 309 312	310 309 312	310 309 312	310 309 312	309 310 312	309 310 312	308	317	316 308	316 308	309 310 317	317 315 309	317 316 309	308 316 317	308 316 317	308 316 317	309 310 312
321 348 347	310 309 312	310 309 312	310 309 312	310 309 312	309 310 312	309 310 312	347	308 347	347 308	347 308	309 310 347	347 308	347 308	347 308 -L	347 308	347 308	309 310 312
400–405 410–420 414	310 309 312	310 309 312	310 309 312	310 309 312	309 310	410† 309††	309 310	309 310	309 310	309 310	310 309	310 309	308 310	309 310	309 310	309 310	309 310 312
416 416SE	310 309	310 309	310 309	310	410–15†	410–15† 309†† 310††	309 310	309 310 312	309 310 312	309 310 312	309 310 312	310 309 312	309 310 312	309 310 312	308 310 312	309 310 312	309 310 312
501 502	310	310	310	502† 310††	310	310	310 309	310 309	310 309	310 309	310 309	310 309	310 309	310 309	310 309	310 309	310 312 309
430 431	310 309	310 309	430–15† 310†† 309††	310	310	310 309	310 309	310 309	310 309	310 309	310 309	310 309	310 309	310 309	310 309	310 309	310 309 312
430F 430FSE	310 309	410 15†	310 309	310 309	310 309 312	310 309 312	309 310 312	309 310 312	310 309 312	310 309 312	310 309 312	310 309 312	310 309 312	310 309 312	310 309 312	310 309 312	310 309 312
442 448	309 310	309 310 312	310 309 312	310 309 312	310 309 312	310 309 312	310 309 312	310 309 312	310 309 312	310 309 312	310 309 312	310 309 312	310 309 312	310 309 312	310 309 312	310 309 312	310 309 312

† Preheat

†† No Preheat Necessary

Bold numbers indicate first choice, light numbers indicate second and third choice. This choice can vary with specific applications and individual job requirements.

Figure 7-4 Filler metal selector guide for joining different types of stainless to the same type or another type of stainless. (Courtesy of Thermacote Welco)

UTP DESIGNATION	AWS/SFA5.4 COVERED	AWS/SFA5.9 TIG AND MIG	DESCRIPTION AND APPLICATIONS
6820	E 308-16	ER 308	For welding conventional 308 type SS.
68 Kb	E 308-15		Low hydrogen coating.
6820 Lc	E 308 L-16	ER 308L	Low carbon grade, prevents carbide precipitation adjacent to weld.
308L Fe Hp	E 308 L-16		Fast depositing for maintenance and production coating.
68 LcHL	E 308 L-16		High-performance electrode with rutile-acid coating, core wire alloyed. For stainless and acid-resisting CrNi steels.
68 LcKb	E 308 L-15		Low-carbon electrode for stainless, acid-resisting CrNi-steels.
6824	E 309-16	ER 309	For welding 309 type SS and carbon steel to SS.
6824 Kb	E 309-15		Special lime-coated electrode for corrosion and heat-resistant 22/12 CrNi-steels.
6824 Lc	E 309 L-16	ER 309L	Same as 309, but with low carbon content.
6824 Nb	E 309 Cb-16		Corrosion and heat resistant 22/12 CrNi-steels.
6824 MoNb	E309MoCb-16		Corrosion and heat resistant 22/12 CrNi-steels.
309L Fe Hp	E 309 L-16		High deposition rate, easy to use.
6824 Mo Lc	E 309 L-16		For welding similar and dissimilar SS.
68H	E 310-16	ER 310	For high temperature service and cladding steel.
6820 Mo	E 316-16	ER 316	For welding acid resistant Stainless steels.
6820 Mo Lc	E 316L-16	ER 316L	Low carbon grade. Prevents intergranular corrosion.
68 TI Mo	E 316 L-16		Most efficient type. For maintenance and production. **High Performance.**
68 MoLcHL	E 316 L-16		High-performance electrode with rutile-acid coating, core wire alloyed, for stainless and acid-resisting CrNi-Mo-steels.
68 MoLcKb	E 316L-15		Low-carbon electrode for stainless and acid-resisting CrNiMo-steels.
317 Lc Titan	E 317 L-16	ER 317L	Deposit resist sulphuric acid corrosion.
317LFe Hp	E 317 L16		Fast melt off rate, excellent for overlays, easy to use. **High Performance.**
68 Mo	E 318-16		Versatile stainless all position electrode.
320 Cb	E 320-15		For welding similar acid resistant ss.
3320 Lc	E 320		Is a rutile-coated electrode for welding in all positions except vertical down.
6820 Nb	E 347-16	ER 347	Stabilized grade, prevents carbide precipitation.
347 FeHp	E 347-16		High performance stainless steel electrode of class E 347-16 for welding stabilized Cr Ni alloys.
66	E 410-15		Low-hydrogen electrode for corrosion and heat-resistant 14% Cr-steels.
1915 HST			Low-hydrogen fully austenitic electrode with 0% ferrite content.
1925			Extremely corrosion resistant to phosphoric and sulfuric acids.
68 Hcb	E 310 Cb	ER 310 Cb	For high heat applications, and joining steels to stainless steel.
2535 NbSn			Electrode is a lime-type special electrode and is used for surfacing and joining heat resistant base metals, especially cast steel.
E 330-16	E 330-16		Excellent for welding furnace parts.
6805			For welding of base material 17-4 Ph.
6808 Mo			Rutile lime-type austenitic-feritic electrode with low carbon content is suited for joining and surfacing on corrosion-resistant steels and cast steel type with austenitic-ferritic structure (Duplex-steels).
6809 Mo			Rutile-basic austenitic-ferritic electrode with low carbon content is suited for joining and surfacing on corrosion-resistant steels and cast steel types with an austenitic-ferritic structure (Duplex-steels).

Figure 7-5 Stainless steel electrodes, filler metals, and wires. (Courtesy of UTP Welding Materials, Inc.)

BASE METAL	319 355	43 356	214	6061 6063 6151	5456	5454	5154 5254	5086	5083	5052 5652	5005 5050	3004	1100 3003	1060
1060	4145 4043 4047	4043 4047 4145	4043 5183 4047	4043 4047	5356 4043	4043 5183 4047	4043 5183 4047	5356 4043	5356 4043	4043 4047	1100 4043	4043	1100 4043	1260 4043 1100
1100 3003	4145 4043 4047	4043 4047 4145	4043 5183 4047	4043 4047	5356 4043	4043 5183 4047	4043 5183 4047	5356 4043	5356 4043	4043 5183 4047	4043 5183 5356	4043 5183 5356	1100 4043	
3004	4043 4047	4043 4047	5654 5183 5356	4043 5183 5356	5356 5183 5556	5654 5183 5356	5654 5183 5356	5356 5183 5556	5356 5183 5556	4043 5183 4047	4043 5183 5356	4043 5183 5356		
5005 5050	4043 4047	4043 4047	5654 5183 5356	4043 5183 5356	5356 5183 5556	5654 5183 5356	5654 5183 5356	5356 5183 5556	5356 5183 5556	4043 5183 4047	4043 5183 5356			
5062 5662	4043 4047	4043 5183 4047	5654 5183 5356	5356 5183 4043	5366 5183 5556	5654 5183 5356	5654 5183 5356	5356 5183 5556	5356 5183 5556	5654 5183 4043				
5063	NR	5356 4043 5183	5356 5183 5556	5356 5183 5556	5183 5356 5556	5356 5183 5556	5356 5183 5556	5356 5183 5556	5183 5356 5556					
5086	NR	5356 4043 5183	5356 5183 5556	5356 5183 5556	5356 5183 5556	5356 5183 5554	5356 5183 5554	5356 5183 5556						
5154 5254	NR	4043 5183 4047	5654 5183 5356	5356 5183 4043	5356 5183 5554	5654 5183 5356	5654 5183 5356							
5654	4043 4047	4043 5183 4047	5654 5183 5356	5356 5183 4043	5356 5183 5554	5554 4043 5183								
5456	NR	5356 4043 5183	5356 5183 5556	5356 5183 5556	5556 5183 5356									
6061 6063 6151	4145 4043 4047	4043 5183 4047	5356 5183 4043	4043 5183 4047										
214	NR	4043 5183 4047	5654 5183 5356											
43 358	4043 4047	4145 4043 4047												
319 355	4145 4043 4047													

Note: First filler alloy listed in each group is the all-purpose choice. NR means that these combinations of base metals are not recommended for welding.

Figure 7-6 Recommended filler metals for joining different types of aluminum to the same type or a different type of aluminum. (Courtesy of Thermacote Welco)

Aluminum Bare Welding Rods and Electrodes

ER1100

ER1100 aluminum has the lowest percentage of alloy agents of all of the aluminum alloys and melts at 1,215° F. The filler wire is also relatively pure. ER1100 produces welds that have good corrosion resistance, high ductility, and tensile strengths ranging from 11,000 to 17,000 psi. The weld deposit has a high resistance to cracking during welding. This wire can be used with OFW, GTAW, and GMAW. Preheating to 300° to 350° F is required for GTA welding on plate or pipe 3/8″ and thicker to ensure good fusion. Flux is required for OFW. ER1100 aluminum is commonly used for items such as food containers, food processing equipment, storage tanks, and heat exchanges. ER1100 can be used to weld 1100 and 3003 grade aluminum.

ER4043

ER4043 is a general-purpose welding filler metal. It has 4.5 to 6.0% silicon added, which lowers its melting temperature to 1,155° F. The lower melting temperature helps promote a free-flowing molten weld pool. The welds have high ductility and a high resistance to cracking during welding. This wire can be used with OFW, GTAW, and GMAW. Preheating to 300° to 350° F is required for GTA welding on plate or pipe 3/8″ and thicker to ensure good fusion. Flux is required for OFW. ER4043 can be used to weld on 2014, 3003, 3004, 4043, 5052, 6061, 6062, 6063, and cast alloys 43, 355, 356, and 214.

ER5356

ER5356 has 4.5 to 5.5% magnesium added to improve the tensile strength. The weld has high ductility but only an average resistance to cracking during welding. This wire can be used for GTAW and GMAW. Preheating to 300° to 350° F is required for GTAWelding on plate or pipe 3/8″ and thicker to ensure good fusion. ER5356 can be used to weld on 5050, 5052, 5056, 5083, 5086, 5154, 5356, 5454, and 5456.

ER5556

ER5556 has 4.7 to 5.5% magnesium and 0.5 to 1.0% manganese added to produce a weld with high strength. The weld has high ductility and only average resistance to cracking during welding. This wire can be used for GTAW and GMAW. Preheating to 300° to 350° F is required for GTAWelding on plate or pipe 3/8″ and thicker to ensure good fusion. ER5556 can be used to weld on 5052, 5083, 5356, 5454, and 5456.

Magnesium Alloys

The joining of magnesium alloys by welding is possible without a fire hazard because the melting point of magnesium is 1202° F (651° C) to 858° F (459° C), well below its boiling point, where magnesium could start to burn.

ER AZ61A

The ER AZ61A filler metal can be used to join most magnesium-wrought alloys. This filler has the best weldability and weld strength for magnesium alloys AZ31B, HK31A, and HM21A.

ER AZ92A

The ER AZ92A filler metal can be used on cast alloys, Mg-Al-Zn and AM 100A. This filler metal has a somewhat higher resistance to cracking.

REVIEW QUESTIONS

1. What groups have developed filler metal identification systems?
2. What is the AWS classification system based on?
3. What is one of the most important decisions the welder will make?
4. What types of general information about filler metals may be given by different filler metal manufacturers?
5. Define tensile strength.
6. What chemicals alloys:
   a) contaminate the weld metal?
   b) increase tensile strength in the weld metal?
   c) increase corrosion resistance?
   d) reduce creep?
7. What should CE be used for?
8. What welding parameters should be used with a metal that has a CE of more than .60%?
9. What welding parameters should be used with a metal that has a CE of .40 to .60%?
10. What do the following filler metal designations stand for?
    a) E
    b) ER
    c) RG
    d) IN
11. What do the following filler metal designations stand for?
    a) R
    b) EC
    c) B
    d) RB

12. What is the AWS specification for carbon steel filler metals?
13. What does the S to the right of tensile strength indicate in AWS filler metal number ER70S?
14. What is the purpose of the deoxidizers in ER70S-2?
15. How can stainless steel be stabilized?
16. What stainless steel would be used for food service equipment?
17. What are 309 stainless steels used for? Give examples.
18. What stainless steel would have low creep at high temperatures?
19. What stainless steel would be used for high temperatures, high strength, and low creep?
20. Which aluminum melts at 1,215° F?
21. Which magnesium filler metal has the best weldability and weld strength?

CHAPTER 8

WELDING SYMBOLS

INTRODUCTION

Welding symbols on drawings show a welder exactly what welding is needed. The welding symbol is a shorthand method to provide the welder with all of the required information to make the correct weld. Learning to interpret welding symbols will enable the skilled welder to place the correct weld in the correct locations. Welds that are not located correctly may result in greater welding expense, fewer serviceable weldments, and possibly even unsafe parts.

WELD JOINT DESIGN

A variety of factors must be considered when selecting any joint design for a specific joint weldment. Each factor, considered alone, would result in a part that might not be able to be fabricated. For example, magnesium is very susceptible to post-weld stress, and the U-groove works best for thick sections.

Joints must be designed so that they allow parts to be welded together to best distribute the stresses, reduce distortion, and provide for the greatest strength. Forces applied to a joint are tensile, compression, bending, torsion, and shear (Figure 8-1). A weldment's ability to withstand these forces depends on both joint design and weld integrity.

The basic parts of a weld joint design that can be changed include the following:

- Joint type — The type of joint is determined by the way the joint members come together (Figure 8-2).
- Edge preparation — The faying surfaces of the mating members that form the joint are shaped for a specific joint. This preparation may be the same on both members of the joint, or each side can be shaped differently (Figure 8-3).
- Joint dimensions — The depth and/or angle of the preparation and the joint spacing can be changed to make the weld (Figure 8-4).

PLATE WELDING POSITIONS

The ideal welding position for most joints is the flat position, which usually allows for the largest possible weld pool to be controlled. As a general rule, the larger the weld bead that can be produced in a single pass, the faster the joint can be completed.

Not all weldments can be positioned so that all of the welding can be made in the flat position. Some joints on a weldment may have to be produced in the out-of-positions, which refers to all welds that are not produced in the flat position.

Some applications lend themselves very successfully to welds made in a position other than flat. For example, some welds made in very thin metal sections may be more easily controlled in the vertical position.

The AWS has divided plates into four basic positions for groove (G) and fillet (F) welds as follows:

- Flat 1G or 1F — Welding is performed from the upper side of the joint, and the face of the weld is approximately horizontal (Figure 8-5).

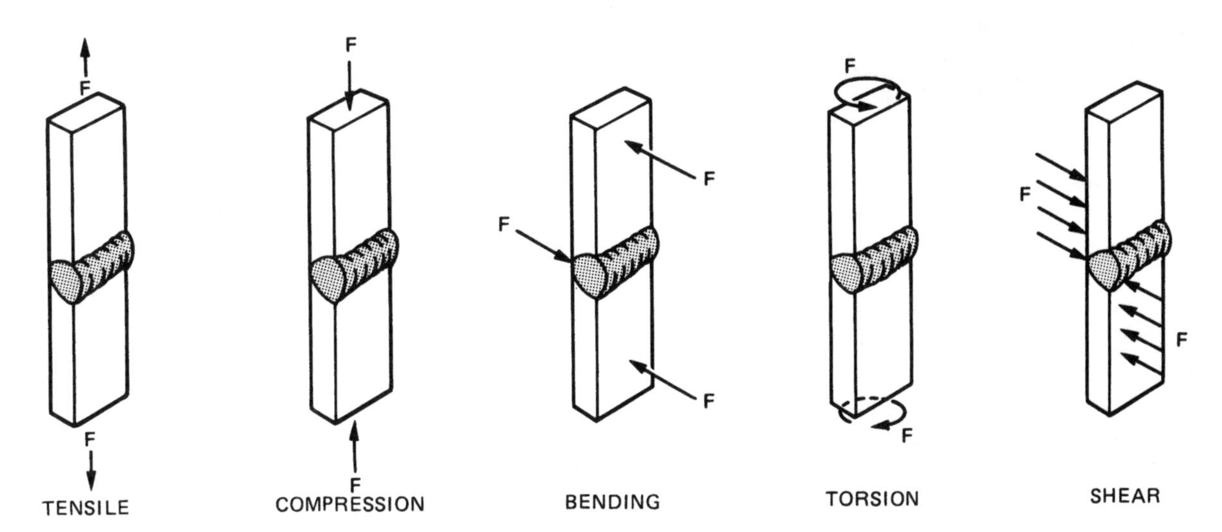

Figure 8-1 Forces on a weld.

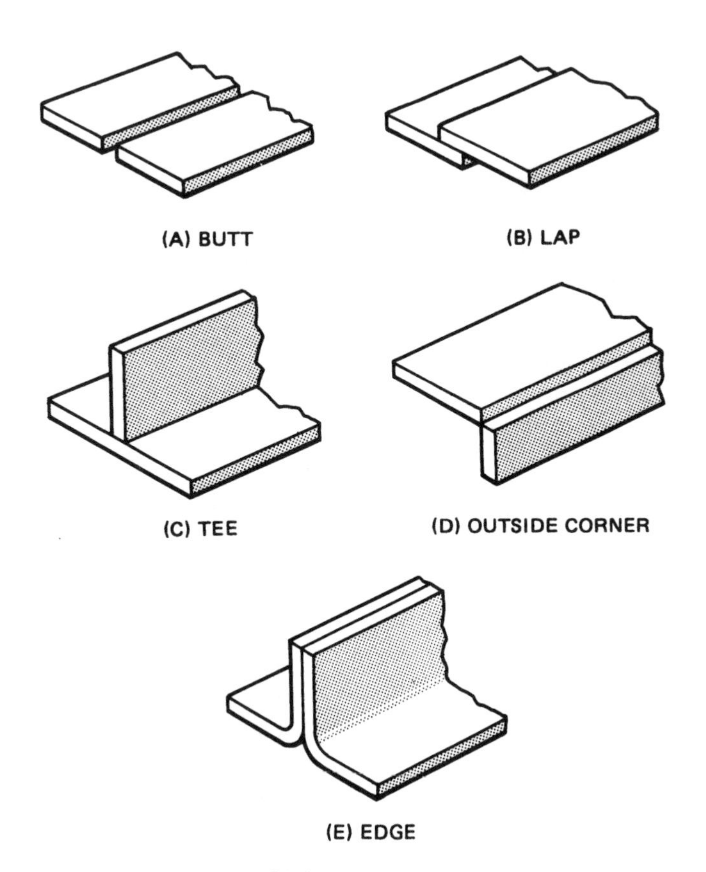

Figure 8-2 Types of joints.

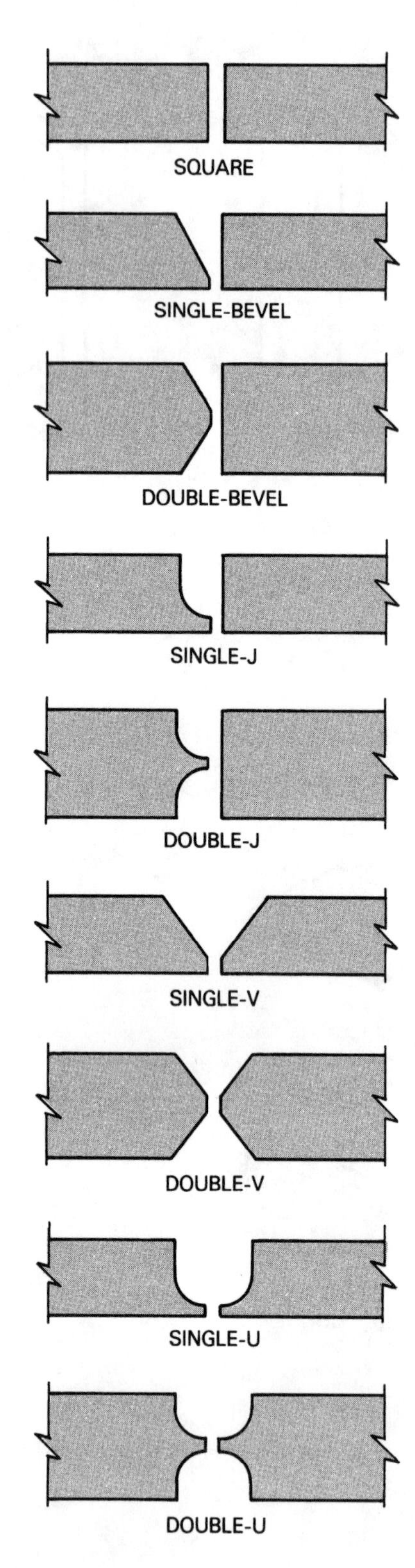

Figure 8-3 Edge preparation.

- Horizontal 2G or 2F — The axis of the weld is approximately horizontal, but the type of the weld dictates the complete definition. For a fillet weld, welding is performed on the upper side of an approximately vertical surface. For a groove weld, the face of the weld lies in an approximately vertical plane (Figure 8-6).
- Vertical 3G or 3F — The axis of the weld is approximately vertical.
- Overhead 4G or 4F — Welding is performed from the underside of the joint.

PIPE WELDING POSITIONS

The AWS Pipe has divided pipe welding into the following five basic positions:

- Horizontal rolled 1G — The pipe is rolled either continuously or intermittently so that the weld is performed within 0° to 15° of the top of the pipe.
- Horizontal fixed 5G — The pipe is parallel to the horizon, and the weld is made vertically around the pipe.
- Vertical 2G — The pipe is vertical to the horizon, and the weld is made horizontally around the pipe.
- Inclined with a restriction ring 6GR — The pipe is fixed in a 45° inclined angle, and a restricting ring is placed around the pipe below the weld groove.

METAL THICKNESS

The thickness of a metal is a major contributing factor in the selection of a joint design. On some thick sections it may be possible to make full penetration welds using a square butt joint (Figure 8-7). However, as the material becomes thicker, some method of edge preparation may be required if 100% joint penetration welds are required (Figure 8-8). The preparation of the edge for welding, grooving, or bubbling may be on either one or both sides of the joint. The various edge shapes for this purpose include beveled, V-groove, J-groove, and U-grooves (Figure 8-9). Some of the factors that determine the particular shape of the edge preparation include the type of metal, thickness, weld position, code or specification, and accessibility.

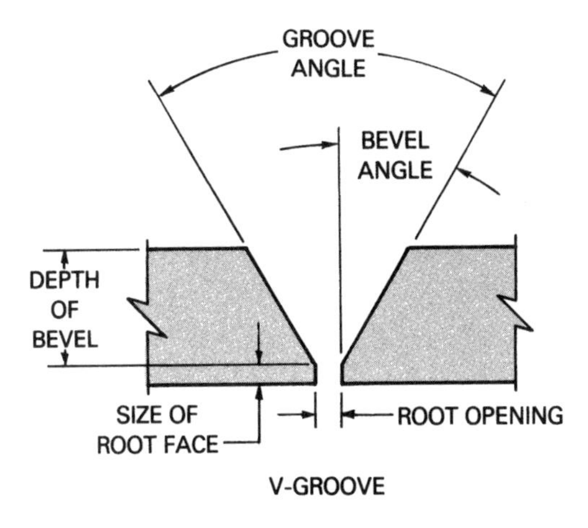

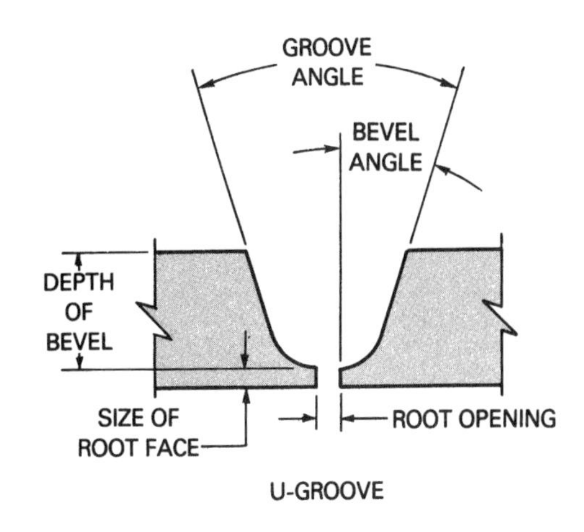

Figure 8-4 Groove terminology.

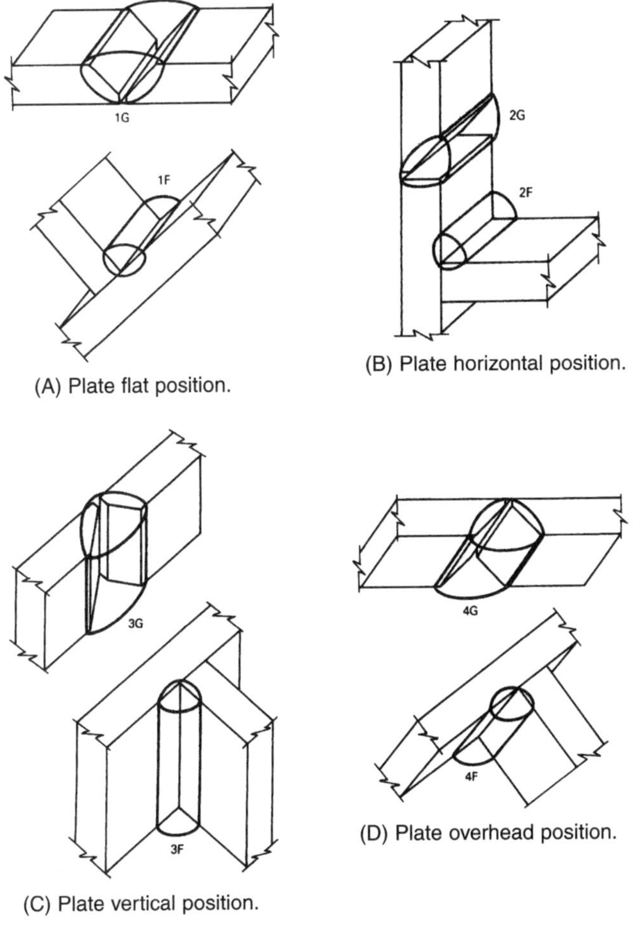

(A) Plate flat position.

(B) Plate horizontal position.

(C) Plate vertical position.

(D) Plate overhead position.

Figure 8-5 Plate welding positions.

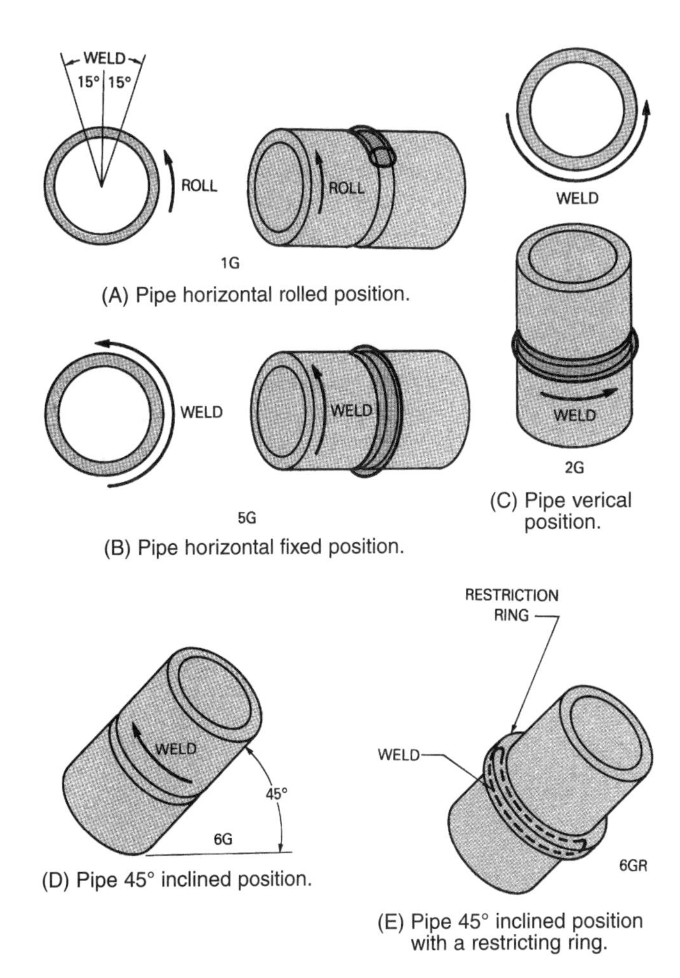

(A) Pipe horizontal rolled position.

(B) Pipe horizontal fixed position.

(C) Pipe verical position.

(D) Pipe 45° inclined position.

(E) Pipe 45° inclined position with a restricting ring.

Figure 8-6 Pipe welding positions.

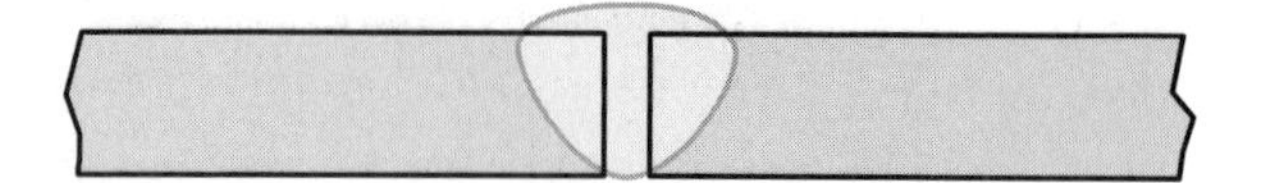

Figure 8-7 Full penetration weld on square butt joint.

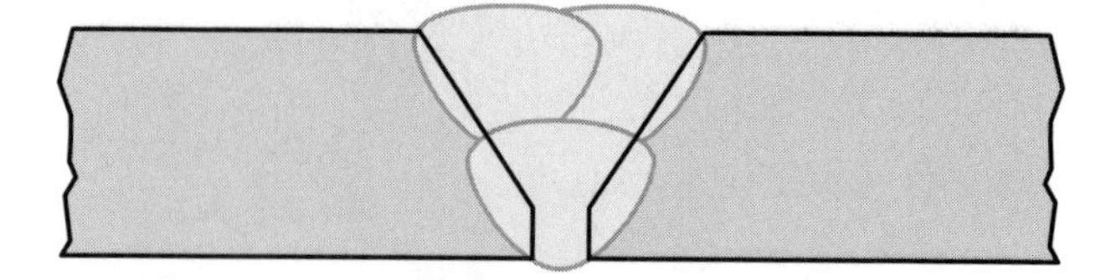

Figure 8-8 Multipass weld on thicker metal section that has been v-grooved for welding.

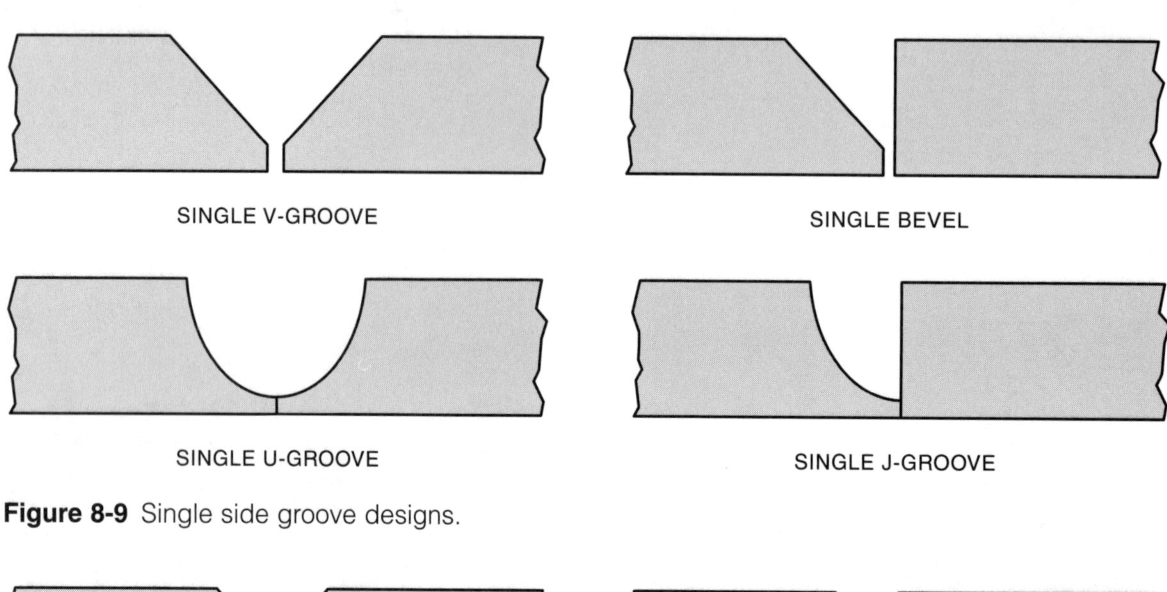

Figure 8-9 Single side groove designs.

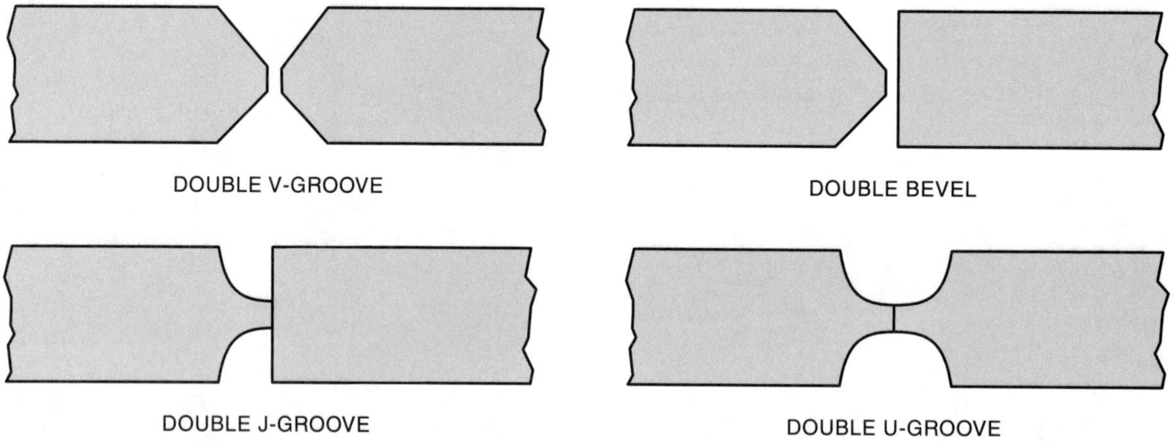

DOUBLE V-GROOVE

DOUBLE BEVEL

DOUBLE J-GROOVE

DOUBLE U-GROOVE

Figure 8-10 Double sided groove designs.

To help provide the required penetration, grooves or bubbles may be cut into either one side or both sides of the joint (Figure 8-10). The edge preparation may be produced by grinding, flame cutting, gouging, sawing, or machining. Bevels and V-grooves are more easily cut on the parts before they are assembled. J-grooves and U-grooves may be cut either before or after assembly (Figure 8-11).

A 90° or squared edge is the only preparation used for lap joints. No strength can be gained by grooving the metal's edge for this joint (Figure 8-12).

If grooving is required, the choice to groove one or both sides of the plate is most often determined by joint design, position, code, and application. Plate in the flat position is usually grooved on only one side unless it can be repositioned so that the back side can be welded in the flat position also.

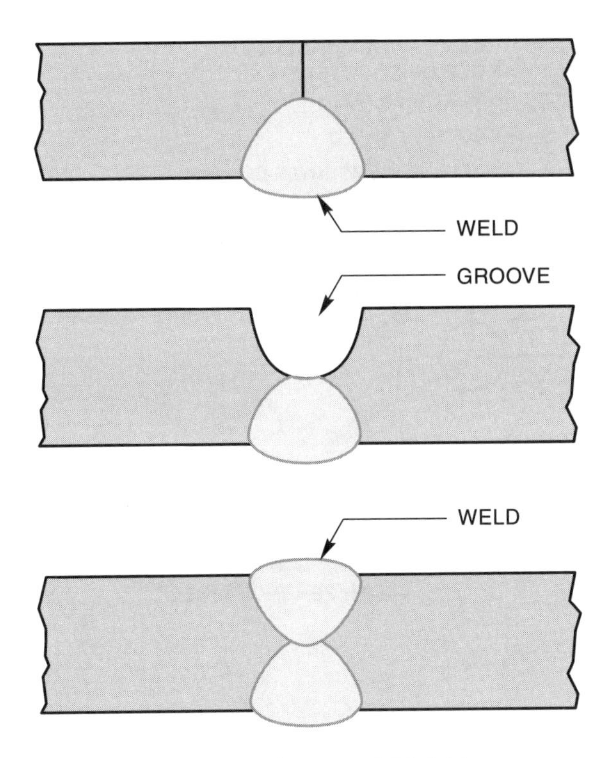

Figure 8-11 Back-gouging a weld joint to ensure 100% joint penetration.

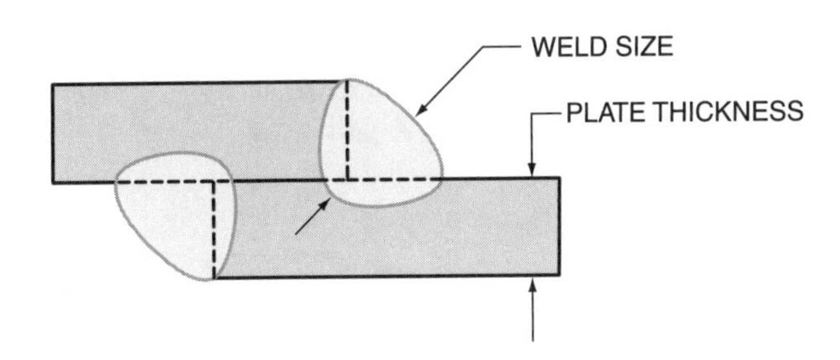

Figure 8-12 The weld's strength is greater than the plate's strength if the weld size is equal to or greater than the plate thickness.

CODE OR STANDARD REQUIREMENTS

The type, depth, angle, and location of the groove is usually determined by a code or standard that has been qualified for the specific jobs. Organizations such as the American Welding Society, American Society of Mechanical Engineers (ASME), and the American Bureau of Ships (ABS) are a few of the agencies that issue such codes and specifications. The most common code or standards are the AWS D1.1 and the ASME Boiler and Pressure Vessel (BPV) Section IX. The joint design for a specific set of specifications often must be prequalified, which means the joints have been tested and found reliable for the weldments for specific applications. The joint design can be modified, but the cost of having the new design accepted under the applicable standard is often prohibitive.

WELDING SYMBOLS

Welding symbols enable a designer or engineer to indicate clearly to the welder important detailed information regarding the weld. Information about the weld that can be included in the symbol includes details such as length, depth of penetration, height of reinforcement, groove type, groove dimensions, location, process, filler metal, strength, number of welds, weld shape, and surface finishing. Without the symbol, such detailed information would require lengthy written instructions to be included with the drawing.

The AWS has established standards for welding symbols. Because of the complexity and diversity of these symbols, only the more common ones have been included in this chapter. Detailed information regarding all of the welding symbols is available in a booklet published by the AWS and American National Standards (ANS) entitled *Standard Symbols for Welding, Brazing, and Nondestructive Examination* (ANSI/AWS A2.4).

All of the various components that make up a welding symbol are located on a horizontal reference line (Figure 8-13). An arrow extends from one end of the reference line and points to the location on the drawing where the weld is to be performed. A tail may be included on the opposite end of the reference line from the arrow. This tail is added to the symbol when it is necessary to include specific information (Figure 8-14). Specifications that may be found in the tail include filler metal specifications, number of weld passes, and pre- or postheat temperatures.

TYPES OF WELDS

Welds can be classified as follows: fillets, grooves, flange, plugs or slots, spot or projection, seam, back or backing, and surface. A specific symbol has been designed for each of these weld types and can be included on the reference line. All of the basic symbols are shown in Figure 8-15.

Although some of the symbols closely resemble the weld shape they indicate, they are only symbols

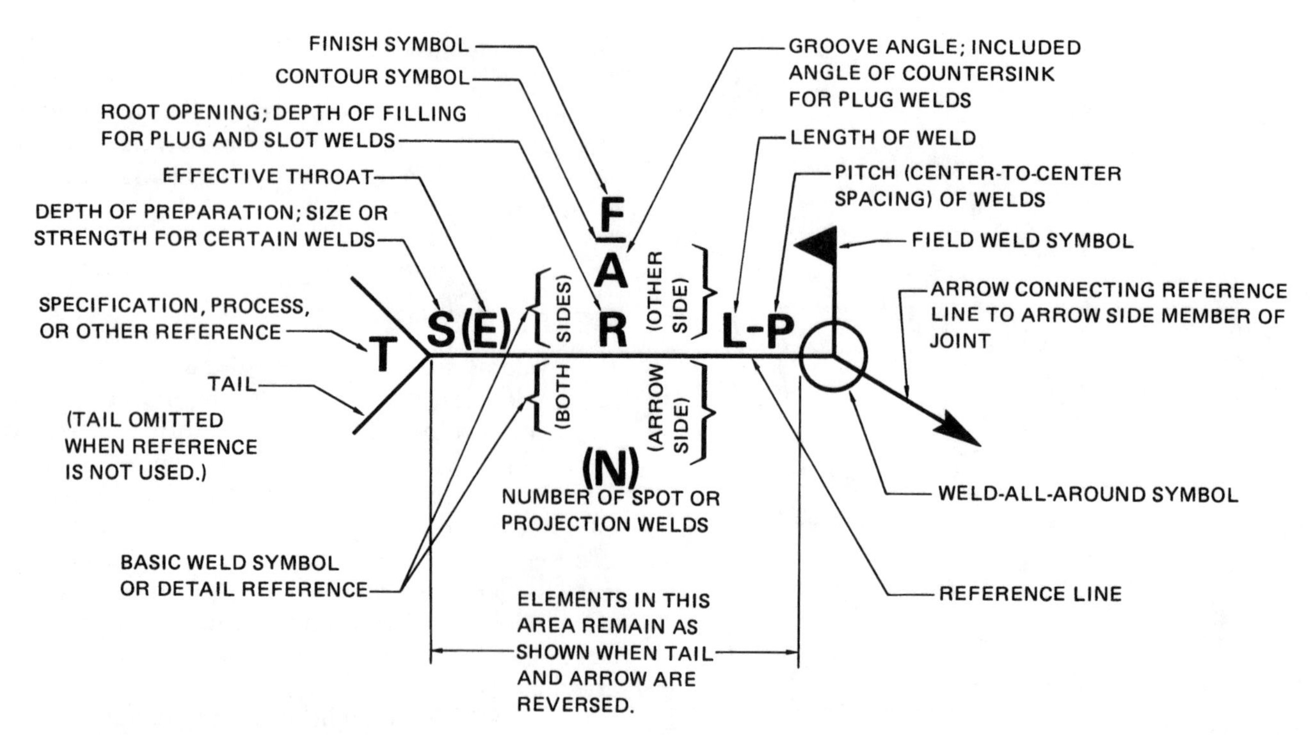

Figure 8-13 Standard location of elements of a welding symbol. (Courtesy of the American Welding Society)

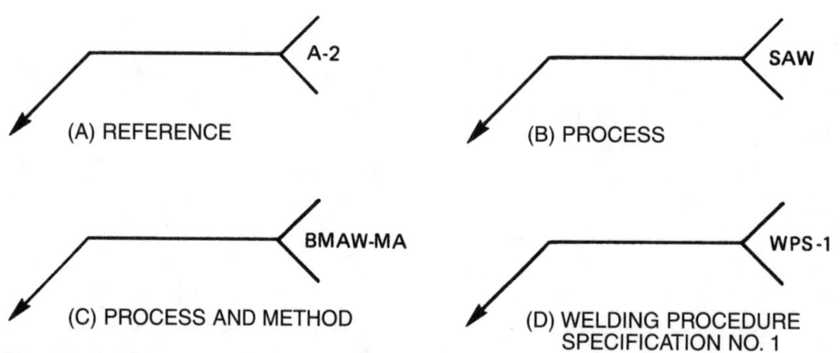

Figure 8-14 Locations of specifications, processes, and other references on weld symbols.

and not pictorial representations of the weld. For example, the fillet weld symbol—a triangular shape—is always drawn with the vertical line on the right side and the sloped line on the left, irrespective of the weld's actual configuration (Figure 8-16).

WELD LOCATION

The reference line is always drawn horizontally. Any symbol located on the top of this reference line is referred to as being an *other side* symbol. Any symbol located below the reference line is referred to as an *arrow side* symbol. Both these terms refer to the side of the joint that the arrow touches (Figure 8-17).

Arrow side and other side symbols as related to the reference line do not change regardless of the direction the arrow points. If the weld is to be deposited on the arrow side of the joint (near side), the proper weld symbol is placed below the reference line (Figure 8-18).

If the weld is to be deposited on the other side of the joint (far side), the weld symbol is placed above the reference line.

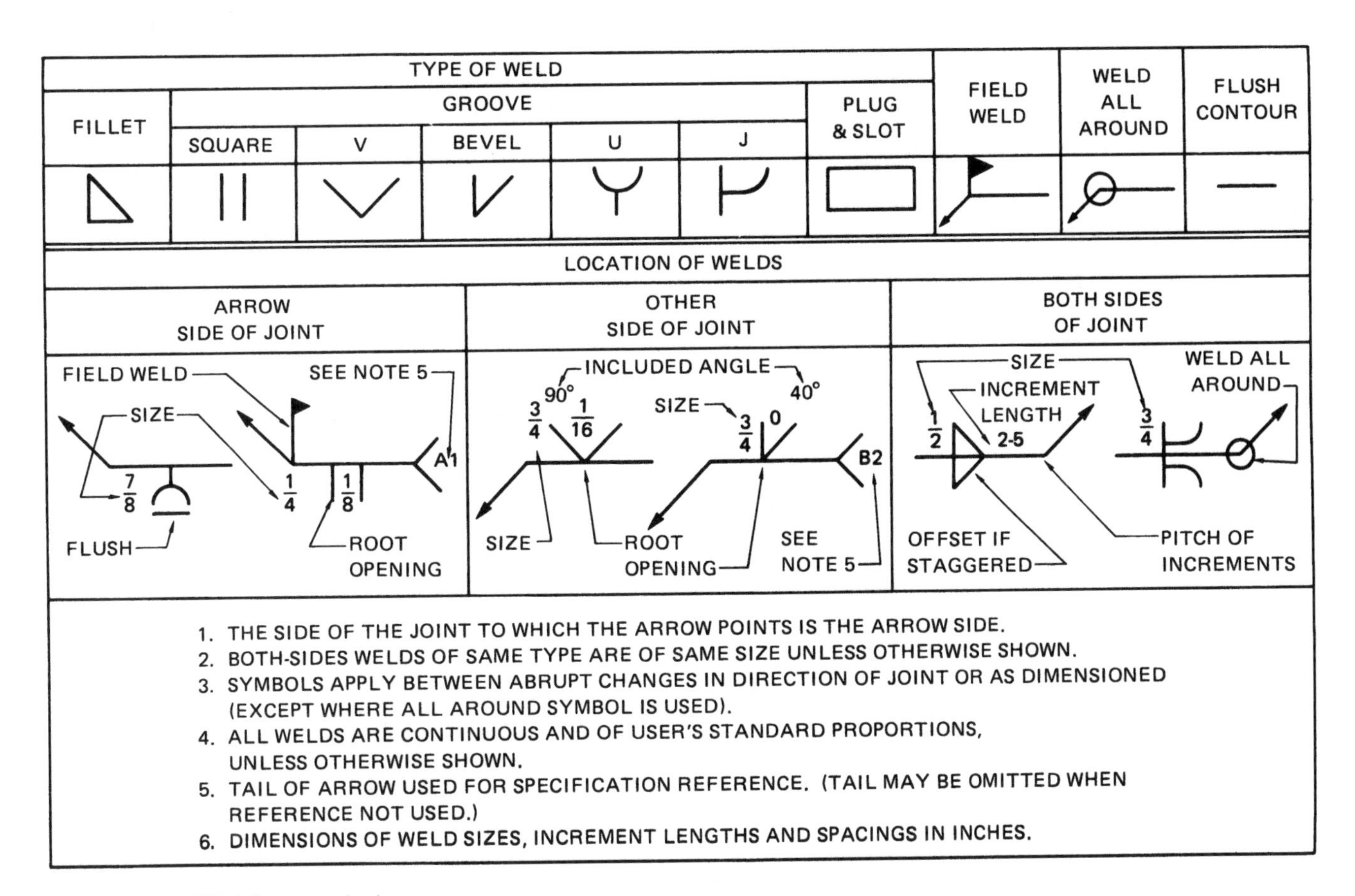

Figure 8-15 Welding symbols.

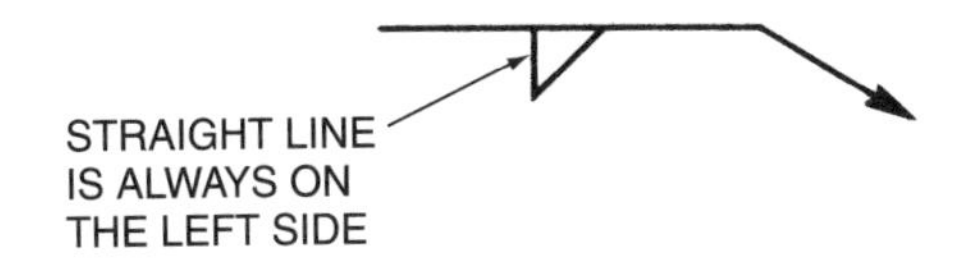

Figure 8-16 Fillet welding symbol.

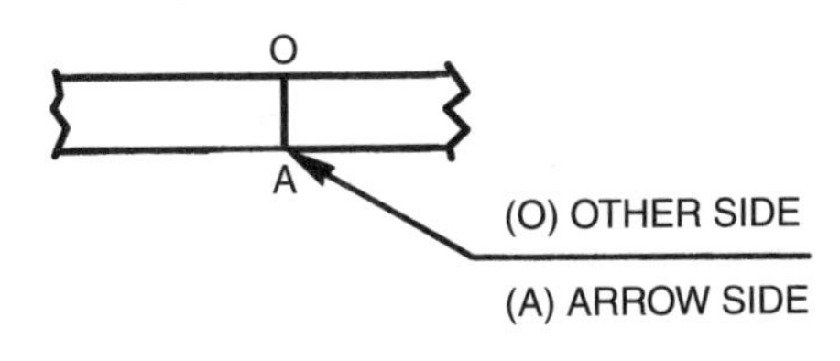

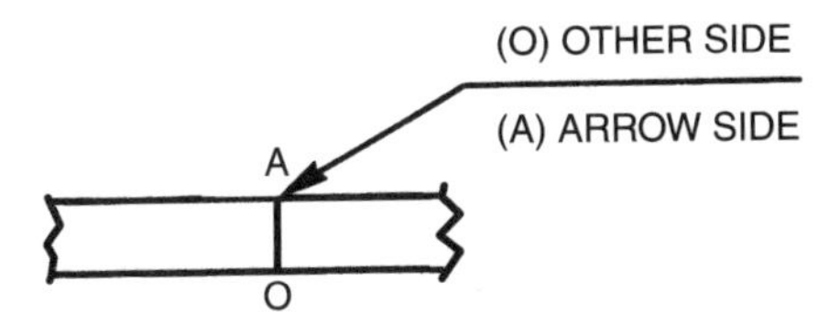

Figure 8-17 Reference line side designation.

If a weld is to be deposited on both sides of the same joint, then the weld symbol would appear both above and below the reference line.

When the same weld is to be performed at different locations, the arrow line may be divided and extended to each of those points (Figure 8-19).

The surface of the joint that the arrow actually touches is called the arrow side of the joint. The opposite side of the joint the arrow contacts is called the other side of the joint. On a drawing, when the joint is illustrated by a single line, it indicates that the weld joint extends outward behind the line (Figure 8-20). That surface represented by the line is considered to be the arrow side of the joint, and the unseen back side is considered to be the other side of the joint.

FILLET WELDS

Fillet weld dimensions are shown to the left of the welding symbol that they refer to (Figure 8-21). If there is to be a weld on both sides of the joint, the fillet welding symbol would appear both above and below the reference line; however, if both fillet welds

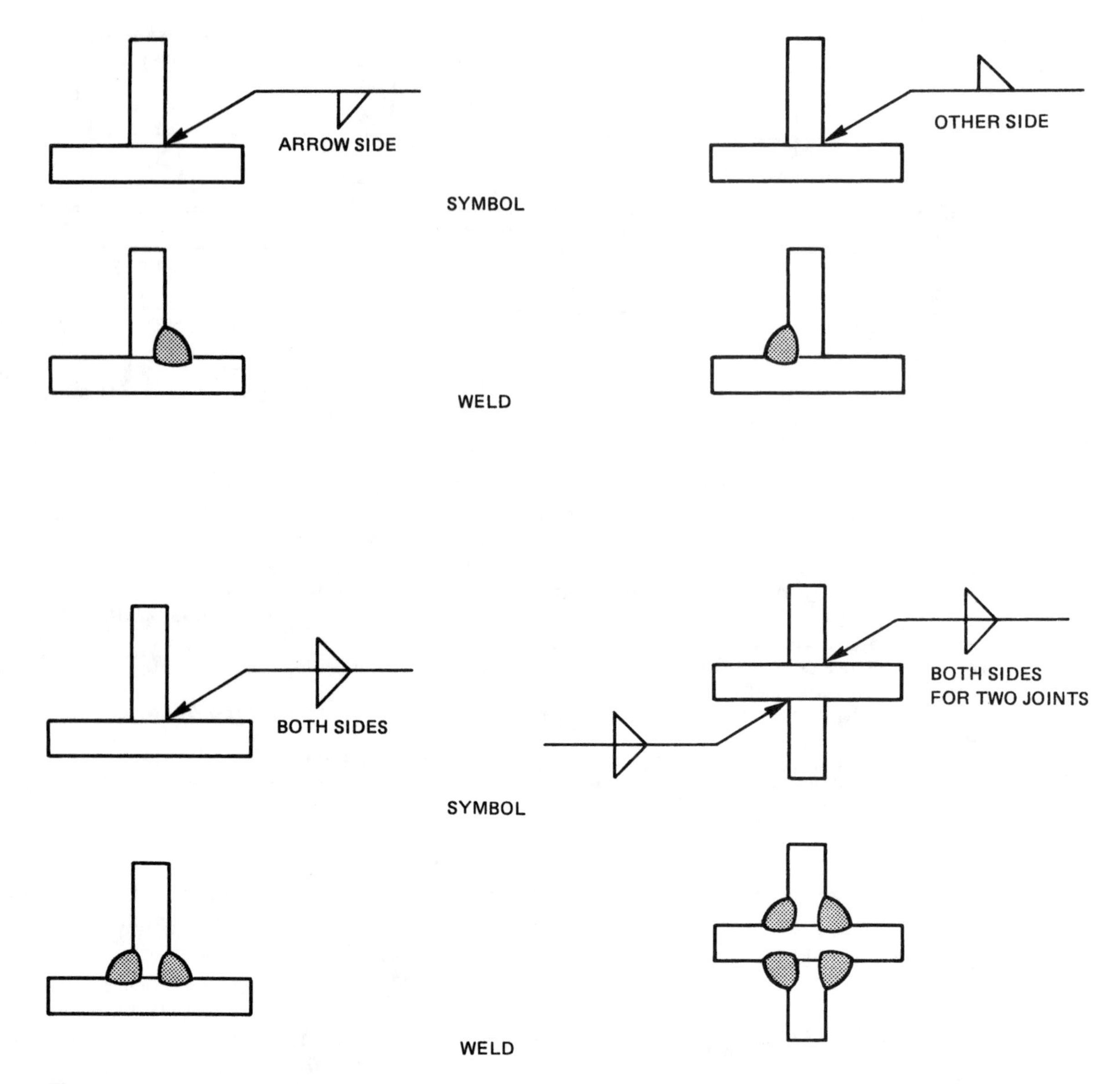

Figure 8-18 Designating weld locations. (Courtesy of the American Welding Society)

are to be the same size, the dimension may be placed on only one side of the reference line. When both sides of the joint are to be welded with different size fillet weld, the dimensions for each size fillet weld would appear on the appropriate side of the reference line.

The size of a fillet is given as the length of its leg. The dimension may be in fractions of an inch, decimals of an inch, or metric measurements and appear in parentheses to the left of the weld symbol. The leg of a fillet weld is the distance from the root measured horizontally and/or vertically to the toe of the weld (Figure 8-22).

The length of a fillet weld may also be included as part of its dimensioning. This information can be displayed in a number of ways, either on the welding symbol or on the drawing (Figure 8-23). When the length is indicated on the welding symbol, it is shown to the right of the symbol (Figure 8-24). When the dimension is shown on the drawing, the arrow line intersects the appropriate dimension line that extends between two extension lines locating the weld's length (Figure 8-25).

For a variety of reasons, a fillet weld may not be made continuous along a joint. Intermittent welds are often used to control distortion and reduce the tendency for crack propagation as well as reducing weld cost by increasing production speed and producing consumable supplies (Figure 8-26). The

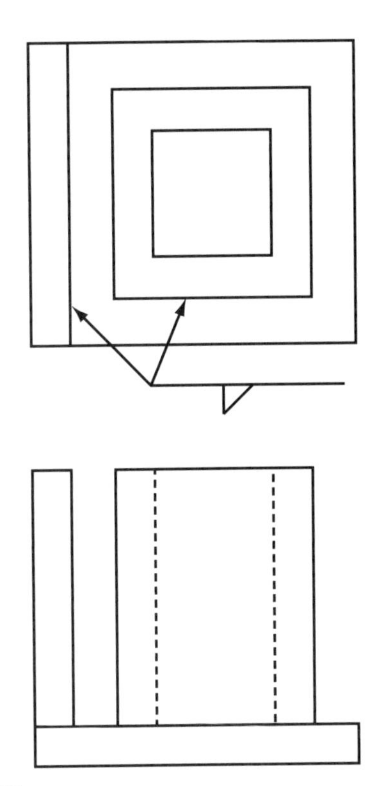

Figure 8-19

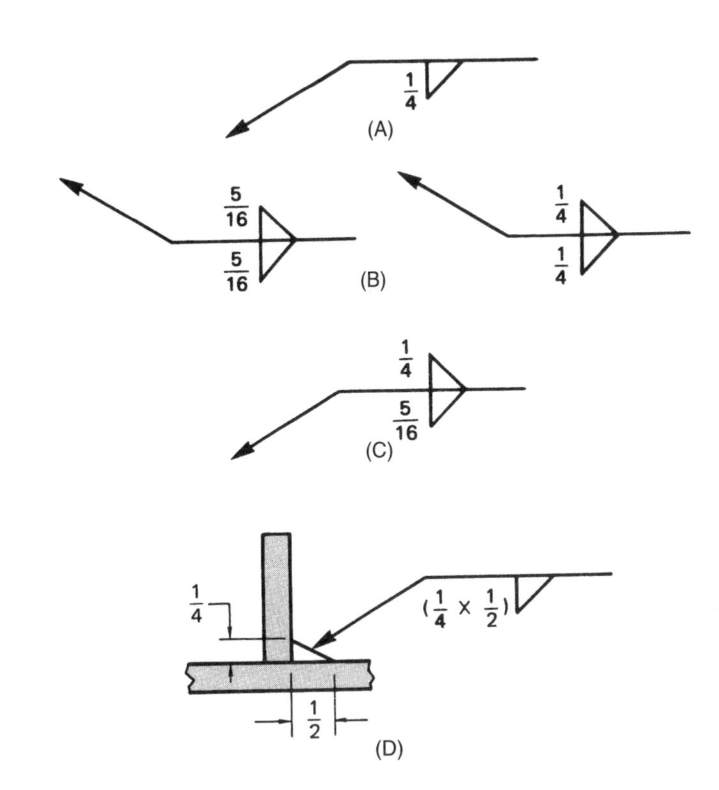

Figure 8-21 Dimensioning the fillet weld symbols.

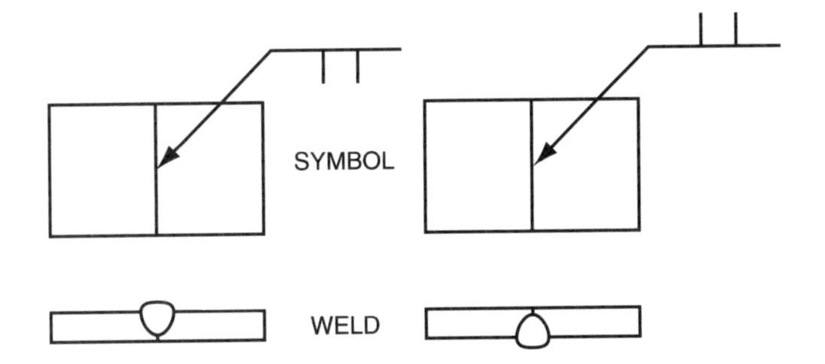

Figure 8-20 Weld location significance.

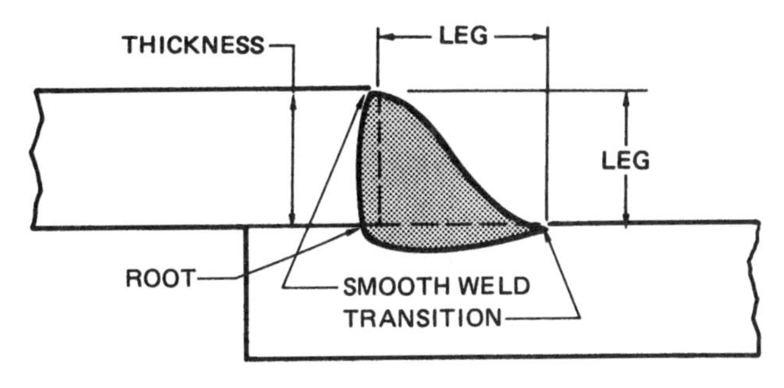

Figure 8-22 The legs of a fillet weld generally should be equal to the thickness of the base metal.

dimensions for an intermittent weld are given as the length and pitch. The length is the actual length of each individual weld. Thus a length indicating 3 inches would be a series of 3-inch-long welds made in the joint (Figure 8-27). The pitch is the distance from the center of one weld to the center of the next. Often the pitch is assumed to be the distance between each of the welds; however, this is incorrect. To properly dimension an intermittent weld, both the length and pitch must be included. The first number represents the length, and the second number represents the pitch (Figure 8-28).

A combination of weld lengths can be incorporated on a single part. It may be necessary for an ini-

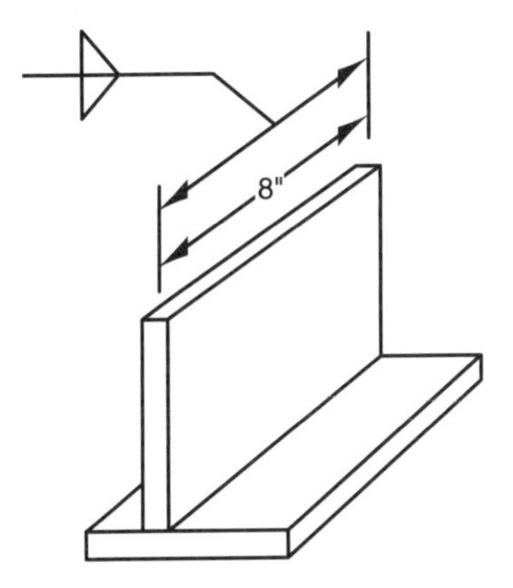

Figure 8-23

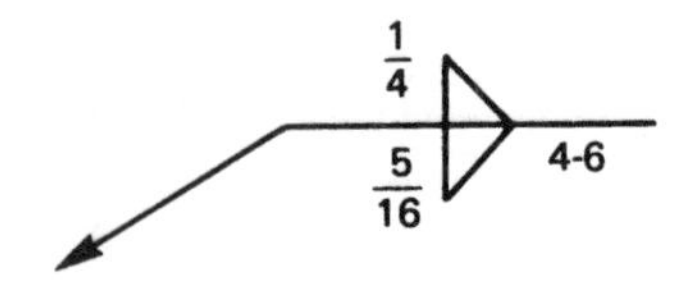

Figure 8-24 Dimensioning the fillet weld symbol.

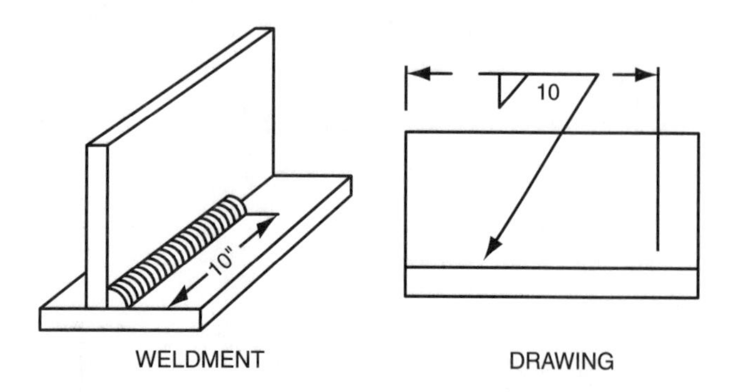

Figure 8-25

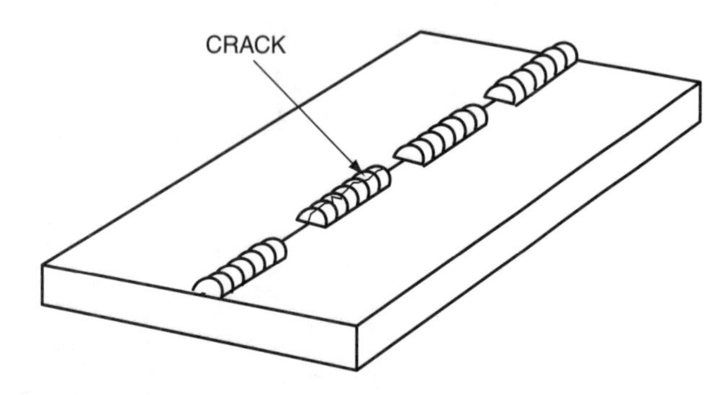

Figure 8-26 A crack that passes through one weld will not easily pass through the next weld.

Figure 8-27

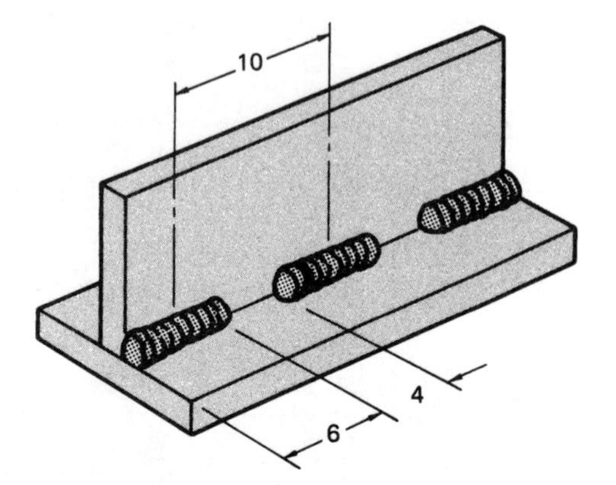

Figure 8-28 Dimensioning intermittent fillet welds.

tial weld to be made from the corner or edge some distance along the joint and then intermittent welds to be made across the remainder of the joint. Figure 8-29 illustrates a joint using this method of weld configuration.

PLUG WELDS

To make a plug weld, you must first cut a hole through the top plate exposing a small portion of the back plate's surface through the hole. The weld is then made by striking an arc on the back plate, establishing a molten weld pool and slowly adding filler metal until the weld's bead surface is even with the top plate's surface (Figure 8-30). The welding symbol for a plug weld is a small rectangular box located on the reference line (Figure 8-31). Dimensioning information for the plug weld includes size, pitch, depth, and angle. The size refers to the diameter of the hole

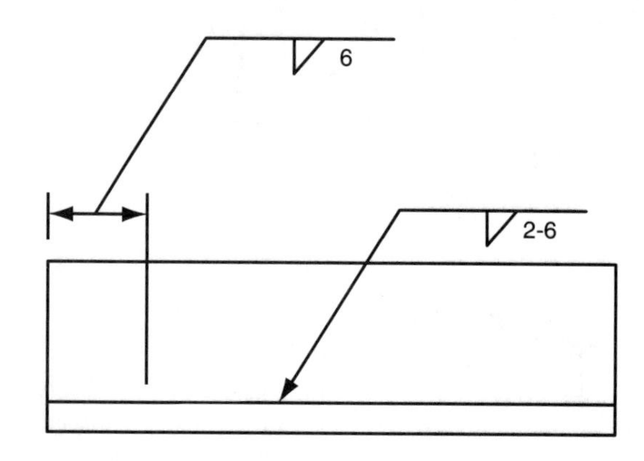

Figure 8-29

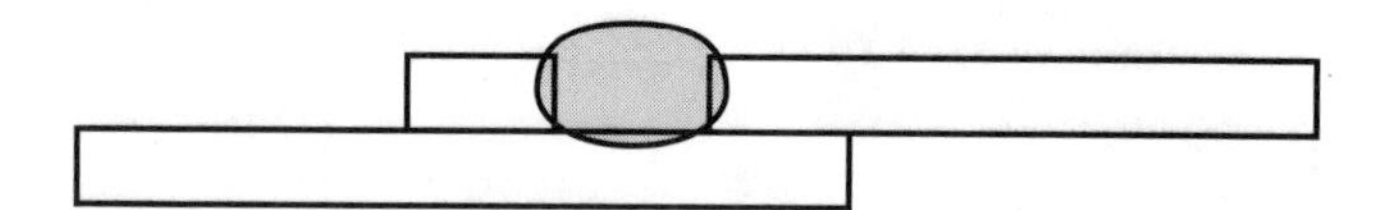

Figure 8-30 Plug weld.

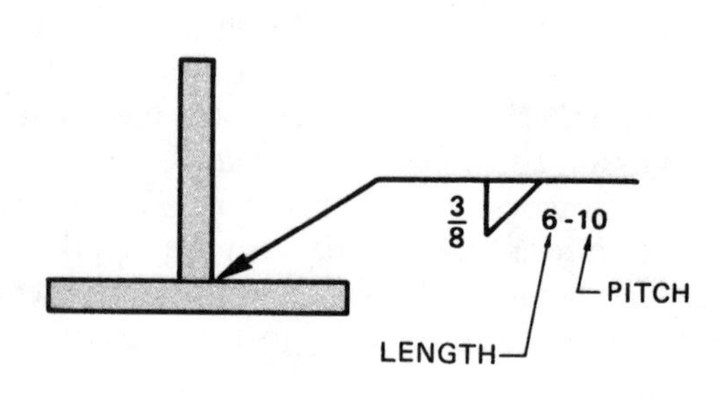

in the top plate. The pitch indicates the distance between the center of one plug weld and the center of the next. For gas tungsten arc welds, the depth is usually the same as the thickness of the top plate; however, the welds may occasionally not be filled flush with the top surface. That is the purpose of the depth dimension. The sides of a plug weld may have an angle other than 90°. This allows for better accessibility of the base plate (Figure 8-32).

SPOT WELDS

GTA spot and plug welds are similar in appearance to the finished weld and application. Both of these techniques allow plates to be fused along an overlapping seam (Figure 8-33).

A GTA spot weld is produced by establishing a molten weld pool on the top plate that is large enough to allow for fusion to occur on the surface of the back

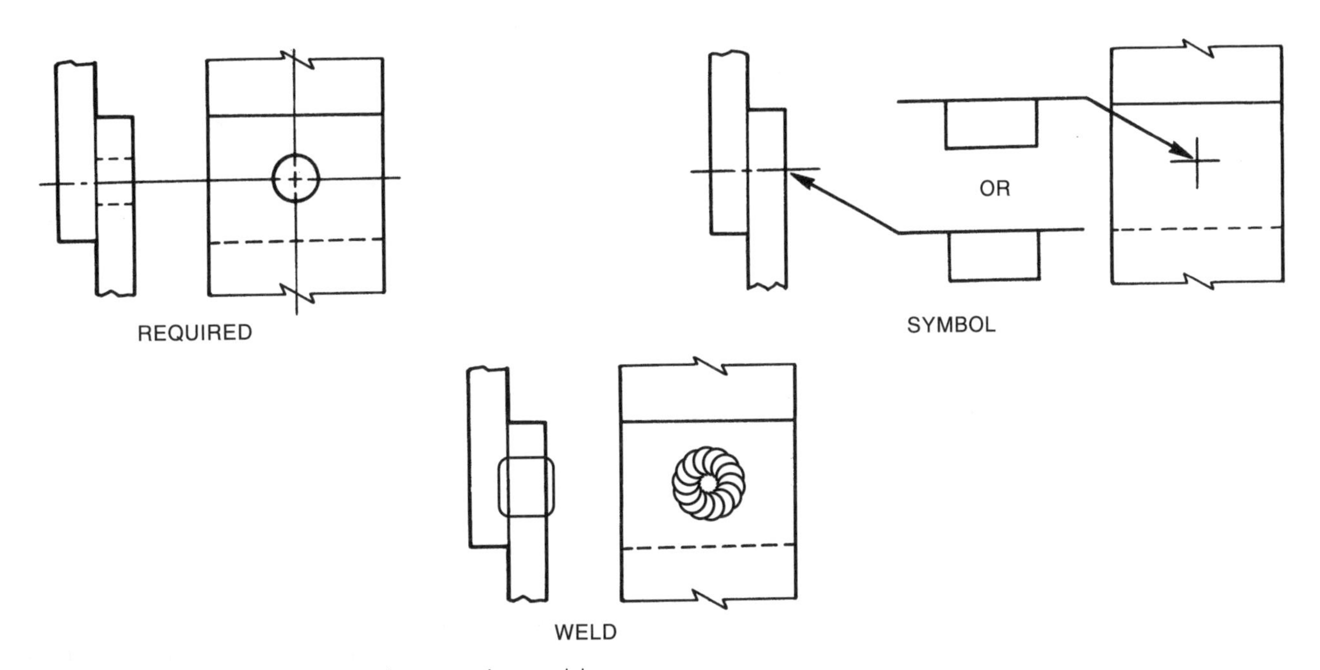

Figure 8-31 Applying dimensions to plug welds.

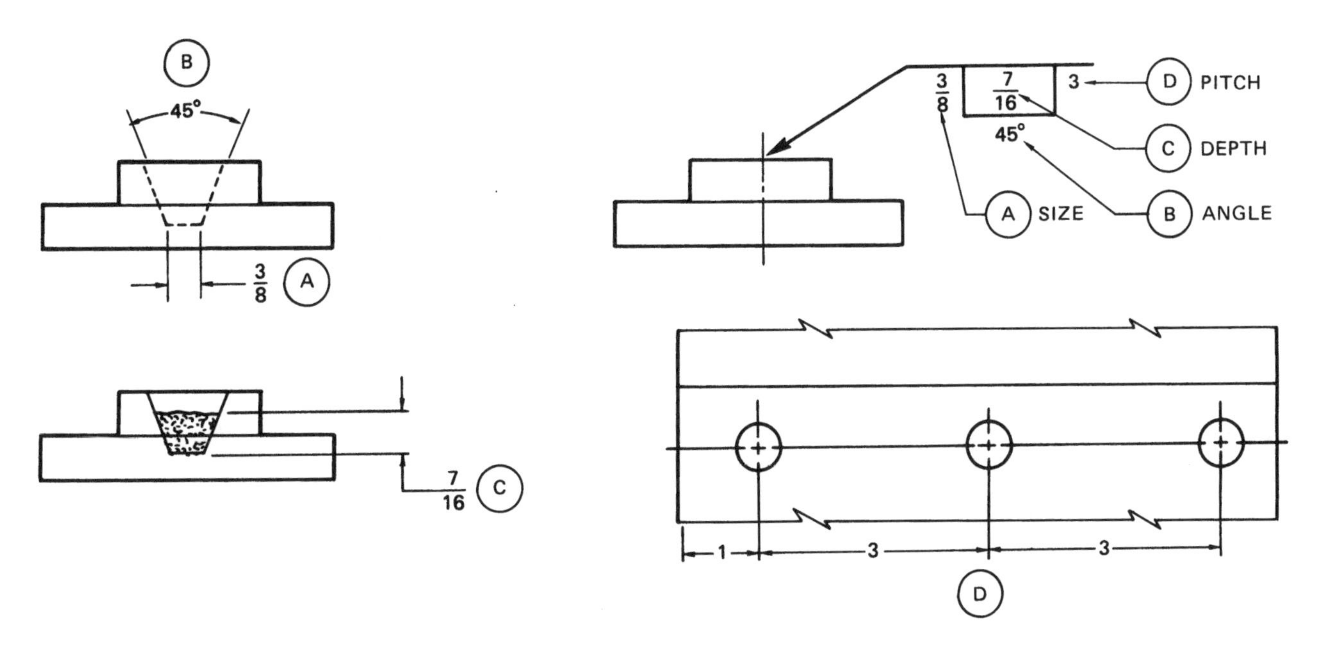

Figure 8-32

plate. GTA spot welds made in this manner will have the spot welding symbol: a circle placed on the appropriate side of the reference line and the letters *GTAW* placed in the tail of the weld symbol. The size of a spot weld may be dimensioned as either the diameter of the weld or as its strength in either pounds (newtons) per square inch per weld or pounds per square inch per linear inch of weld (Figure 8-34). Additionally, the pitch or center-to-center spacing of the weld may be given or the number of welds that are to be produced.

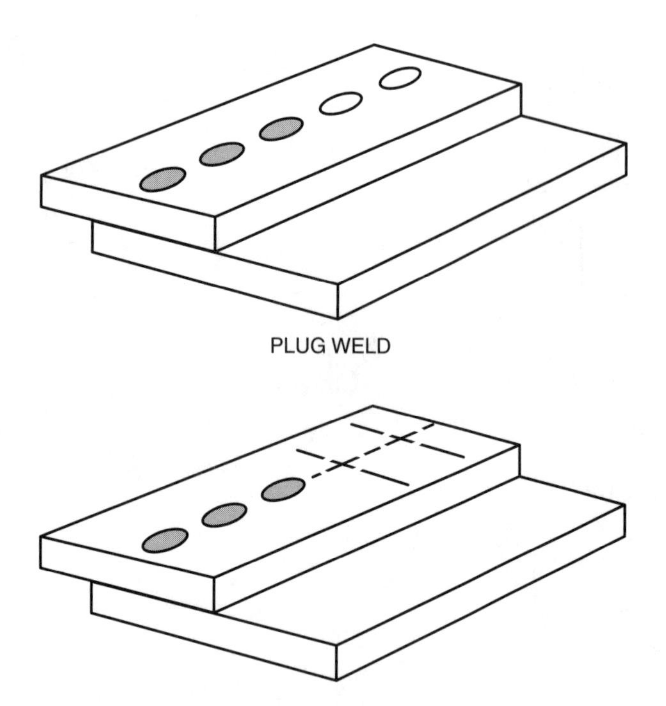

Figure 8-33 The completed plug and spot GTA welds look the same.

GROOVE WELDS

To ensure that the proper depth and penetration have been obtained, some joints are grooved before welding. There are several different types of grooves, which can be made in one or both plates and on one or both of the joint's sides. Grooves are most often used to increase the joint's strength without reducing its flexibility.

A groove can be cut into the base metal in a variety of ways. A groove may be produced by using oxyfuel cutting torches, air carbon arc cutting, plasma arc cutting, machining, or sawing.

The various types of groove welds are classified as follows:

- Single-groove and symmetrical double-groove welds that extend completely through the members being joined. No size is included on the weld symbol (Figure 8-35).
- Groove welds that extend only part way through the parts being joined. The size as measured from the top of the surface to the bottom (not including reinforcement) is included to the left of the welding symbol (Figure 8-36).
- The size of groove welds with a specified effective throat is indicated by showing the depth of groove preparation with the effective throat appearing in parentheses and placed to the left of the weld symbol (Figure 8-37). The size of

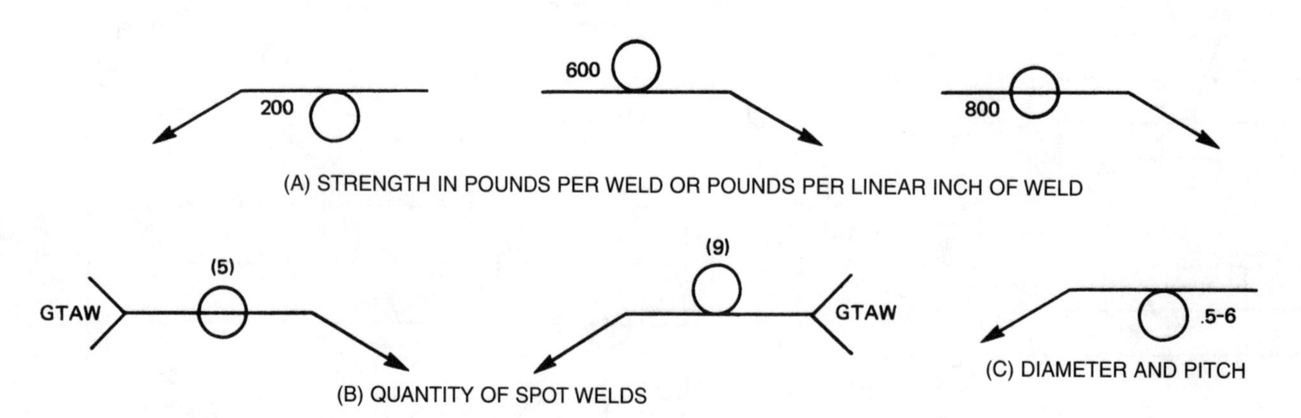

Figure 8-34 Designating strength and number of spot welds.

square groove welds is indicated by showing the root penetration. The depth of chamfering and the root penetration is read in that order from left to right along the reference line.

- The root face's main purpose is to minimize the burn-through that can occur with a feather edge. The size of the root face is important to ensure good root fusion (Figure 8-38).
- The size of flare groove welds is considered to extend only to the tangent points of the members (Figure 8-39).
- The root opening of groove welds is the user's standard unless otherwise indicated. The root opening of groove welds, when not the user's standard, is shown inside the weld symbol (Figure 8-40).

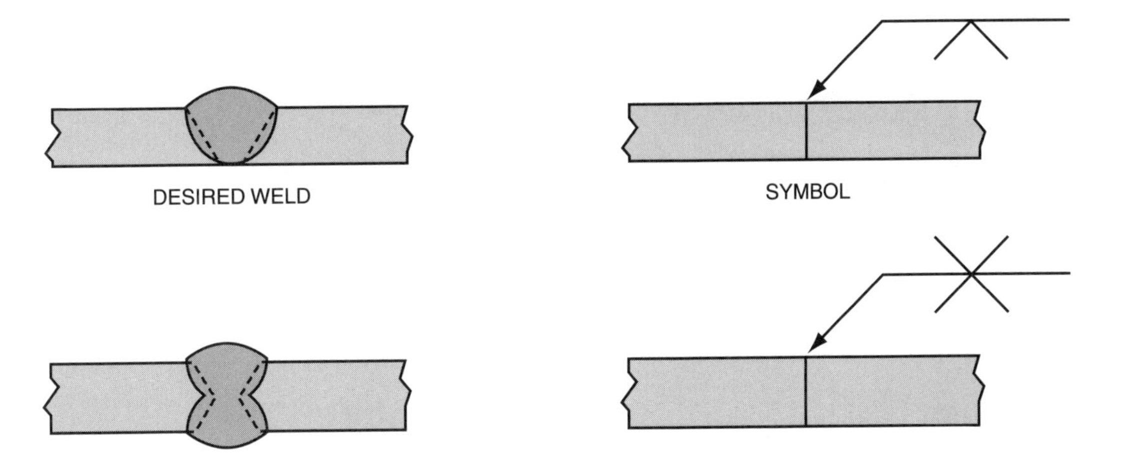

Figure 8-35 Designating single- and double-groove welds with complete penetration. (Courtesy of the American Welding Society)

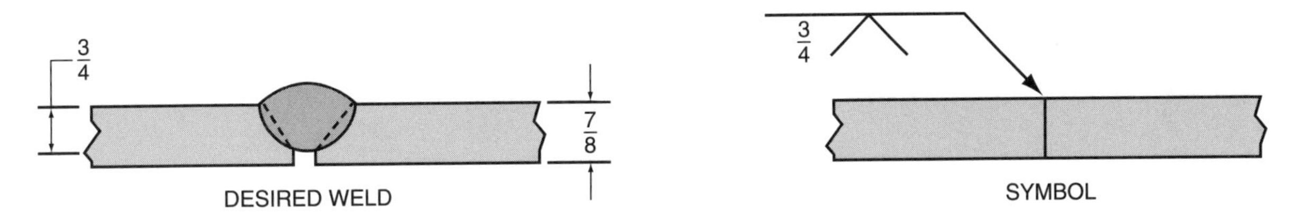

Figure 8-36 Designating the size of grooved welds with partial penetration. (Courtesy of the American Welding Society)

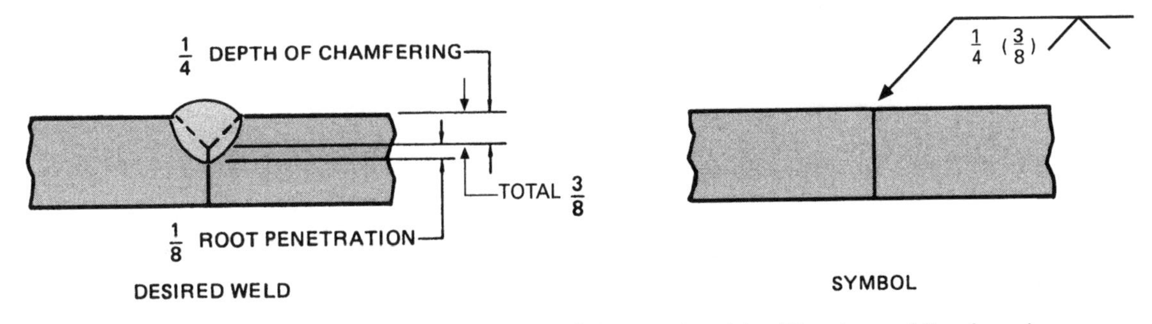

Figure 8-37 Showing size and root penetration of grooved welds. (Courtesy of the American Welding Society.)

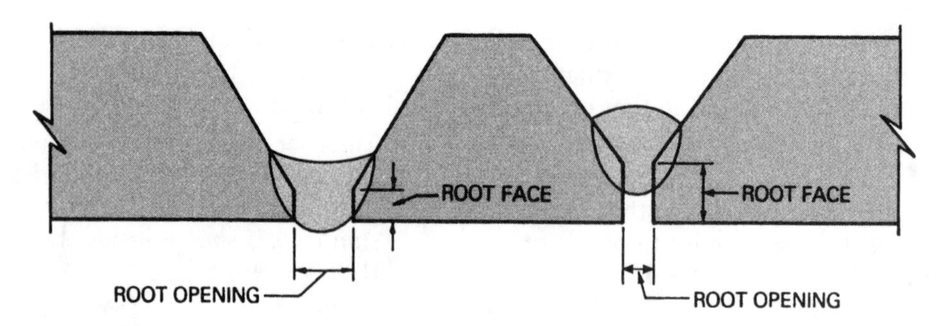

Figure 8-38 Root opening effect on the root weld.

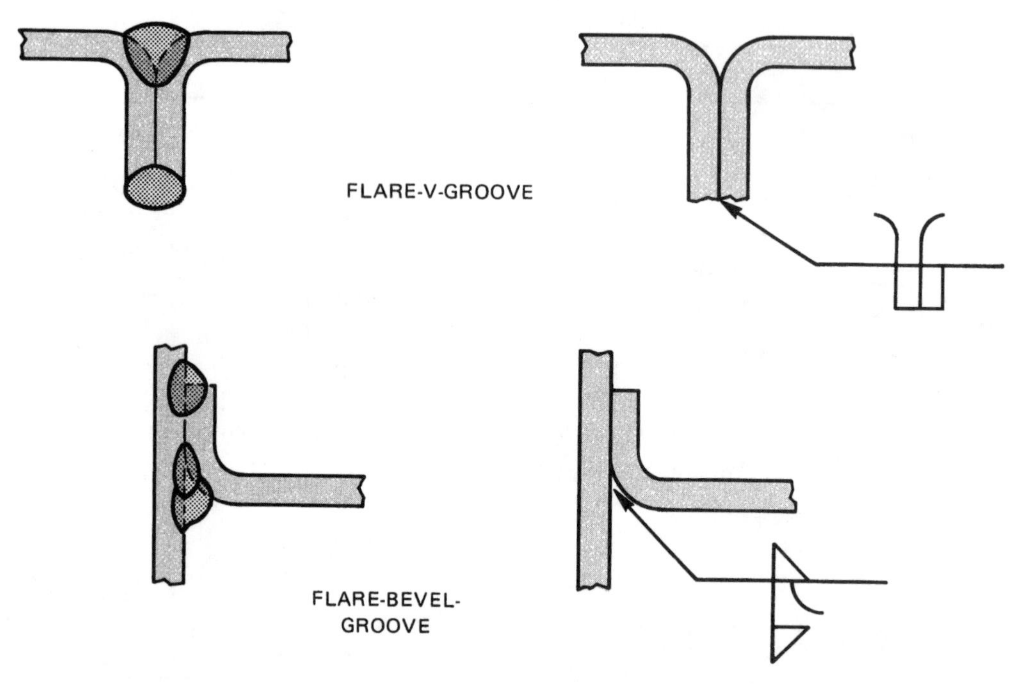

Figure 8-39 Designating flare-V- and flare-bevel-groove welds. (Courtesy of the American Welding Society)

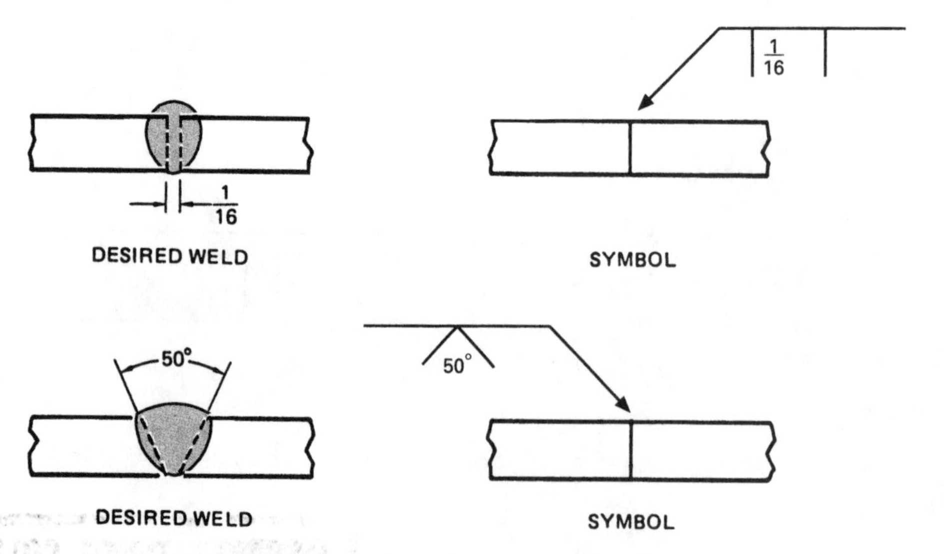

Figure 8-40 Designating root openings and included angle for groove welds. (Courtesy of the American Welding Society)

BACKING

A backing (strip) is sometimes placed on the root side of a joint. The purpose of this backing strip for GTAWelding is to both control the root face and prevent burn-through. The backing strip can be made of the exact same material as the base metal, or it may be made from a material very similar to the filler metal. A backing strip may be used on butt joints, tee joints, outside corner joints, and pipe joints (Figure 8-41).

Depending on the purpose of the finished product and any appropriate code or standard, the backing strip may be allowed to remain on the finished weld, or it may be removed. If the backing is to be removed, the letter *R* is placed in the backing symbol. The backing strip may be removed because it is often a source of stress concentration and a crevice into which trapped moisture can cause corrosion.

FLANGED WELDS

Flanged welds are sometimes used with thin-gauge sheet metal. Before welding begins, the edge of the sheet metal is bent to form a small flange (Figure 8-42). This flange controls joint distortion and burn-through and may serve as the filler metal for the completed weld. Flanged welds are most often performed in butt joints and outside corner joints.

- Edge flange welds are shown by the edge flange weld symbol.
- Corner flange welds are indicated by the corner flange weld symbol.
- Dimensions of flange welds are shown on the same side of the reference line as the weld symbol and are placed to the left of the symbol (Figure 8-43). The radius and height above the point of tangency are indicated by showing

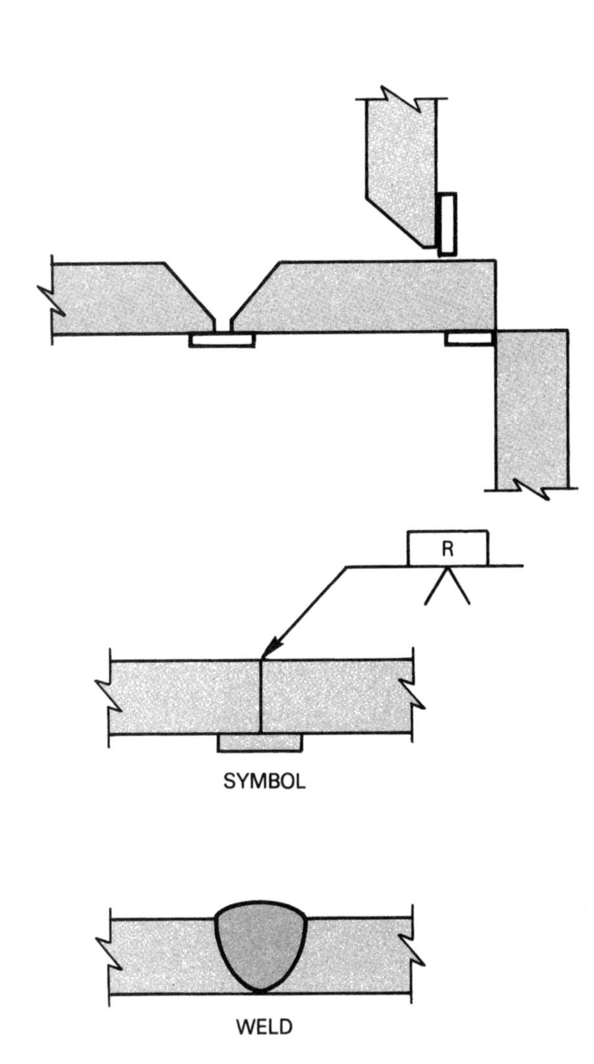

Figure 8-41 Backing strips.

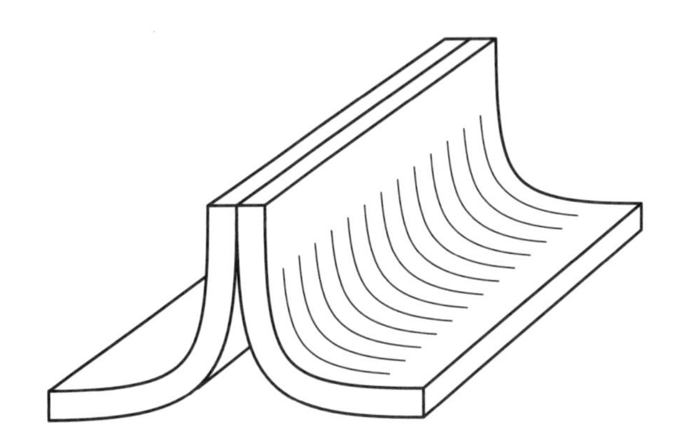

Figure 8-42 The edge of the metal is prebent before welding.

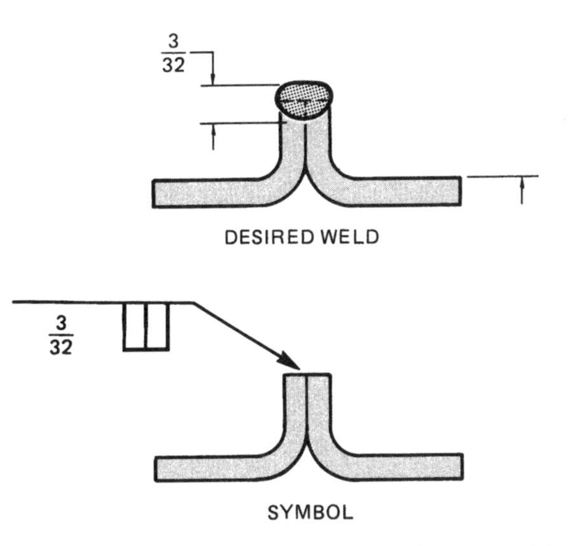

Figure 8-43 Applying dimensions to flange welds. (Courtesy of the American Welding Society)

both the radius and the height separated by a plus sign.

- The size of the flange weld is shown by a dimension placed outward from the flanged dimensions.

NONDESTRUCTIVE TESTING SYMBOLS

Because GTA welds are frequently used in applications where extremely high-quality welds are required, they are frequently inspected before, during, and after the actual welding. It is necessary, therefore, for a welder to become familiar with the nondestructive testing (NDT) standardized symbols, which use the same basic reference line and arrows as the welding symbol (Figure 8-44).

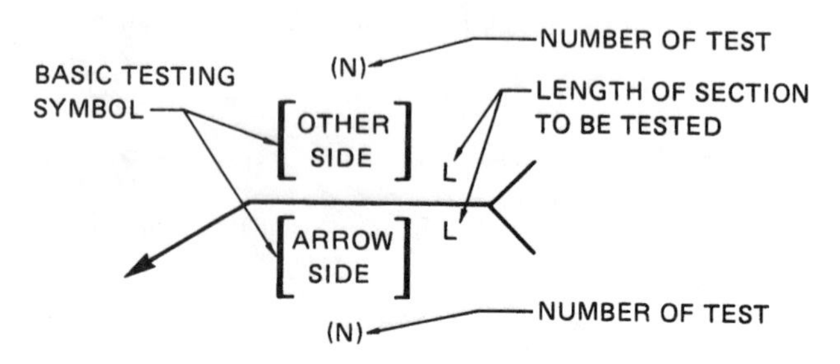

Figure 8-44 Basic nondestructive testing symbol.

The symbol for the particular type of NDT to be performed is shown on the reference line. As with the welding symbol, the significance of having the NDT symbol above or below the reference line refers to arrow or other side of the joint. Symbols that are above the line indicate other side; symbols that are located below the reference line indicate arrow side; and those that are located on the reference line indicate that there is no preference for which side of the joint will be tested (Figure 8-45). Some tests may be performed on both sides of the weld surface; therefore, the NDT symbol will appear above and below the reference line to indicate this.

In some cases two or more tests may be performed on the same joint. The first test usually is less expensive and serves as a screening test to determine whether more expensive testing should follow. By doing this the weld can frequently be repaired and brought up to standard before the complete bank is performed. Figure 8-46 shows several methods that are commonly used to combine NDT weld test symbols.

Dimensional information that may be included with NDT test symbols includes the length of weld to be tested and the number of locations the test is to be performed. The length dimension may be given either to the right of the test symbol or can be shown by using extension lines and the arrow line (Figure 8-47).

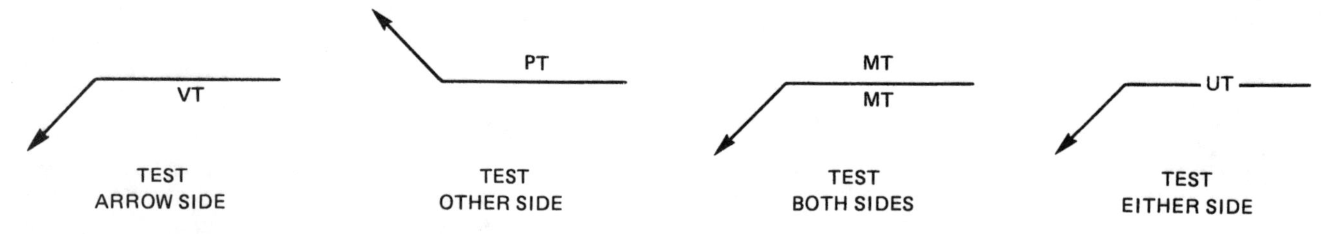

Figure 8-45 Testing symbols used to indicate what side is to be tested.

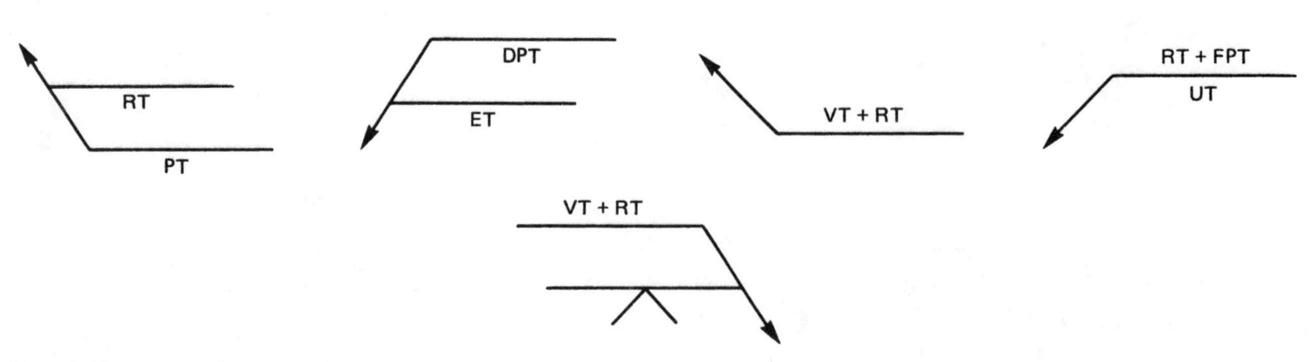

Figure 8-46 Methods of combining testing symbols.

The number of tests to be made is given in parentheses above or below the test symbol (Figure 8-48).

Both welding symbols and NDT symbols can be combined to form a single symbol (Figure 8-49). The combining of symbols can serve both to alter the welder of an impending test that will be performed post-welding and to emphasize the importance of a particular weld's quality.

Because an X ray is in essence the shadow of a weld captured on film, the angle used to expose the film is important (Figure 8-50). If the X ray is taken to vertically through a joint, no depth perception is available to the technician reading the X ray to determine exactly where discontinuities may have appeared within the weld. If the angle is too sharp, the discontinuity showing on the film may also be skewed excessively, making size determination difficult. The engineer or designer of the weldment can include the desired exposure angle as part of the radiographic testing symbol.

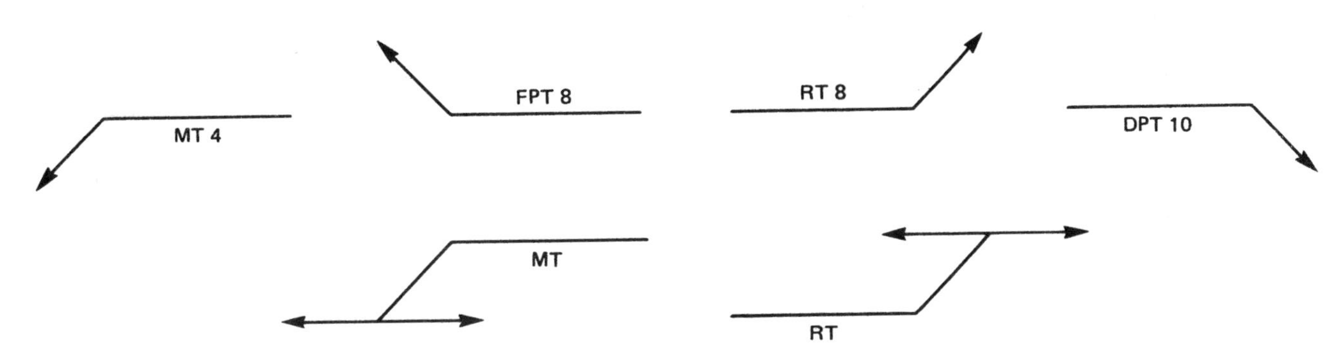

Figure 8-47 Two methods of designating the length of weld to be tested.

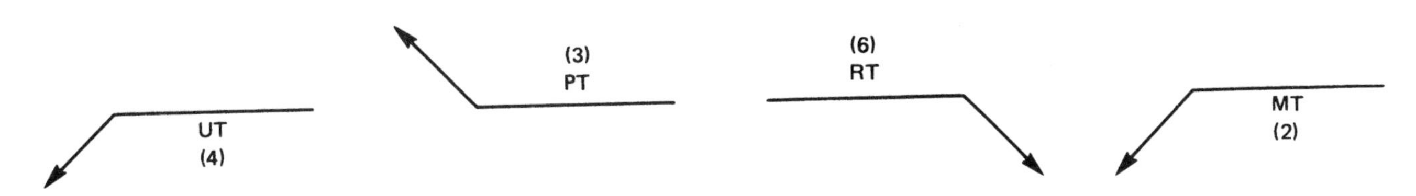

Figure 8-48 Method of specifying the number of tests to be made.

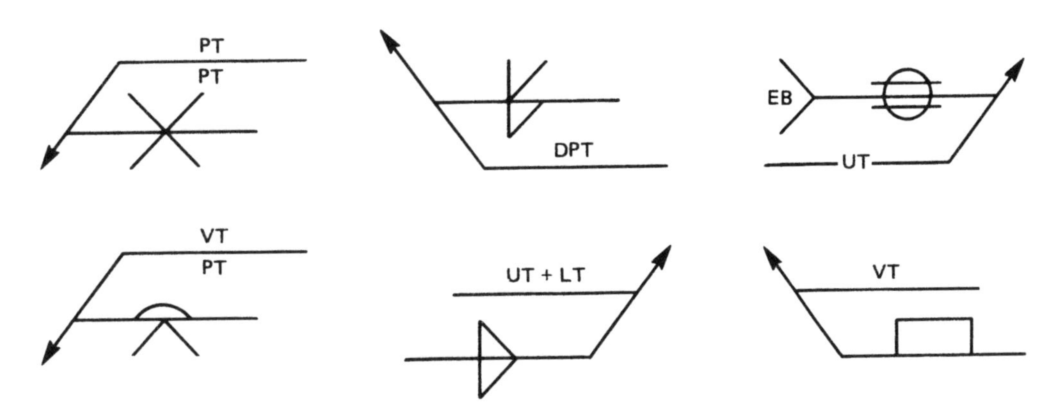

Figure 8-49 Combination welding and nondestructive testing symbols.

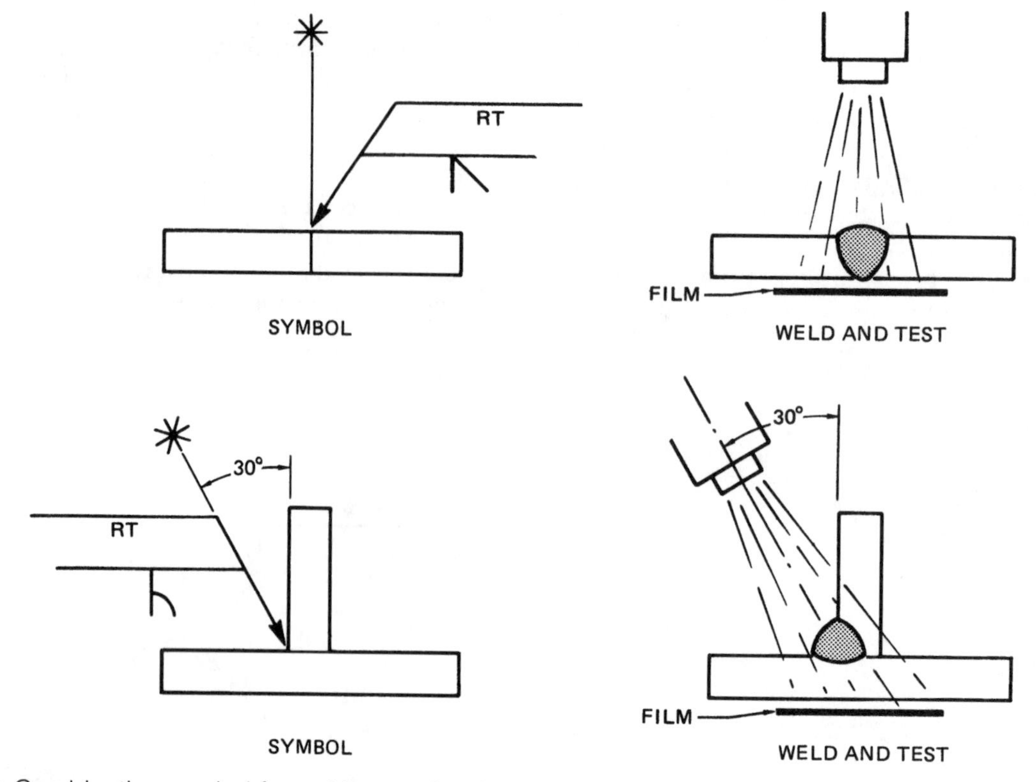

Figure 8-50 Combination symbol for welding and variation location for testing.

REVIEW QUESTIONS

1. What stresses must a weld joint withstand?
2. Sketch and label five edge preparations that are used for welding joints.
3. Why is it usually better to make a weld in the flat position?
4. List four basic positions for groove and fillet welds.
5. Sketch a weld on a pipe in the 1G position.
6. What are some of the factors that determine the shape of the edge preparation?
7. What is a prequalified joint?
8. Why are welding symbols used?
9. What type of information can be included in the welding symbol?
10. Why is a tail added to the basic welding symbol?
11. What are different classifications of welds that a symbol can indicate?
12. How is the reference line always drawn?
13. How are the dimensions for a fillet weld given?
14. What dimensions can be given for a plug weld?
15. What two units show the minimum shear strength of a spot weld?
16. How can the groove be cut on the edge of a plate?
17. What type of joints may joints be used on?
18. How is the removal of backing strip noted on a welding symbol?
19. How are flanged edges formed?
20. Sketch two NDT symbols that illustrate different methods of indicating multiple test requirements for the same section of weld.

CHAPTER 9

BLUEPRINT READING

INTRODUCTION

Mechanical drawings have been called the universal language because they are produced in a similar format worldwide. There are only a few differences in the arrangement of the views on a drawing (Figure 9-1). These slight differences in arrangement are minor, and once you understand the procedure for placing the various views on the drawing, you will have no difficulty interpreting them correctly. Such interpretation is facilitated by the fact that the method for representing the view on both types of drawings is the same as the method of dimensioning.

A set of drawings contains all of the information required to produce a particular part, fabrication, or finished product. This set of drawings may consist of a single sheet with one or more drawings or multiple sheets that may be required for larger, more complex parts. The pages may include a title page, pictorial, assembly drawing, detailed drawing, and various views (Figure 9-2).

In addition to the specifications and dimensions, other information may be included, such as a bill of materials and a title block. The bill of materials will list the materials, including grades of metal,

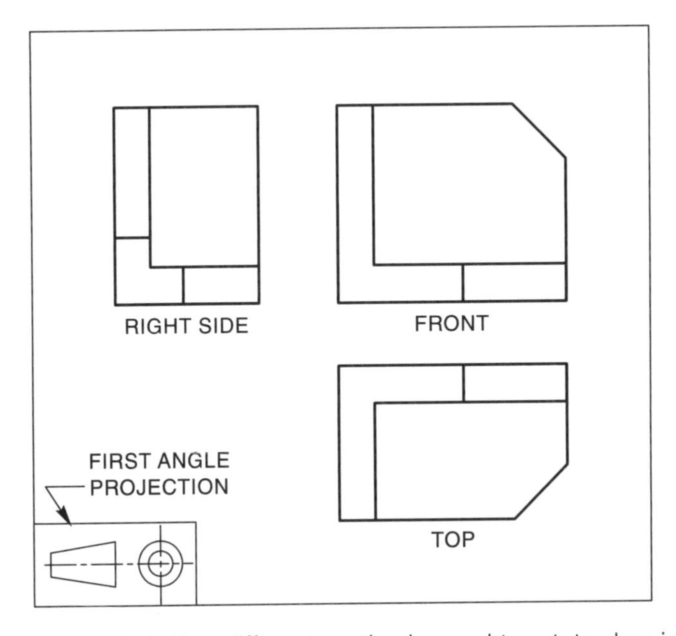

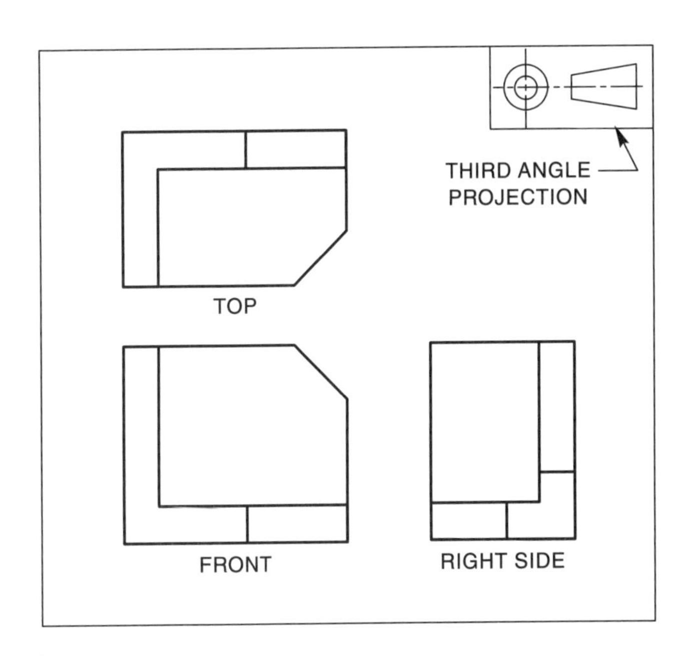

Figure 9-1 Two different methods used to rotate drawing views.

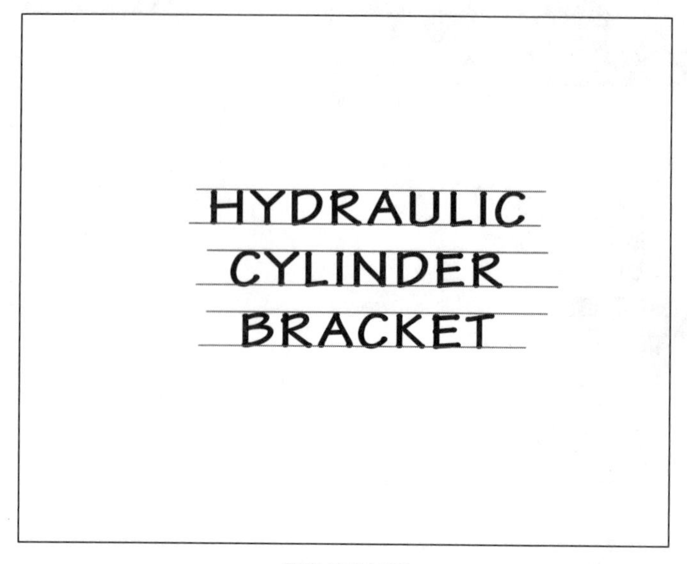

TITLE PAGE

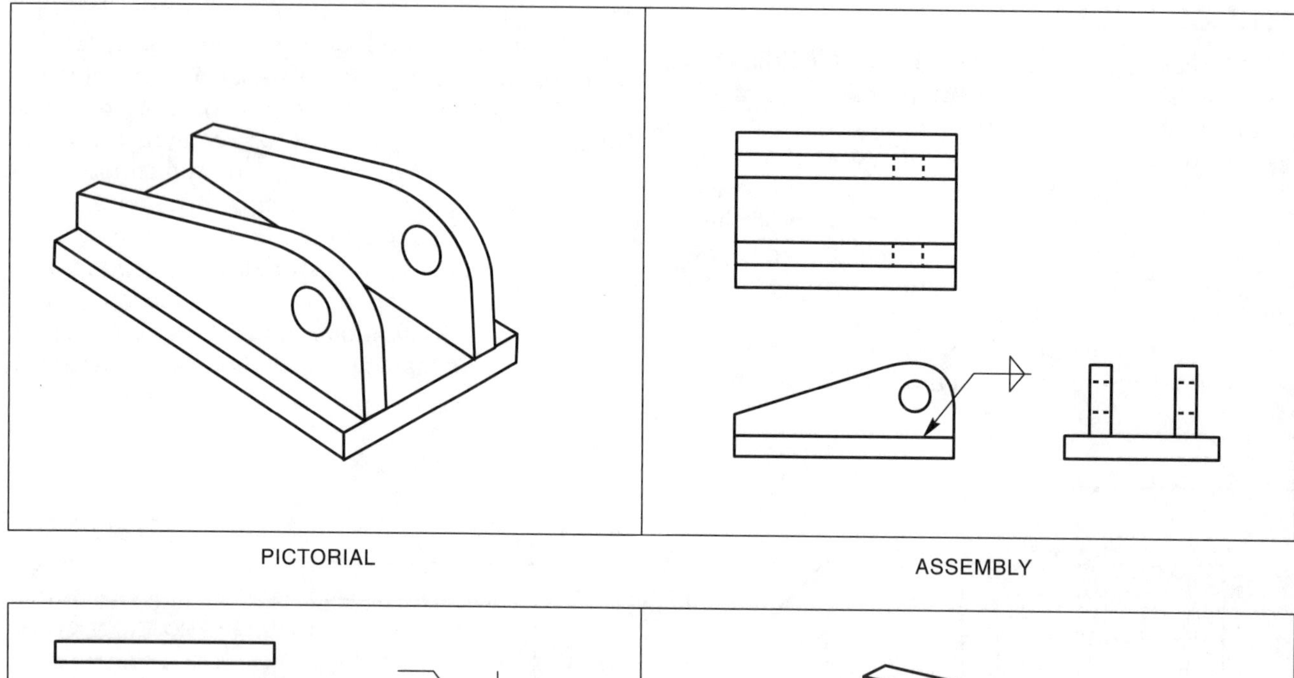

PICTORIAL

ASSEMBLY

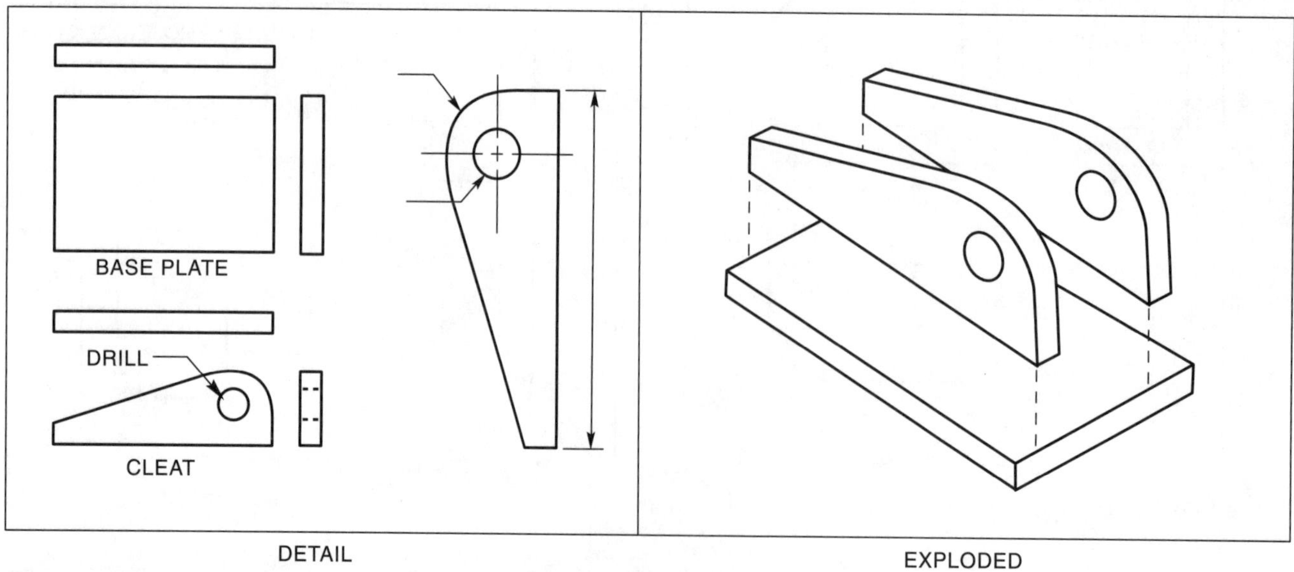

DETAIL

EXPLODED

Figure 9-2 Drawings that can make up a set of drawings.

BILL OF MATERIALS				
Part	Number Required	Type of Material	Size Standard Units	Size SI Units
Base	1	Hot Roll Steel	1/2" × 5" × 8"	12.7 mm × 127 mm × 203.2 mm
Cleat	2	Hot Roll Steel	1/2" × 4" × 8"	12.7 mm × 101.6 mm × 203.2 mm

Figure 9-3 Typical Bill of Materials.

thicknesses, and sizes for each separate component or item needed to build the weldment (Figure 9-3).

The *title box* will normally appear in one corner or on one edge of the drawing. It contains general information about the part and specific information about the drawing. General information about the part includes its name, the name of the individual or business that ordered the part, tolerances, and so on. Information regarding the drawing may include the scale of the drawing, the date of the drawing, the creator of the drawing, the number of each page, the number of pages in the set, and, if this is a revised drawing, the effective date of the revision.

LINES

The drawing consists of a variety of types of lines, each of which has a specific meaning. The lines are collectively known as the *alphabet of lines* (Table 9-1 and Figure 9-4). A description of the most commonly used lines and their purpose is as follows:

- Object line — The object line is a solid, dark bold line that shows the shape of an object, including the outline, holes, slots, the edges of different pieces that make up the part, and (on flat parts) any corner created by a change in angle. On round or circular parts, the object line illustrates the maximum diameter or radius. This is true even though the part may look flat as seen in one view (Figure 9-5). In such a case, you must look at another view to determine the exact shape of the object.
- Hidden line — A hidden line is a medium weight series of short dashes that can range from approximately 1 1/4″ to 3/8″ in length; the space between the dash can be from 1/16″ to 1/8″. The dashes and spaces on the drawing should be consistent. Hidden lines represent the same thing that object lines represent, except that the surface of extent of the curb surface occurs below or behind a solid surface. Hidden lines therefore would represent notches, grooves, holes, and so on that are to be made in the back or bottom of the part that would not be visible when looking at the object from this direction.
- Center line — The center line is a thin, fine broken line made up of longer line sections with shorter dashed line sections (Figure 9-6). The long line section may be from 1/2″ to several inches in length. The short dashed sections may range from 1/4″ to 1/2″ in length. Spacing between each of these lined sections may range from 1/8″ to 1/4″. The center line shows the centers of holes, curves, or symmetrical parts. Center lines may also be extended for dimensioning (Figure 9-7).
- Extension line — The extension line is a thin fine line that is drawn as an extension of an object line. There must be a 1/16″ to 1/8″ gap between the end of the extension line and the object line it extends from (Figure 9-8). Extension lines are used for dimensioning purposes and should extend 1/8″ to 1/25″ beyond the

TABLE 9-1

ALPHABET OF LINES

Line Type	Description	Purpose
Object Line	Solid bold line	To show the intersection of surfaces or the extent of a curved surface.
Hidden Line	Broken medium line	To show the intersection of surfaces or the extent of a curved surface that occurs below the surface and hidden from view.
Center Line	Fine broken line made up of longer line sections on both sides of a short dashed line	To show the center of a hole, curve, or symmetrical object.
Extension Line	Extension lines (fine line) extending from near the surface of the object	Extension lines extend from an object line or hidden line to locate dimension points.
Dimension Line	Dimension lines (medium line) extending between extension lines or object lines	Dimension lines touch the extension lines and/or object lines that represent the points being dimensioned
Cutting Plane Line	Bold broken lines with arrowheads pointing in the direction of the cut surface	These lines extend all the way across the surface that is being imaginarily cut. The arrowhead ends point in the direction in which the cut surface will be shown in the sectional drawing.
Section Lines (Steel, Cast Iron)	Series of fine lines drawn at an angle to the object lines. The line angle usually changes from one part to another. The cast iron section lines are used universally for most sections.	Used to indicate a surface that has been imaginarily cut or broken. The spacing and pattern can be used to indicate the type of material that is being viewed.
Leader or Arrow Line	Medium line with an arrowhead at one end	Leader and arrow lines are used to locate points on the drawing to which a specific note, dimension, or welding symbol refers.
Long Break Line	Bold straight line with intermittent zigzag	To indicate that a portion of the part has not been included in the drawing either to conserve space or because the omitted portion was not significant to this specific drawing.
Short Break Line	Bold freehand irregular line	Used for the same purposes as the long break above except on parts not wide enough to allow the long break lines with their zigzags to be used clearly.

last dimension line. Extension lines may be used to represent the extent of a curved surface or circular object (Figure 9-9).

- Dimension lines — Dimension lines are medium-weighted lines that locate the exact spots between which the dimension refers (Figure 9-10). They may extend from an extension line or object line. The exact point they refer to is identified by placing an arrow dot, or slash on the dimension line (Figure 9-11). The dimension line may have a short break in the middle where the dimension is placed, or the dimension may be written just above or below the dimension line (Figure 9-12).
- Cutting plane line — The cutting plane line is a bold line made up of a series of long and short dashed lines. The long dashed line can range from 1/2″ to 1″ in length, and the short dashed line ranges from 1/4″ to 3/8″ in length. Spaces between the line ranges from 1/8″ to 1/4″. Each end of the line has a short line drawn at a right angle to the main line with arrows at their end

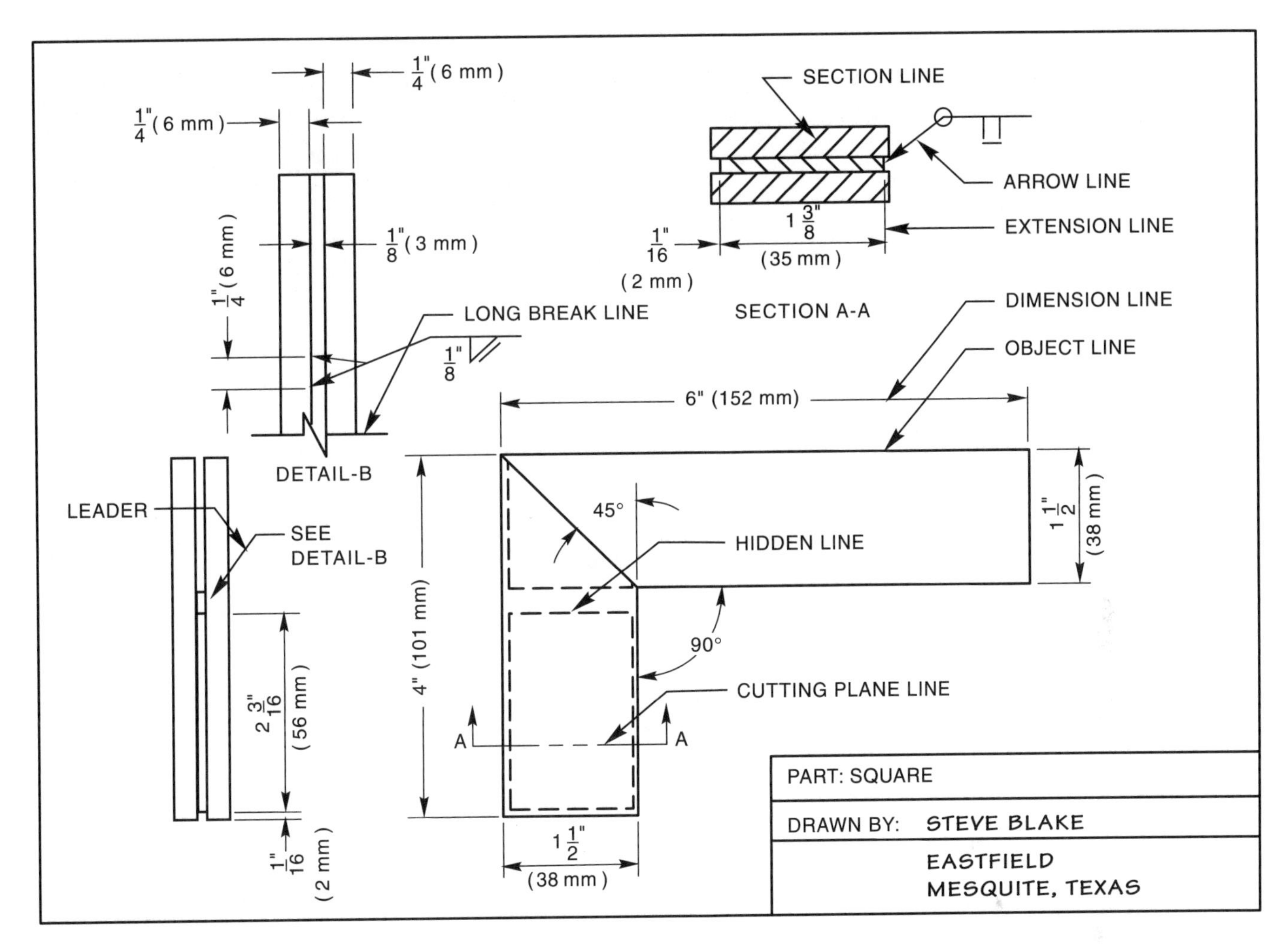

Figure 9-4

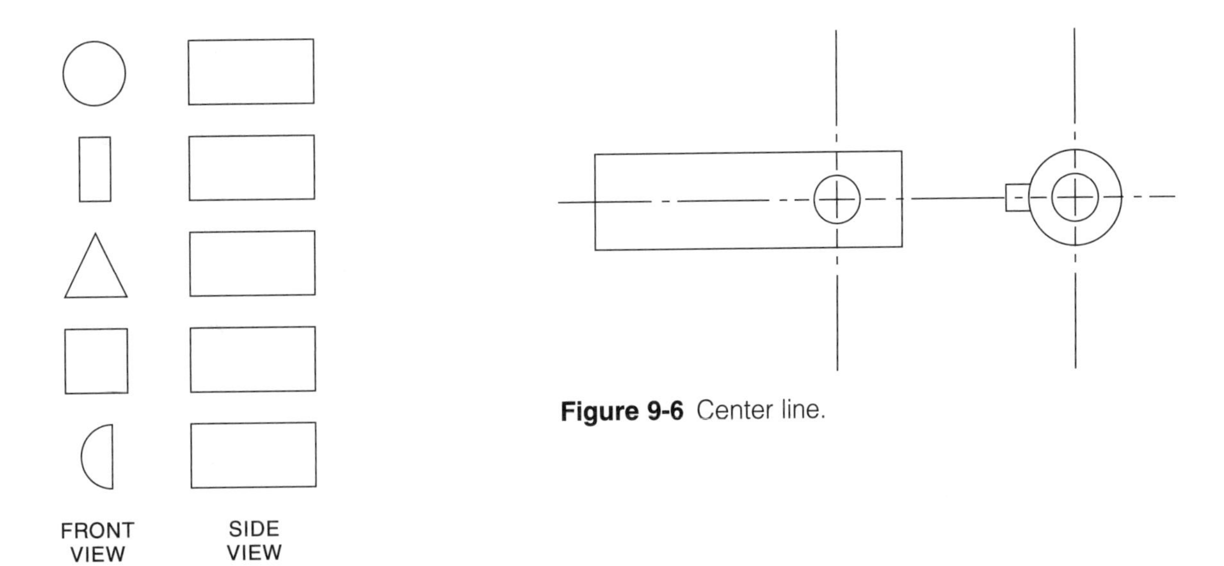

Figure 9-5 All of the side views would look the same for each of these different front views.

Figure 9-6 Center line.

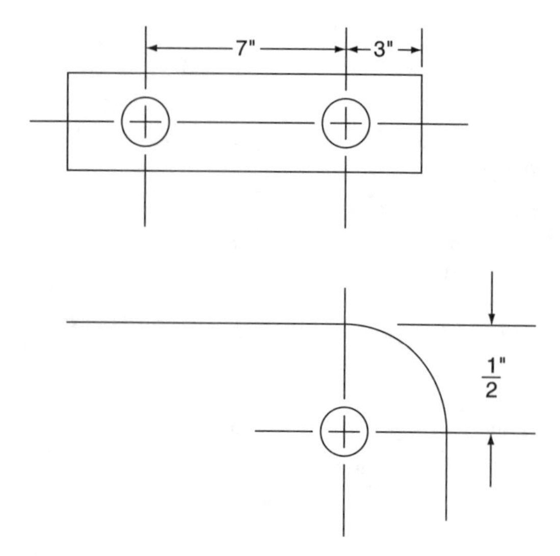

Figure 9-7 Using center lines for dimensioning.

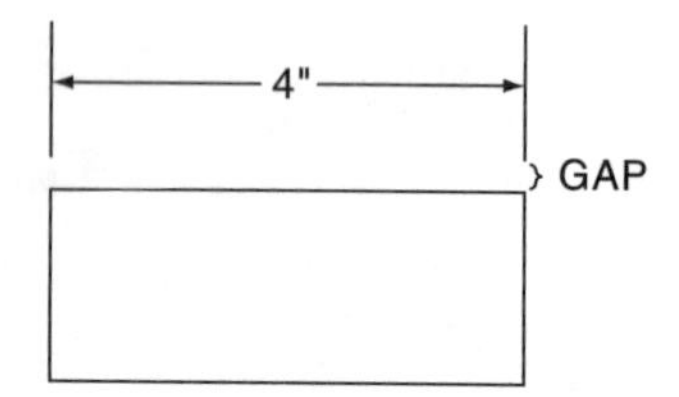

Figure 9-8 Extension line spacing.

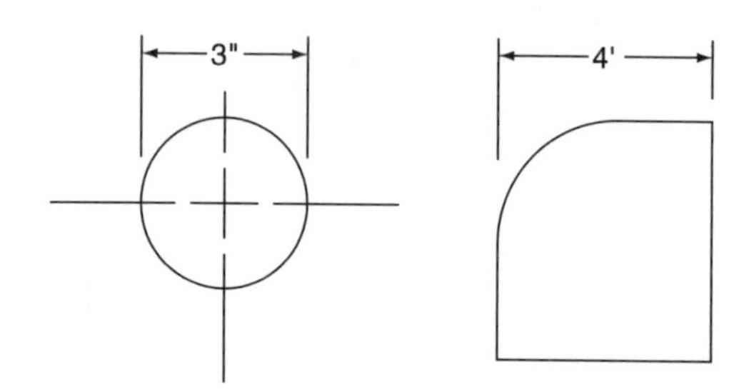

Figure 9-9 Extension lines.

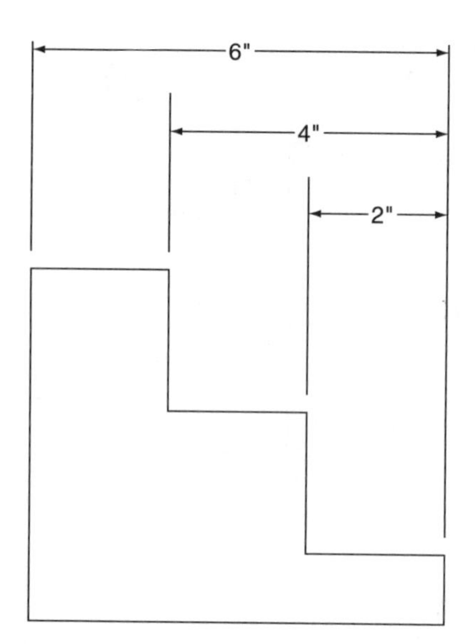

Figure 9-10 Dimension lines.

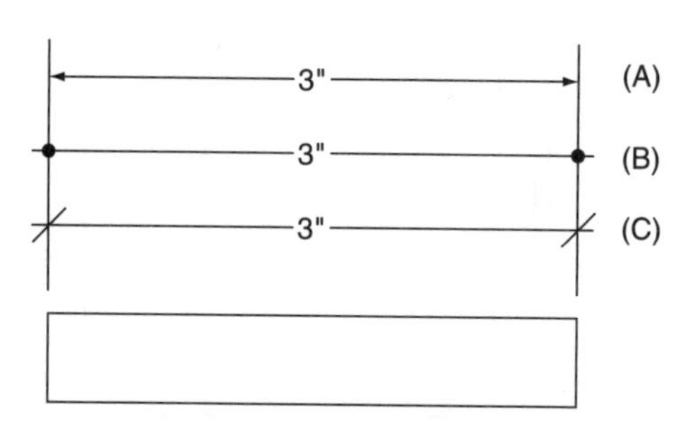

Figure 9-11 The end of a dimension line can have (A) an arrow, (B) a dot, or (C) a slash.

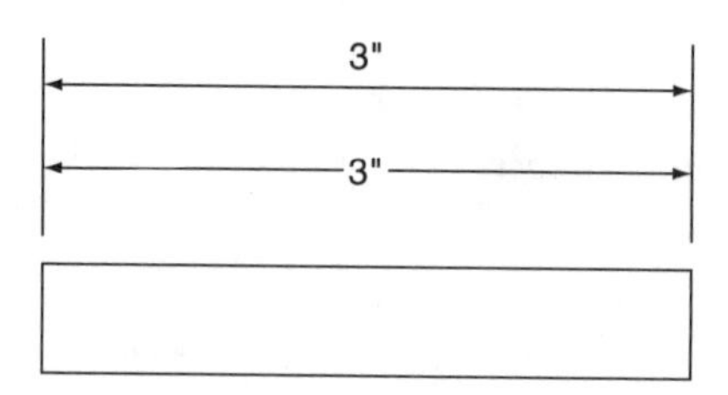

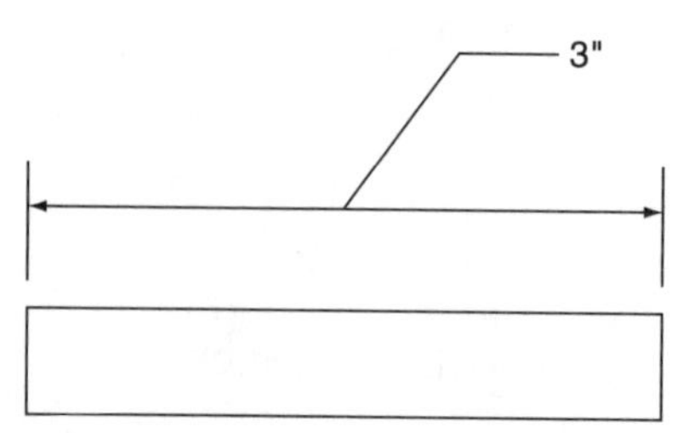

Figure 9-12 Dimensioning

(Figure 9-13). Usually a letter is drawn at each end of the cutting plane line that refers to the detailed drawing its particular cutting plane line locates. Cutting plane lines conceptually open up the inside of a part to show details that could not otherwise be clearly illustrated.

- View line — The view line is drawn exactly the same as the cutting plane line except that it shows surface details and not interior or hid-

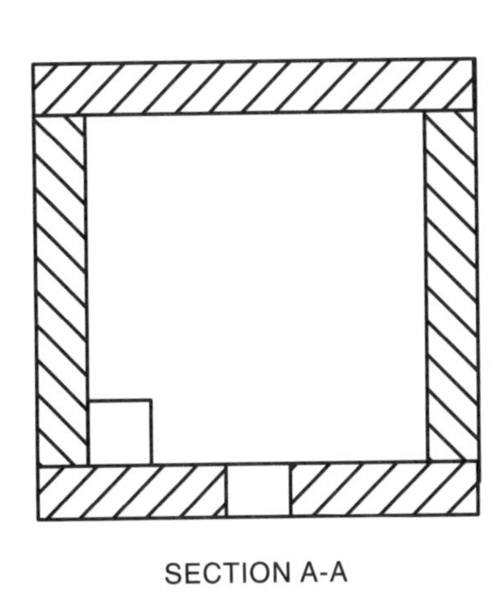

Figure 9-13 Section drawing.

den details. The view line uses letters that refer to a detailed drawing that can be seen clearly only by looking at the object in this specific direction.

- Section lines — Section lines are thin fine lines that are drawn parallel to each other but at an odd angle to the object lines. The angle used to draw the section lines will change from part to part within the sectional drawing (Figure 9-14). A large number of different materials can be represented by the various spacing of section lines (Figure 9-15). However, most draftspersons use the symbols for iron or steel when making sectional drawings, regardless of the actual material the parts are made from. Section lines are used only to illustrate the imaginarily cut surfaces, which are identified by a cutting plane line.
- Leader or arrow lines — These are medium-weighted lines that have an arrowhead drawn on one end. Leader and arrow lines point to a specific location that a note, specification, or dimension refers to. They are used in conjunction with welding symbols to locate the joint or surface to be welded.
- Break lines — There are two types of break lines that may be used on a drawing. One is a bold straight line that has intermittent zigzags drawn along its length (Figure 9-16). The other break line is a bold freehand irregular line (Figure 9-17). Both of these lines are used to illustrate a segment of a part that is not included on a drawing. Break lines are sometimes used to remove central portions of an object when no details are required for the fabrication of the part. Such removals are most often used to allow the draftsperson to fit the part on the drawing page while using a larger scale for the drawing.

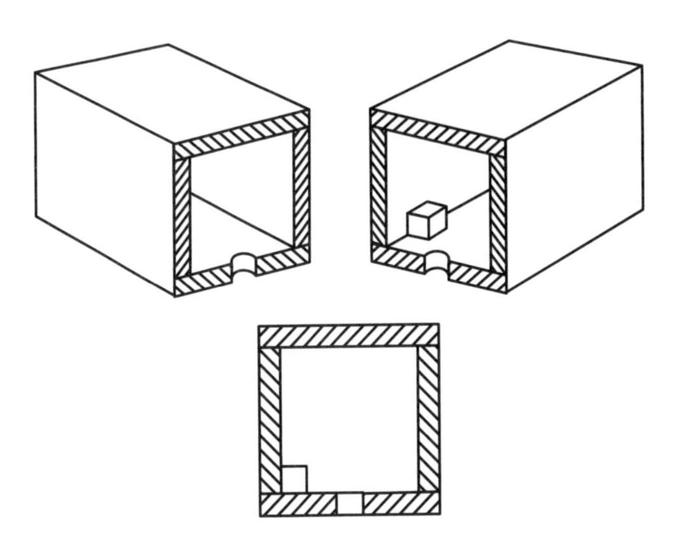

Figure 9-14 Section drawing.

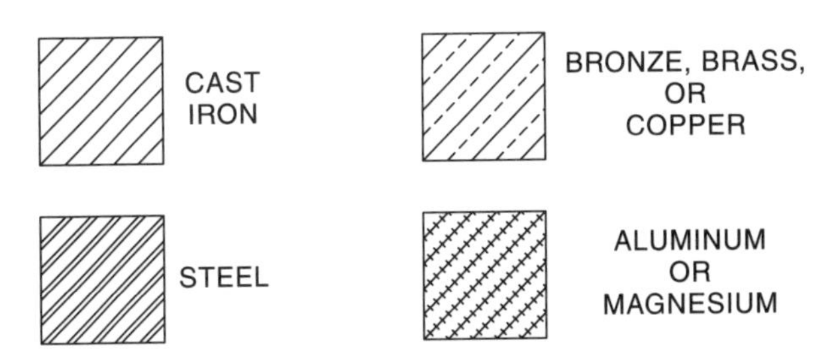

Figure 9-15 Types of section lines.

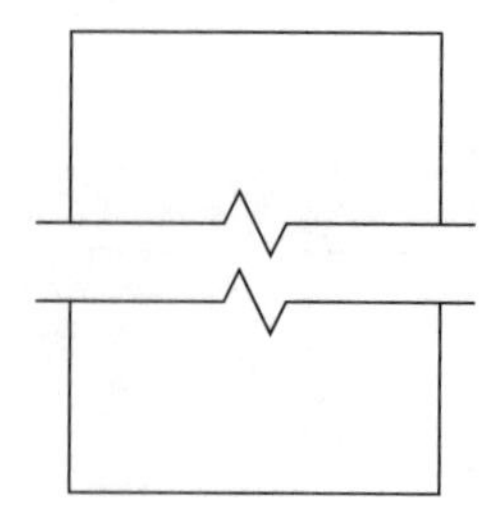

Figure 9-16 Break line.

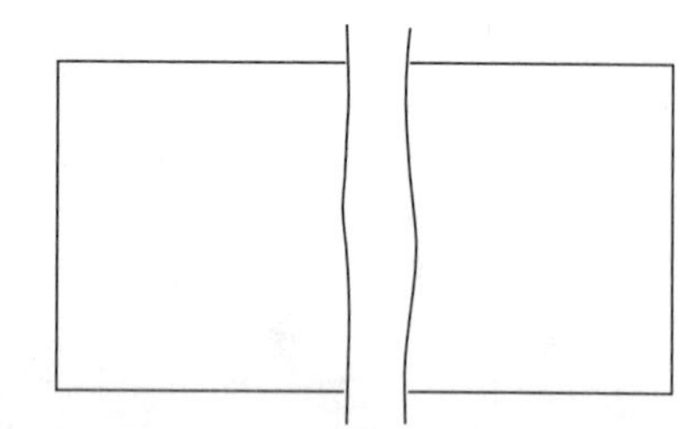

Figure 9-17 Freehand break line.

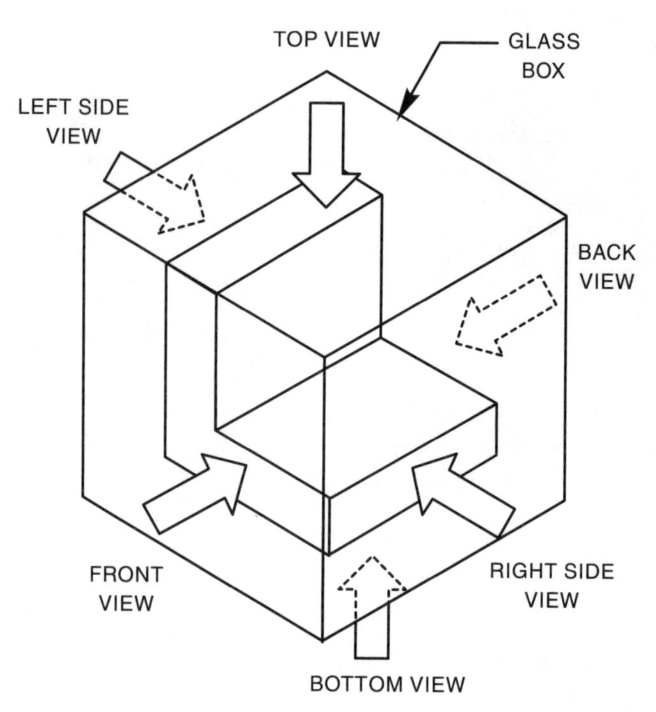

Figure 9-18 Viewing an object as if it were inside a glass box.

TYPES OF DRAWINGS

There are two general categories of drawings used in the welding industry. The majority of them are orthographic projections sometimes referred to as mechanical drawings. These drawings represent what you would see if you were looking through the sides of a glass box at the object (Figure 9-18). The object then appears as if you had traced its shape on the side of the glass box and then unfolded the box, laying it flat. If you were to trace the fixed sides of an object and lay them out as if the box had been unfolded, they would appear as shown in Figure 9-19.

The other type of drawing that may be used to represent that object is called a pictorial. These drawings are more realistic and depict the object as it would appear if you were looking at it. Pictorial drawings can be divided into three major types: isometric, cavalier, and angular prospective (Figure 9-20). Angular prospective drawings are the most realistic and are sometimes used to illustrate to a prospective customer what a finished product would look like. They are seldom used by the welder for actual fabrication.

Projection Drawings

Although up to six views can be included in a projection drawing, seldom are more than three used. For example, when you look at a car, the right and left sides for all practical purposes, would appear on a drawing as mirror images of each other. Therefore, it would not be necessary for the draftsperson to have included both the right and left sides because they would not contain significant details not easily obtained from only one side view. The three most common views used in orthographic projections are the front view, right side view, and top view (Figure 9-21). In some cases, when relatively simple objects are being illustrated, it may be necessary to use only one or two of these views.

The front view is not necessarily the front of the object. The side selected to represent the front view should be the side that best describes the overall shape of the object. As an example, the front view of a car or truck would probably be the side of the vehicle, which would show more about the vehicle than the front. The front of many cars, light trucks, station wagons, and vans all appear similar; therefore, the side view would be necessary to determine type of vehicle it really was.

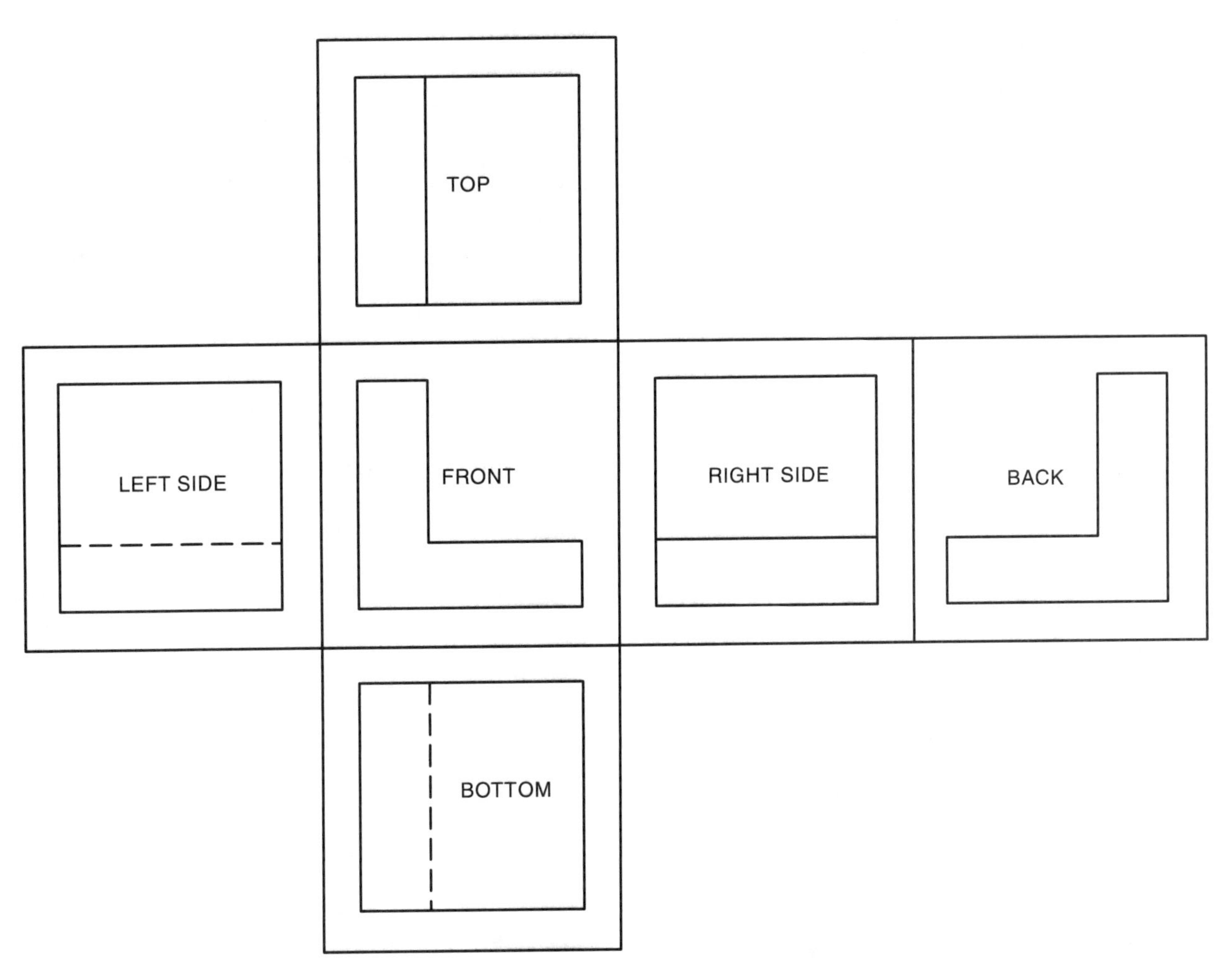

Figure 9-19 The arrangement of views for an object if the glass box were unfolded.

Special Views

Special views may be included on a drawing to help describe the object so it can be made accurately. Special views on some drawings may include the following:

- Sections view — The section view is made as if you were to saw away part of the object to reveal internal details (Figure 9-22). This is done when the internal details would not be as clear if they were shown as hidden lines. Sections can be either fully or just partially across the object. The imaginary cut surface is set off from other noncut surfaces by section lines drawn at an angle on the cut surfaces. On some drawings the type of section line used illustrates the type of material the part was made with. The location at which this imaginary cut takes place is shown using a cutting plane line (Figure 9-23).
- Cut-aways — The cut-away view shows detail within a part that would be obscured by the part's surface. Often a freehand break line outlines the area that has been imaginarily removed to reveal the inner workings.
- Detail views — The detailed view is usually an external view of a specific area on the part. Detail views show small details of an area on a part without having to draw the entire part larger. Sometimes only a small portion of a view has significance, and this area can be shown in a detailed view. The detail view can be drawn at the same scale or larger if needed. By showing only what is needed within the detail, the drawn part can be clearer and not require such a large page.
- Rotated views — A rotated view shows a surface of the part that would not normally be drawn square to any one of the six normal view planes. If a surface is not square to the viewing angle, then lines may be distorted. For example, a circle—when viewed at an angle—looks like an ellipse.

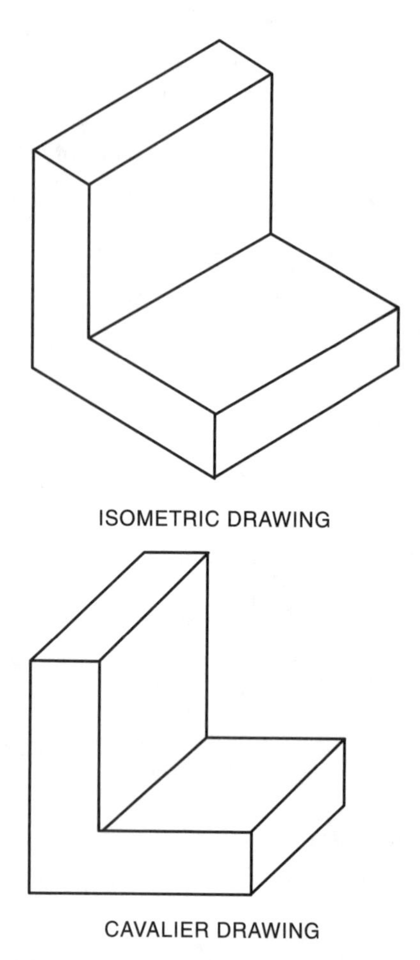

Figure 9-20 Pictorial drawing types.

Dimensioning

Drawings are two-dimensional representations of three-dimensional objects. In other words, on any single view of an orthographic projection, only two dimensions may be illustrated. The length dimension, for example, can be found on the front and top views. The height dimension can be found on the front and right side views. The width dimension can be found on the top and right side views. It is necessary therefore to look at other views in order to locate all of the dimensions required to build an object. Dimensions on a drawing may be given conventionally or from a base line (Figure 9-24). Conventional dimensioning refers to the method in which each dimension is given from a point, and that point's dimension is used as reference to other points. With this type of dimensioning, in order to locate some items on a part, it may be necessary to add together dimensions. For example, in Figure 9-25, to locate the center of the hole, the welder must add together the dimension from the end of the bracket to the brace and then the distance from the brace to the hole.

Base line dimensioning uses an edge, side, or major component as a base line, and all dimensions' distances extend from that single point. With base line dimensioning, however, the welder does not have to add up dimensions to locate points. The drawing may become overly cluttered and thus confusing if a large number of dimensions are given.

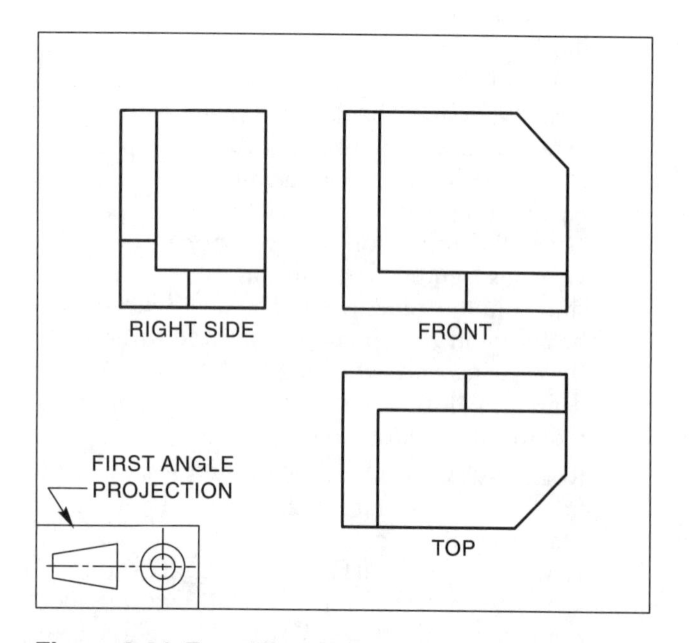

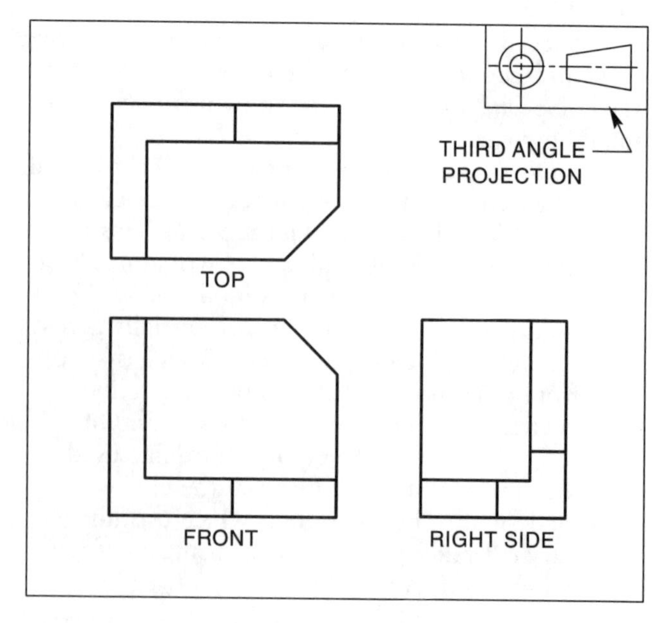

Figure 9-21 Two different methods used to rotate drawing views.

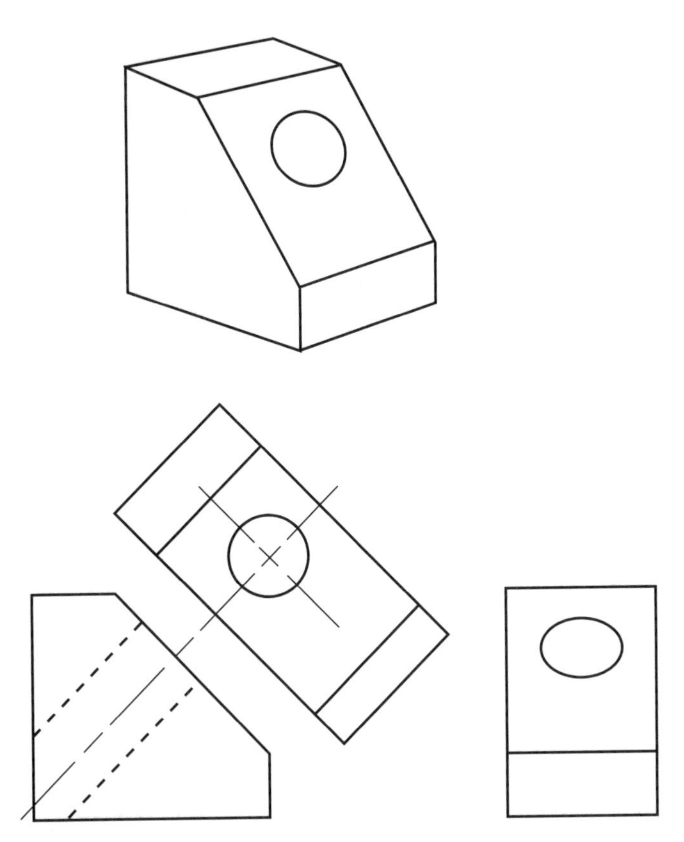
Figure 9-22 Notice that the round hole looks misshapen, elliptical, in the right side view.

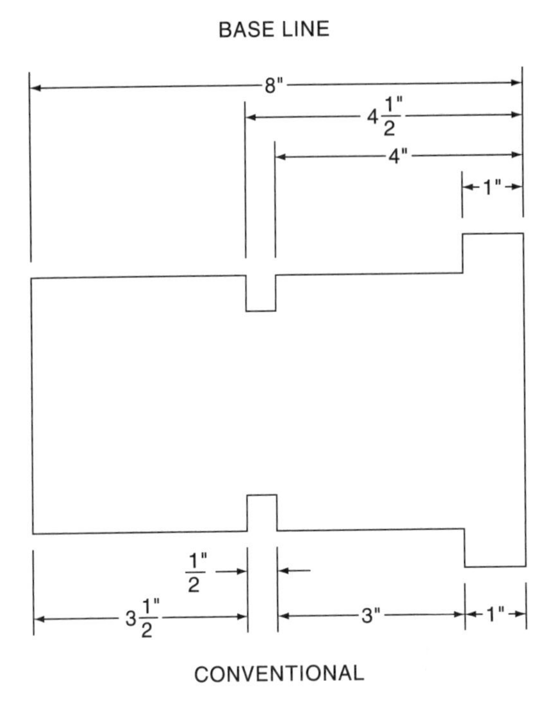

Figure 9-24 Two methods used for dimensioning.

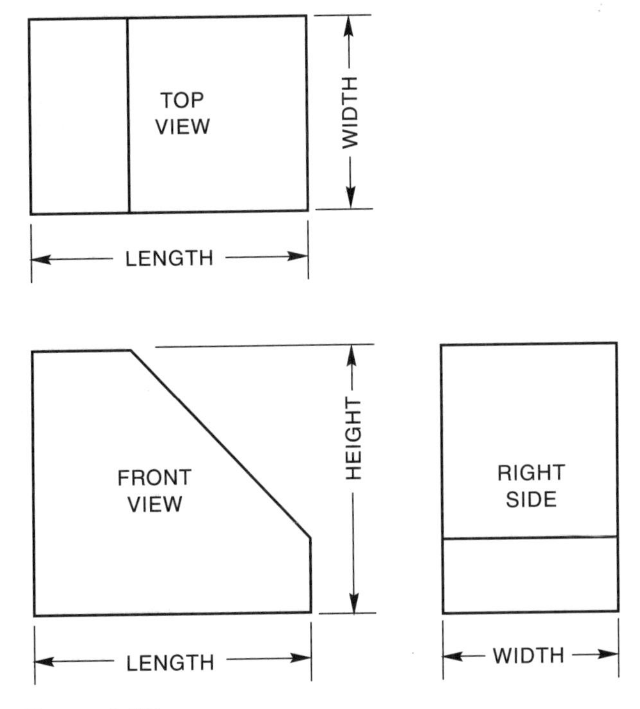

Figure 9-23

In either dimensioning system it is often necessary to locate the dimension distance for any given item on a single view by referencing the item on another view. In Figure 9-26, in order to locate the center of the hole on the front view, you would have to locate the length dimension in the top view and the height dimension in the side view. Such cross-referencing to locate all of the dimensions is common on all drawings. If all of the dimensions of every part and component were shown on each and every view, the drawing would become unnecessarily cluttered. Therefore, only the necessary dimensions required to accurately produce the part are given and usually only once.

If, after a diligent search, you cannot locate a dimension on a drawing, do not try to obtain the dimension by measuring the drawing itself. Even if the original drawing was made very accurately, the paper it is drawn on will change sizes with changes in humidity. Copies of the original drawing will never be the exact same size. The most acceptable way of determining a missing dimension on a drawing is to contact the person responsible for the drawing.

Drawings are often used by more than one person during fabrication, so to avoid possible confusion by these individuals, do not write or do calculations on the drawings. The better care you take of these drawings, the easier it will be for someone else to use them. Always keep drawings clean and well away from any welding.

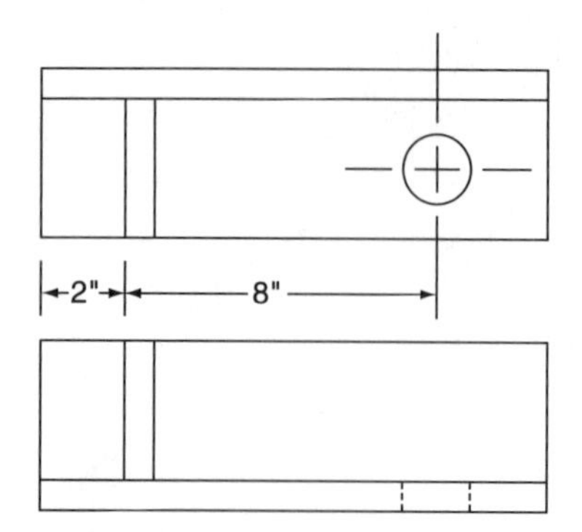

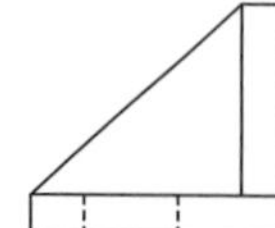

Figure 9-25

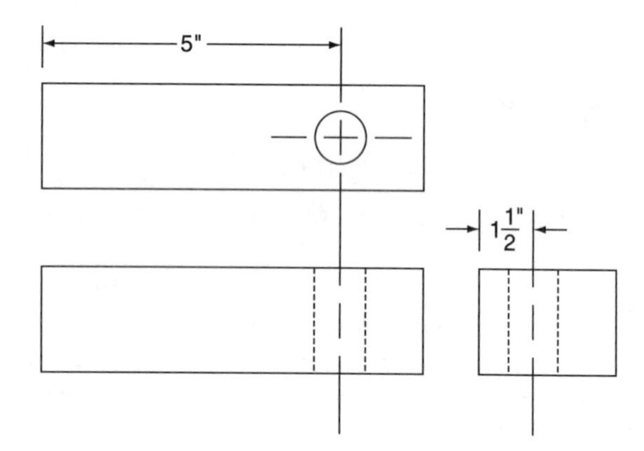

Figure 9-26

REVIEW QUESTIONS

1. What is contained in a set of drawings?
2. What type of pages may be included on a larger set of drawings?
3. What does a bill of materials include?
4. What information can be included in the title box of a drawing?
5. What does an object line represent?
6. What does the center line represent?
7. What are dimension lines used for?
8. How is the view line different from the cutting plane line?
9. What are leader and arrow lines used for?
10. Why are break lines sometime used?
11. How can pictorial drawings be divided?
12. What are the most common orthographic projection views?
13. Why is the front view of an object selected?
14. Why are sections used in drawings?
15. Why are detail views used?
16. What are drawings?
17. In which dimension can you find the length dimension?
18. In what dimension do you find height?
19. In what dimension do you find width?
20. How would you locate the center of the hole in the front view?

CHAPTER 10

LAYOUT AND FABRICATION

INTRODUCTION

Depending on the size of the shop and the type of work being performed, most welders will be required to lay out and fabricate components they are going to be welding. In large shops the laying out and fabricating of virtually all of the components is performed by individuals other than the actual welder. This enables the skilled welder to be more productive by producing more welds. Also in large shops the parts may be preassembled and tacked, ready for welding. Even under these working conditions, from time to time these welders will be asked to lay out and fabricate specialty parts.

Layout refers to the process of transferring the drawing specifications from paper to the metal that they will be fabricated from. Once laid out, the parts may be cut out using any number of processes such as plasma arc cutting, oxyfuel cutting, shearing, machining, or laser cutting.

Fabrication is the assembling of parts into the correct positions as specified by the drawings. Some assemblies may be done by hand; others may use jig fixtures, clamps, or other means to assist in the accurate positioning of each component that will make up the weldment.

To help develop your welding fabrication and layout skills, you can treat each welding practice plate as if it were a part being assembled for production welding.

LAYOUT

In the layout process, lines from drawings, sketches, or concepts within the welder's mind are transferred onto the material that will be cut, bent, grilled, punched, assembled, and welded. During the process of layout, the welder may expect to transfer dimensions and specifications from the drawing to the material to be used for fabrication. Not all layout lines are used to locate cuts to be made on the material. Some layout lines are used to show the positions for locating parts during the assembly. Yet other lines are used only to help in the layout process; these are referred to as construction lines (Figure 10-1). The purpose for each line drawn on the material must be clearly marked to prevent mistakes during fabrication.

The accuracy and squareness of all lines used during the layout process are essential. Fabricated parts can be no more accurate than the lines used during the assembly and fabrication.

Lines may be marked with a soapstone or chalk line; they may be scratched with a metal scribe or punched with a center punch (Figure 10-2). Other items to use for drawing on the material (for example, pencils, pens, felt-tip markers, and paint pens) may contribute to contamination of gas tungsten arc welds. These items are frequently used during the layout and fabrication process of materials that are going to be welded using other processes, such as shielded metal arc welding, flux core welding, and other processes in which fluxes, which are part of the process, can remove such small contaminations that might result from these marks.

Soapstone is available in either flat or round cut plates. In order to draw accurate lines, you must properly sharpen the substone to (Figure 10-3).

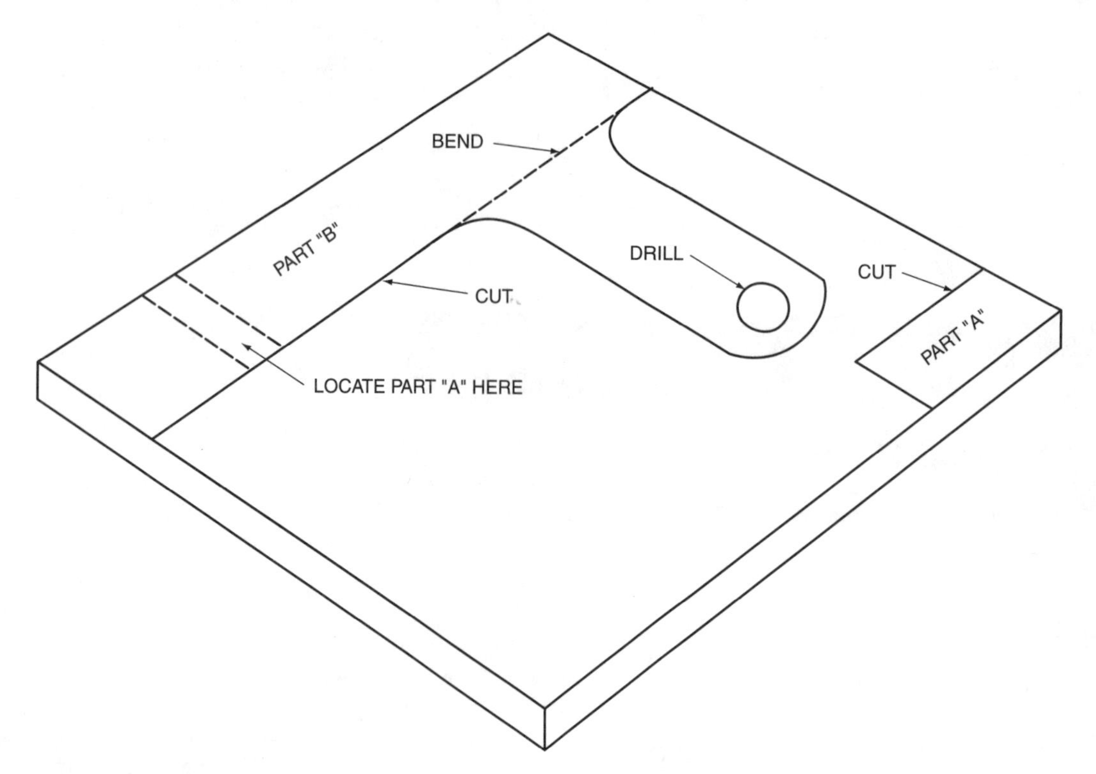

Figure 10-1 Lines and notes used to aid in the cutting out and fabrication of weldments.

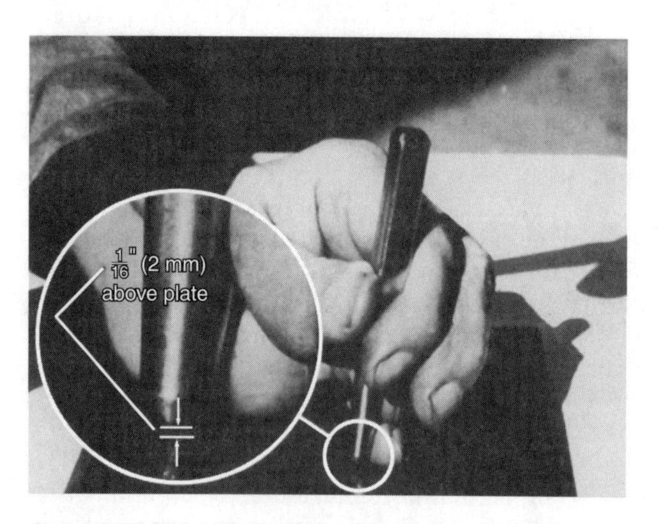

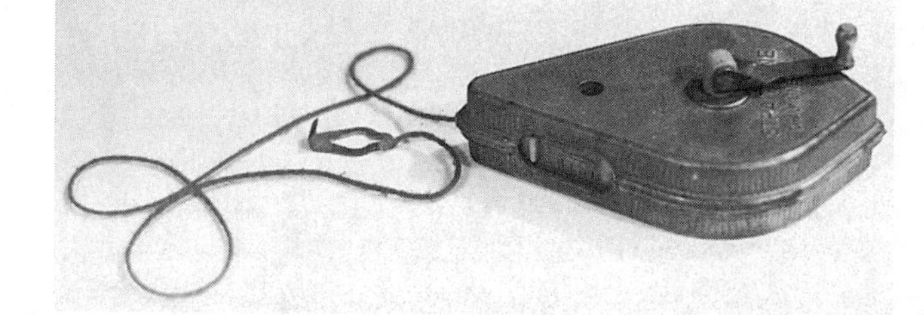

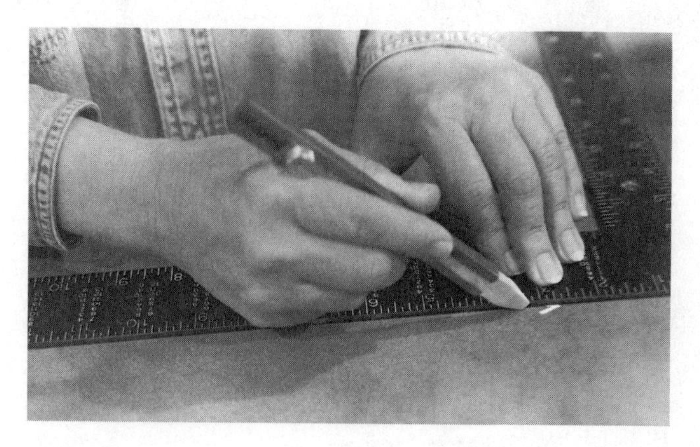

Figure 10-2 Chalk line reel. Holding the punch slightly above the surface allows the punch to be stuck rapidly and moved along a line to mark it for cutting. Using a square to draw a straight line.

A chalk line reel contains powdered chalk and a cotton frame. Chalk lines can make long, straight lines on metal and work best on large jobs. Powdered chalk is available in a variety of colors, including red, green, blue, and white. As the string is pulled from the chalk line, powdered chalk adheres to the string, which becomes straight as it is pulled tight between two marks. Lifting the center of the string and allowing it to snap back against the metal results in a line of powdered chalk sticking to the metal surface (Figure 10-4).

Part Layout

Always start a layout as close as possible to a corner or an edge of the material. This will aid in layout accuracy and help reduce wasted material. Often you can take advantage of a preexisting straight edge on the stock and use it as one surface of the part you are fabricating. An existing edge also aids in aligning tools, such as squares and bevels.

It is often easy to mistakenly cut, bend, or locate a part on the wrong line. In welding shops one person may lay out the parts, while others cut them out or assemble them. However, even if the same person does all of these jobs, it is still possible to use the wrong

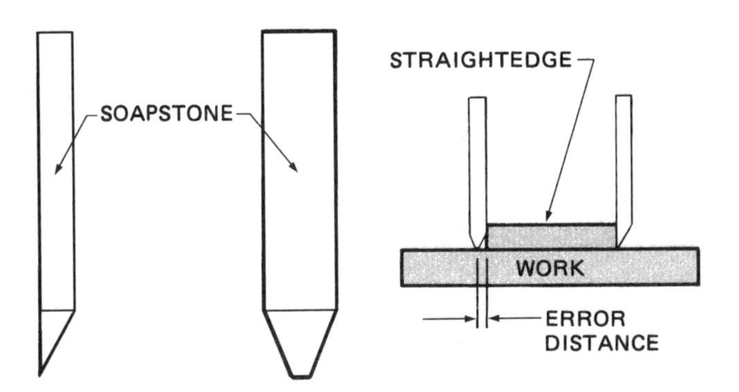

Figure 10-3 Proper method of sharpening a soapstone.

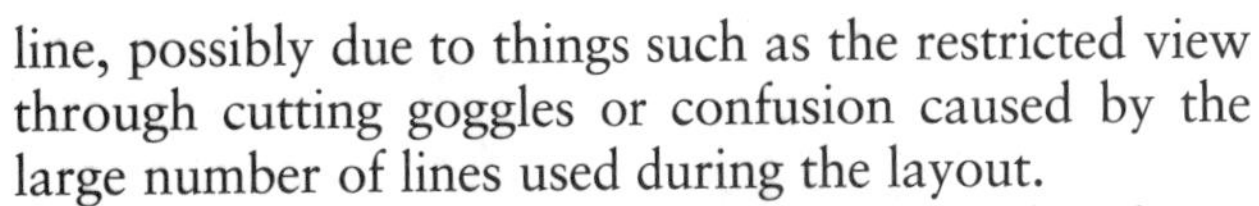

line, possibly due to things such as the restricted view through cutting goggles or confusion caused by the large number of lines used during the layout.

To avoid such mistakes, lines should be identified as to whether they are being used for cutting, bending, drilling, or assembly locations. Common methods used to identify lines such as those that are not to be cut is to mark them with an X. Lines may also be identified by writing directly on the part. Additionally, to ensure the accurate cutting out of parts, sometimes the scrap side where the kerf is to be removed during the cutting process should be identified (Figure 10-5). Any lines that were used for constructing the actual layout lines should be erased completely or clearly marked to avoid confusion.

Some shops have their own shorthand methods for identifying layout lines, or you may develop your own system. If a mistake occurs during cutting out or fabrication, check with your welding shop supervisor to see what corrective steps may be taken. One advantage of most welding assemblies is that such errors may be repairable by welding. There are often pre-qualified procedures established for just such events, so check before deciding to scrap the part.

The process of laying out a part may be affected by the following factors:

- Material shape—Figure 10-6 lists the most common shapes of metal used for fabrication.

(A)

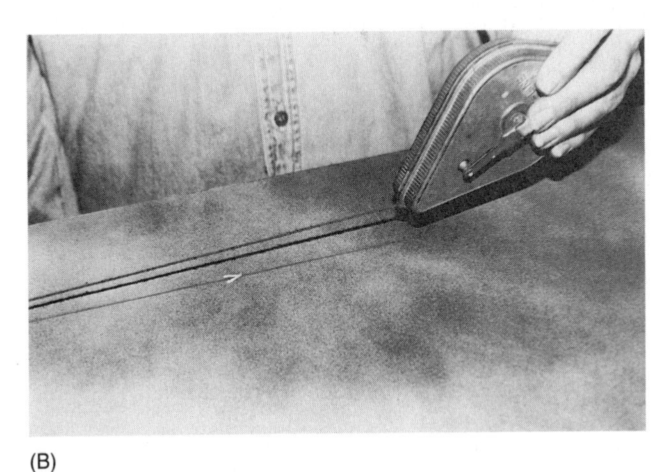

(B)

Figure 10-4 (A) Pull the chalk line tight and then snap it. (B) Check to see that the line is dark enough.

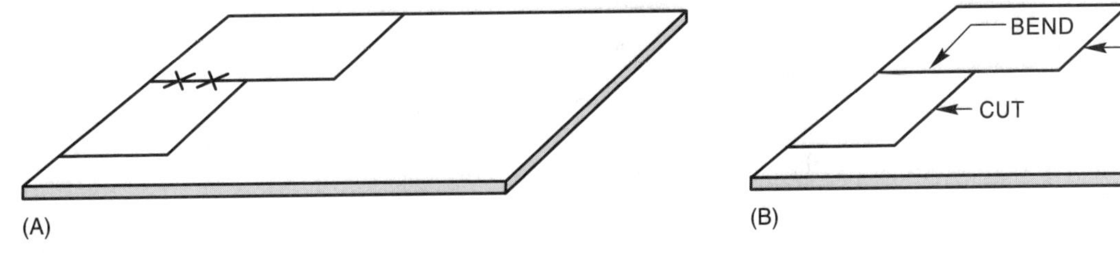

(A) (B)

Figure 10-5 Identifying layout lines to avoid mistakes during cutting.

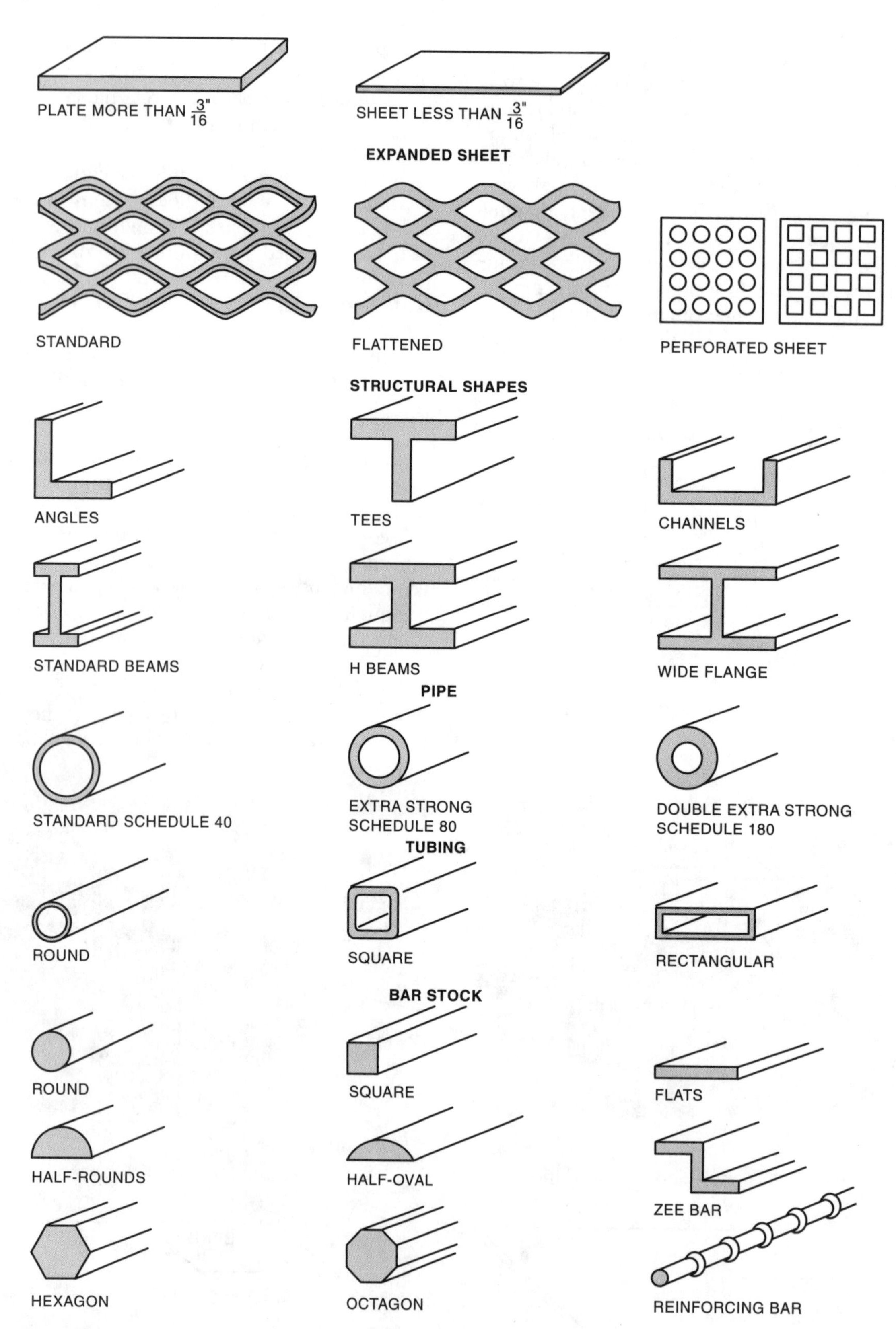

Figure 10-6 Standard metal shapes, most available with different surface finishes such as hot-rolled, cold-rolled, or galvanized.

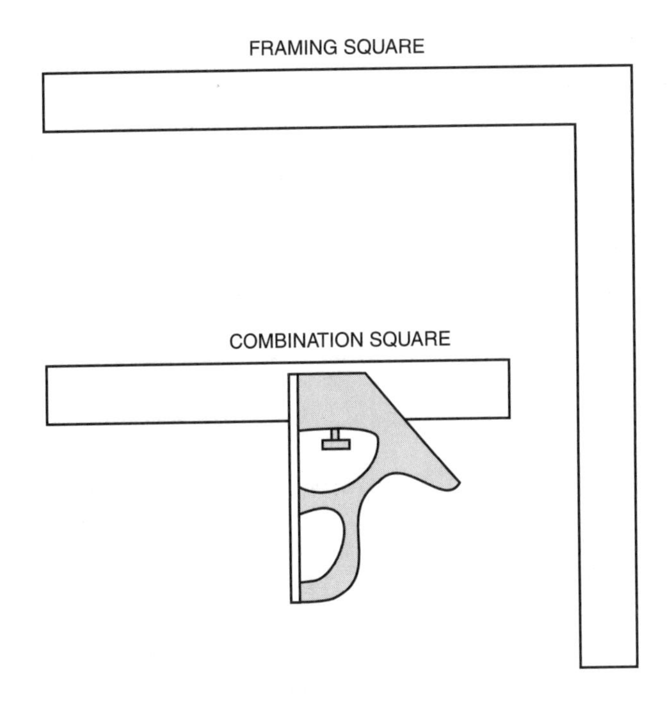

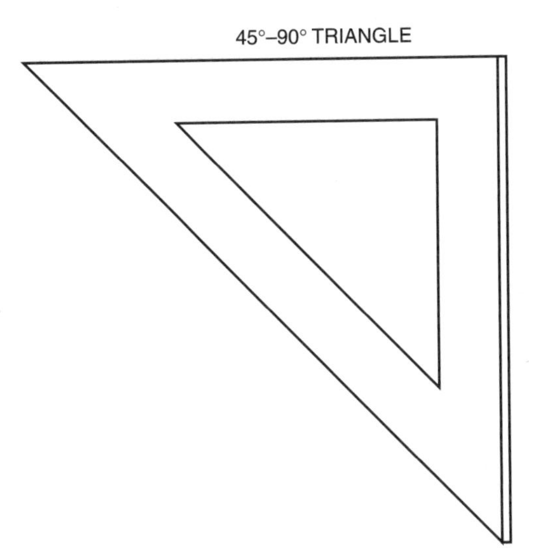

Figure 10-7 Common layout tools.

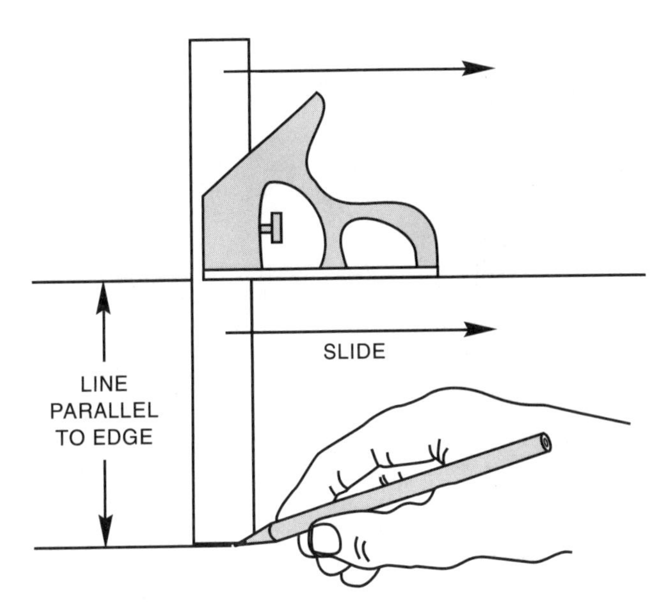

Figure 10-8 Using a combinaiton square to lay out a strip of metal.

Flat stock, such as sheets and plates, are easiest to lay out, and the most difficult shapes to work with are pipes and round tubing.

- Part shape—Parts with square and straight cuts are easier to lay out than are parts with angles, circles, curves, and irregular shapes.
- Tolerance—The smaller or tighter the tolerance that must be maintained, the more difficult the layout.
- Nesting—The placement of parts in a manner that will minimize waste is called *nesting.*

Parts that are square or have straight edges are the easiest to lay out. Square lines can be laid out using tools such as framing squares, combination squares, or tri-squares (Figure 10-7). They may also be made by simply measuring equal distances from existing right-angle corners on the stock.

Straight or parallel cuts can be made by using a combination square and a piece of soapstone. Set the combination square to the correct dimension, and drag it along the edge of a plate while holding the soapstone at the end of the combination square blade (Figure 10-8).

Circles, Arcs, and Curved Lines

Circles, arcs, and curves can be laid out by using either a compass or circle template (Figure 10-9). The diameter is usually the type of dimension given for holes or other such circular parts. The radius is usually the type of dimension given for arcs and curves (Figure 10-10). The center point of the circle, arc, or curve may be located using dimension lines and center lines (Figure 10-11). Curves and arcs that are to be made tangent to another line may be dimensioned with only their radius (Figure 10-12).

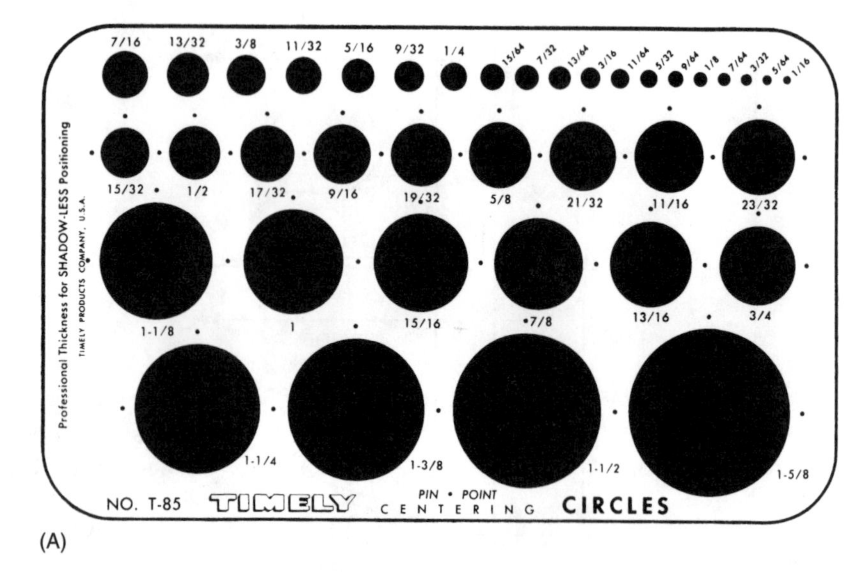

(A)

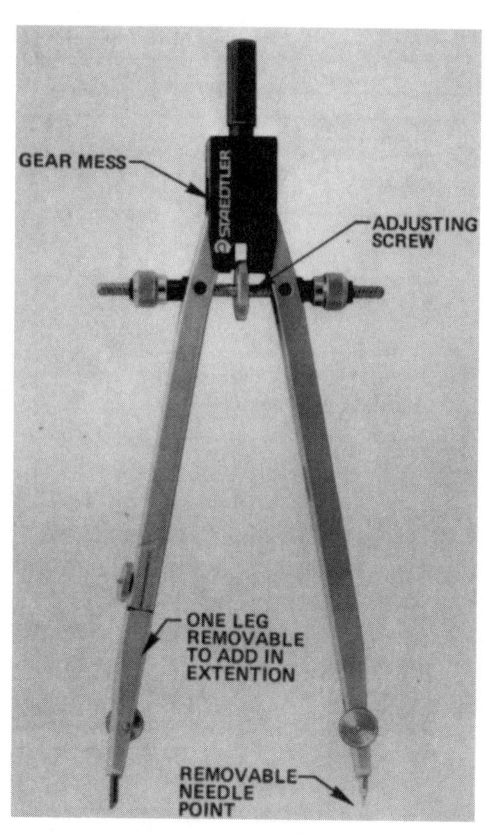

(B)

Figure 10-9 (A) Circle template (Courtesy of Timely Products Co.) (B) Compass (Courtesy of J.S. Staedtler, Inc.)

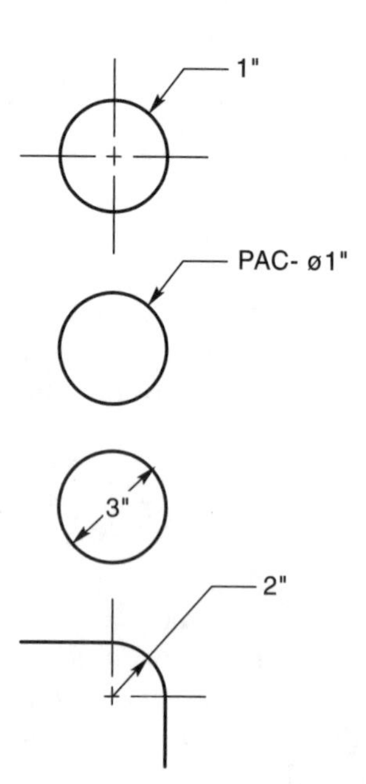

Figure 10-10 Dimensioning circles and radiuses.

Angled Lines

Straight lines that are not drawn at a 90° angle or square can be dimensioned using either their angle or the points between which a line will be drawn (Figure 10-13). Triangles that have 30, 60, 90, or 45, 90 degree angles are available for such layouts odd angles can be located by using a protractor (Figure 10-14).

On large parts angles are more accurately laid out if dimensions are given for their points. Even the slightest angular error as marked using a protractor can result in an inaccurate part if the line is to be extended several feet (Figure 10-15). The interior angles of all triangular parts must add up to 180 (Figure 10-16).

Nesting

Layout parts so they are clustered as tightly as possible, thus eliminating as much scrap as possible, is called *nesting* (Figure 10-17). Oddly shaped parts and those with unusual sizes produce the greatest amount of scrap and are the most difficult to nest. It is often necessary to try several layout possibilities

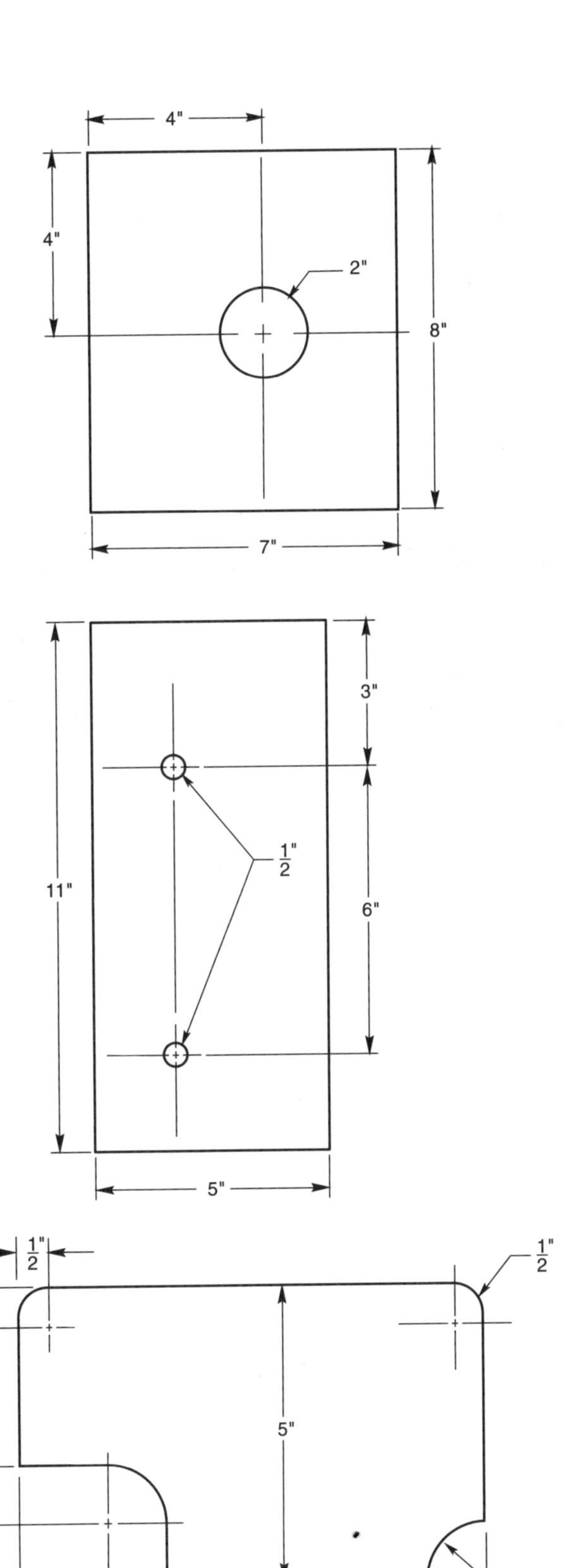

Figure 10-11

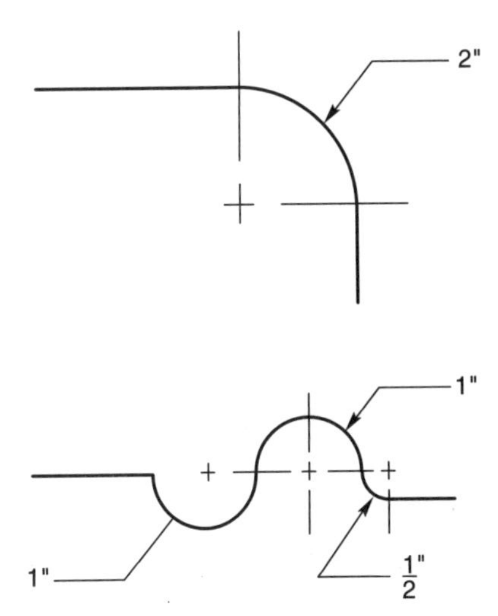

Figure 10-12

before determining which pattern will minimize scrap. Sometimes these trial-and-error efforts can be accomplished using a pencil and paper.

Computerized programs are available to aid in layout. Some automated cutting machines can use these computer-generated layouts to cut out parts.

Manual nesting of parts may require several attempts before you achieve the lowest possible scrap.

Templates and Patterns

Parts can be laid out by tracing an existing part, a template, or a pattern (Figure 10-18). If an existing part is to be used, mark it, and be sure to use the same part for each successive layout. Failure to do so can result in an ever-increasing size as each part is laid out slightly larger than the previous.

To ensure that the parts laid out using templates, patterns, or existing parts are as accurate as possible, make the lines you are drawing as close as possible to the edge (Figure 10-19). When these parts are cut out, the line should be removed with the cut's kerf leaving as little—if any—of the extreme inside edge of the line as possible so that the parts will be the correct size once they are cleaned up (Figure 10-20).

A template can be made in the shape of the parts to be produced. Templates can be useful if the parts are complex or need to fit existing weldment or if there will be a large number of similar parts produced. The advantage of using templates is that, once

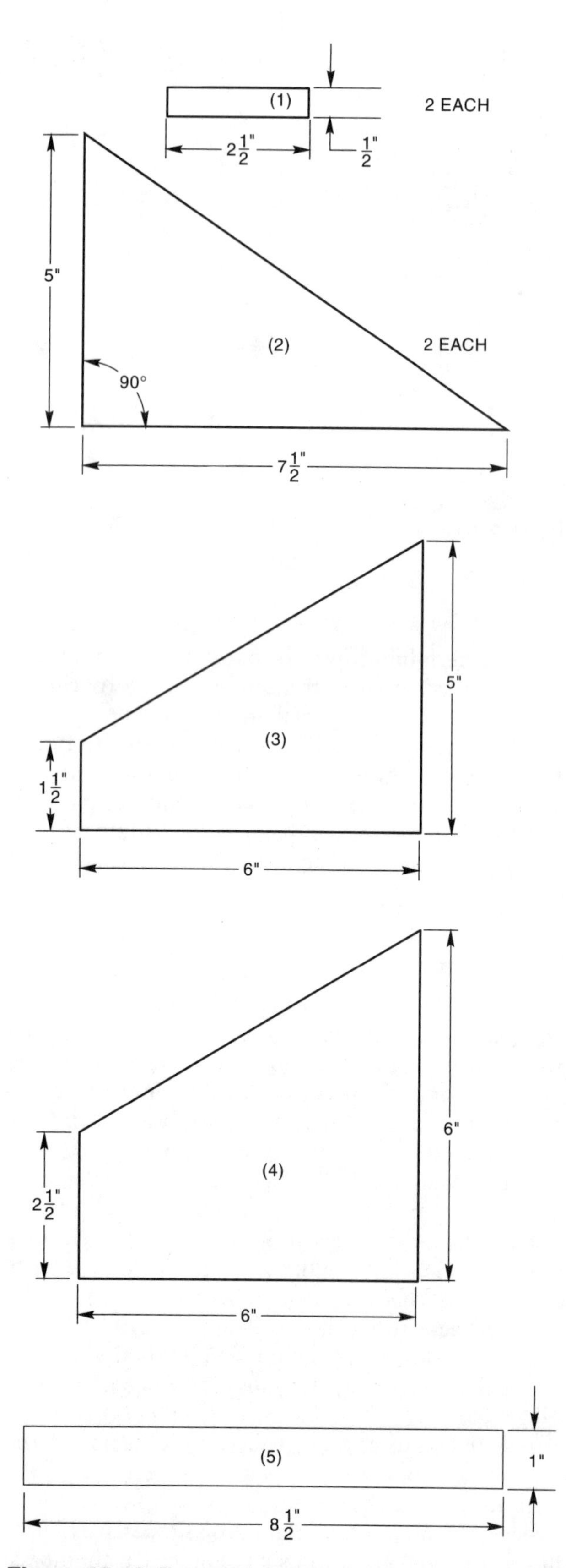

Figure 10-13 Parts to be nested.

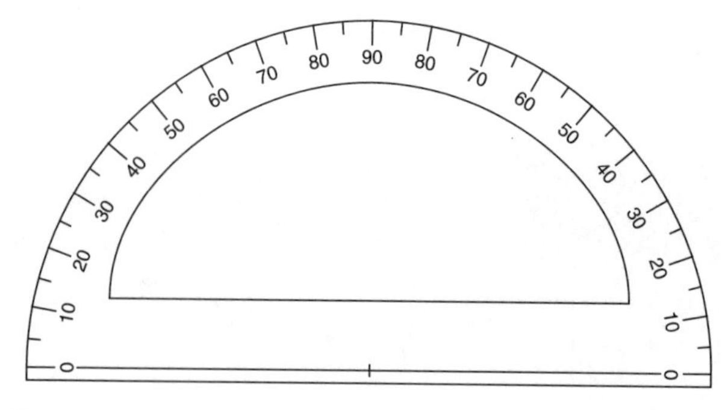

Figure 10-14 Protractor.

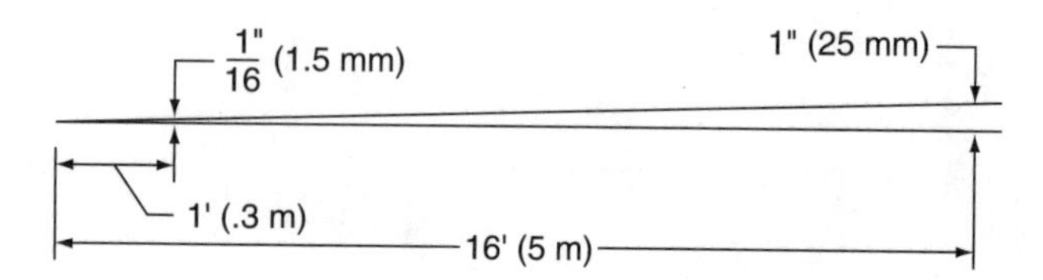

Figure 10-15 A small error at 1' (.3 m) is 16 times greater at 16' (5 m).

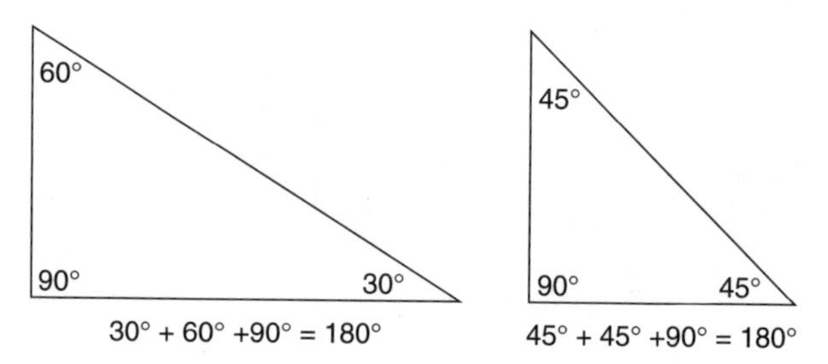

Figure 10-16

they have been produced, they can be used to lay out subsequent parts quickly and accurately. Templates can be made from any sturdy material, such as heavy paper, cardboard paper, wood, or sheet metal. The sturdier the material used for the template, the longer it will last.

Special Tools

Specialty tools have been developed to aid in laying out parts; one such tool is a contour marker (Figure 10-21). These tools, when used correctly, can be used to lay out very accurate parts. Welders require a certain amount of practice to gain experience. Once you have become familiar with the tools, you can lay out an almost infinite variety of joints within the limitations of the tool. An advantage of

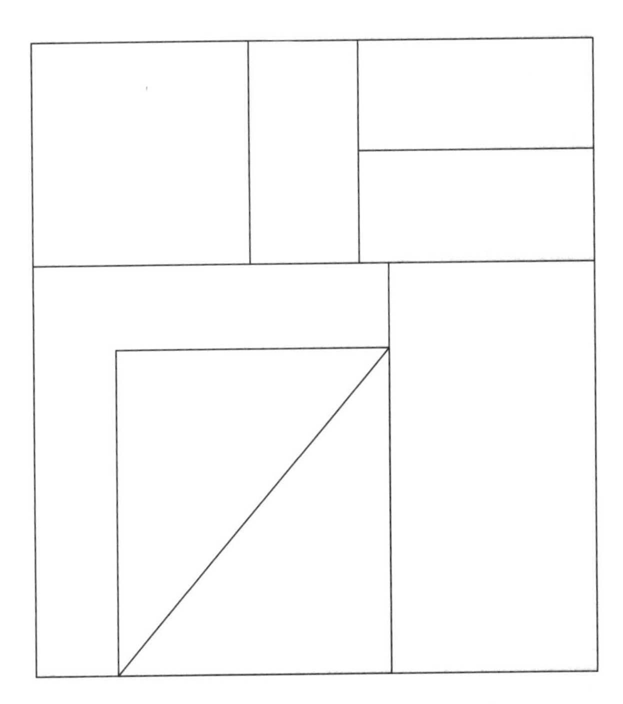

Figure 10-17 All eight of these parts can be cut out of a single piece of stock if they are nested correctly.

(A)

(B)

Figure 10-18 (A) Tracing a part (B) Tracing a template.

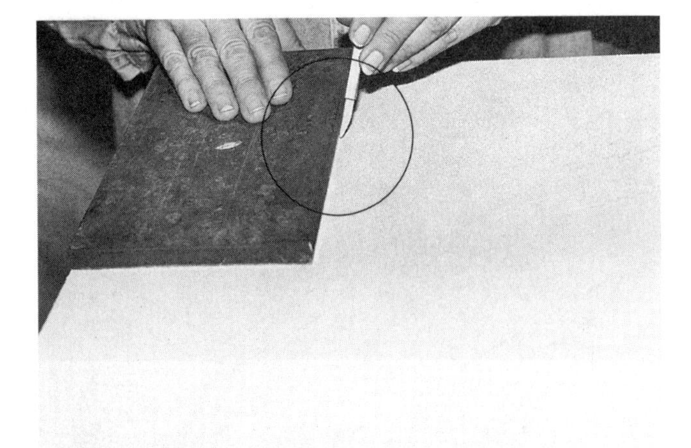

Figure 10-19 Be sure that the soapstone is held tightly against the part being traced.

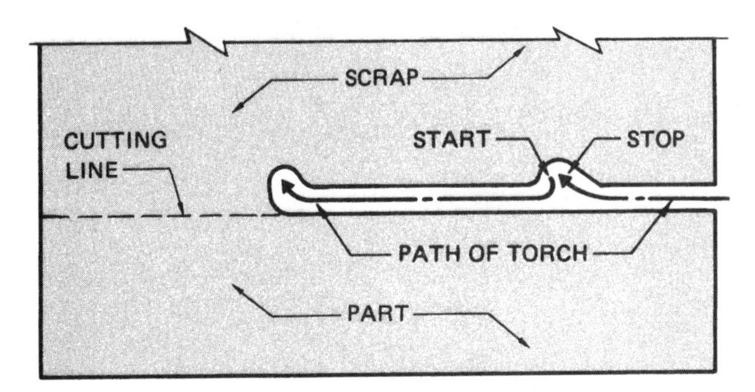

Figure 10-20 Turning out into scrap to make stopping and starting points smoother.

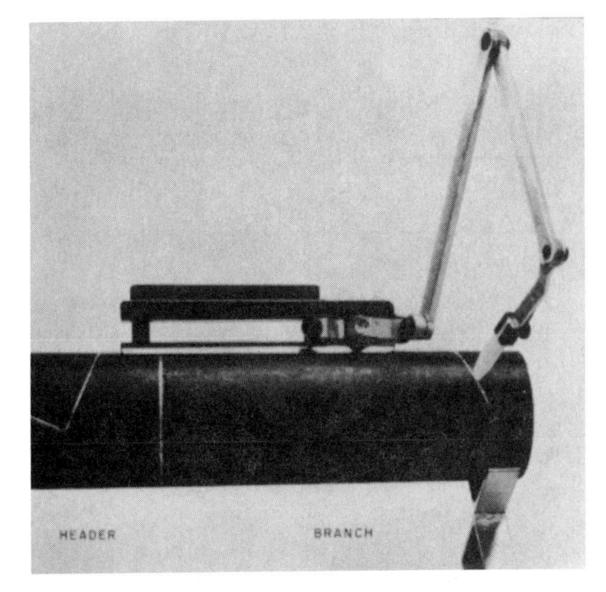

Figure 10-21 Pipe lateral being laid out with contour marker.

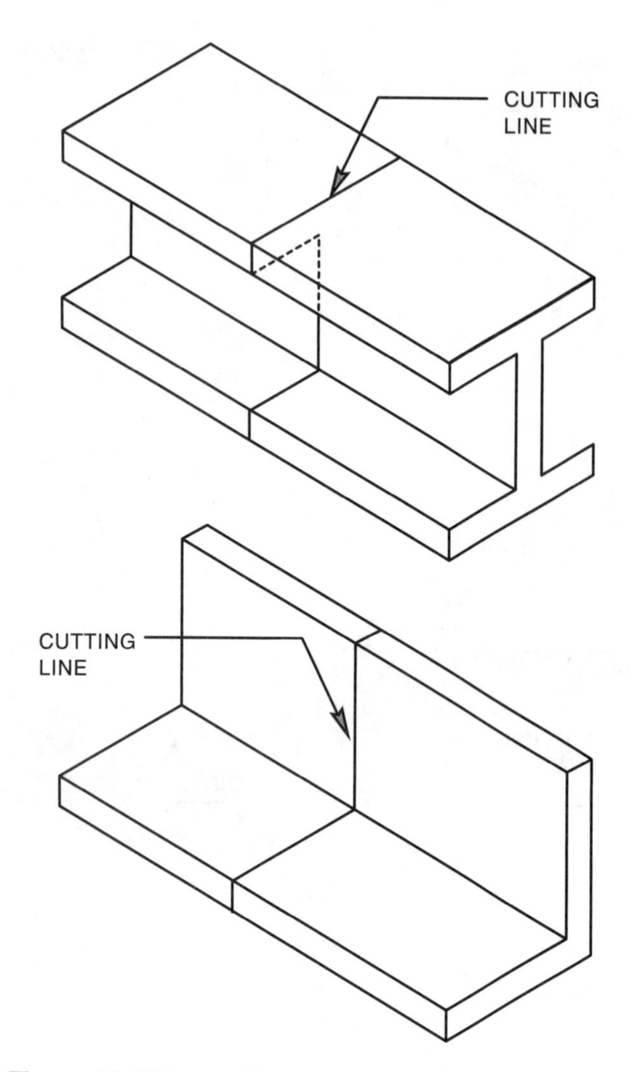

Figure 10-22 Layout lines.

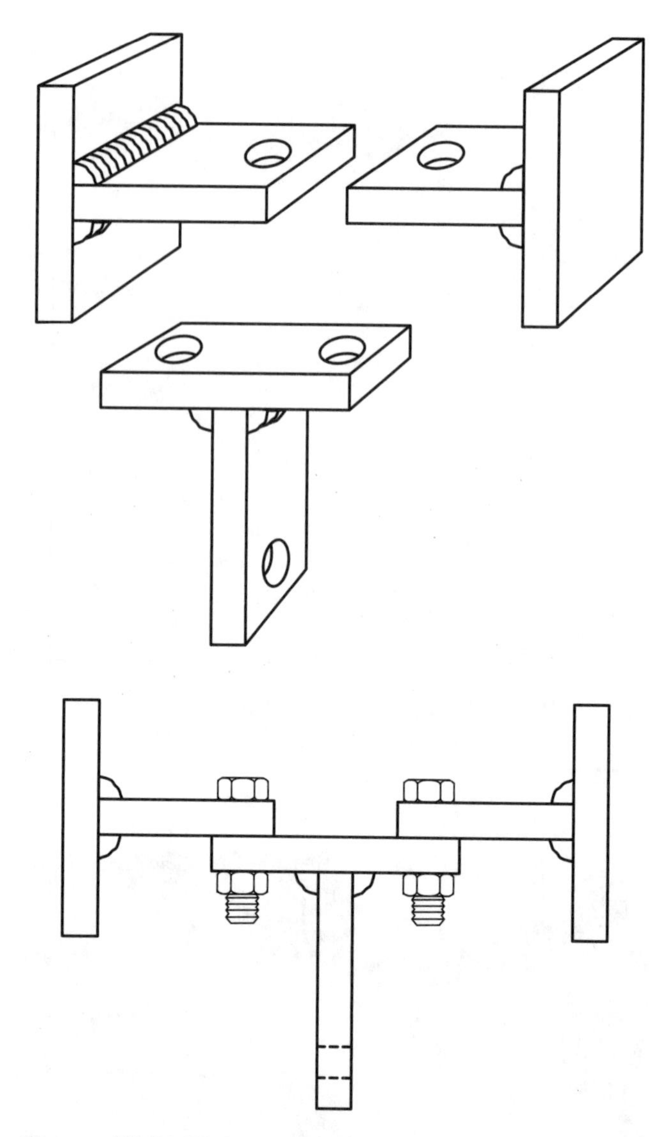

Figure 10-23 Three weldments make up this one bracket.

tools such as the contour marker is that all the sides and edges of a structural shape or pipe can be laid out without relocating the tool (Figure 10-22).

FABRICATION

Fabrication is the process of bringing together the individual pieces, placing them in the correct locations, and welding them to form a completed weldment. A *weldment* refers to an assembly of parts that has been welded together. A weldment may form a completed project itself, or it may be only a single part of a much larger structure. A weldment may consist of two or three pieces such as those used to form a bracket (Figure 10-23), or it may consist of hundreds or even thousands of individual parts that have been assembled into a much larger structure (Figure 10-24). Even the largest welded structures start by placing two parts together.

All welding projects start with a plan, which can range from a simple one that exists only in the mind of the welder to those that have been fully engineered and specified through a complete set of plans, including drawings. Drawings may range from a few lines sketched on a note pad to those created with a computerized drafting program. Throughout their career, welders will have an opportunity to work with every level, type, and style of weldment plans.

Figure 10-24 Large welded oil platform. (Courtesy of Amoco Corporation)

ASSEMBLY

The assembly process brings together all of the various parts and components and places them together securely in their proper locations using clamps, fixtures, or tack welds in preparation for welding. The assembly process requires proficiency in several areas. You must be able to read a set of specifications and interpret drawings in order to properly locate and assemble all of the individual parts of a project.

An assembly drawing has all of the necessary information that will be required to properly locate the various parts of the weldment. A set of assembly drawings shows both graphically and dimensionally the proper location of components. These drawings may include either pictorial or exploded views that make it much easier for the beginning assembler to visualize the completed project. Most assembly drawings, however, contain only two, three, or more orthographic views of the project, which requires a higher level of understanding and visualization by the assembler in order to complete the project (Figure 10-25).

On some projects the welder begins the assembly by placing parts together along their edges (Figure 10-26). On larger, more complex parts, a center line or base line may be drawn from which all other dimension locations are measured or taken (Figure 10-27). On some projects, such as this bearing bracket, dimensions between two points are more critical than others because the assembler will locate other parts within tolerance so that the more critical dimension is maintained (Figure 10-28).

The order of assembling parts may be specified on the drawings so that tolerance and accessibility to make the necessary welds do not become a problem. At other times the sequence of assembly is strictly up to the individual worker. Often there is a single part that may be designated as the base, or central piece, to which all the other parts will be joined.

When starting an assembly, select the largest or most central part, which will become the base for

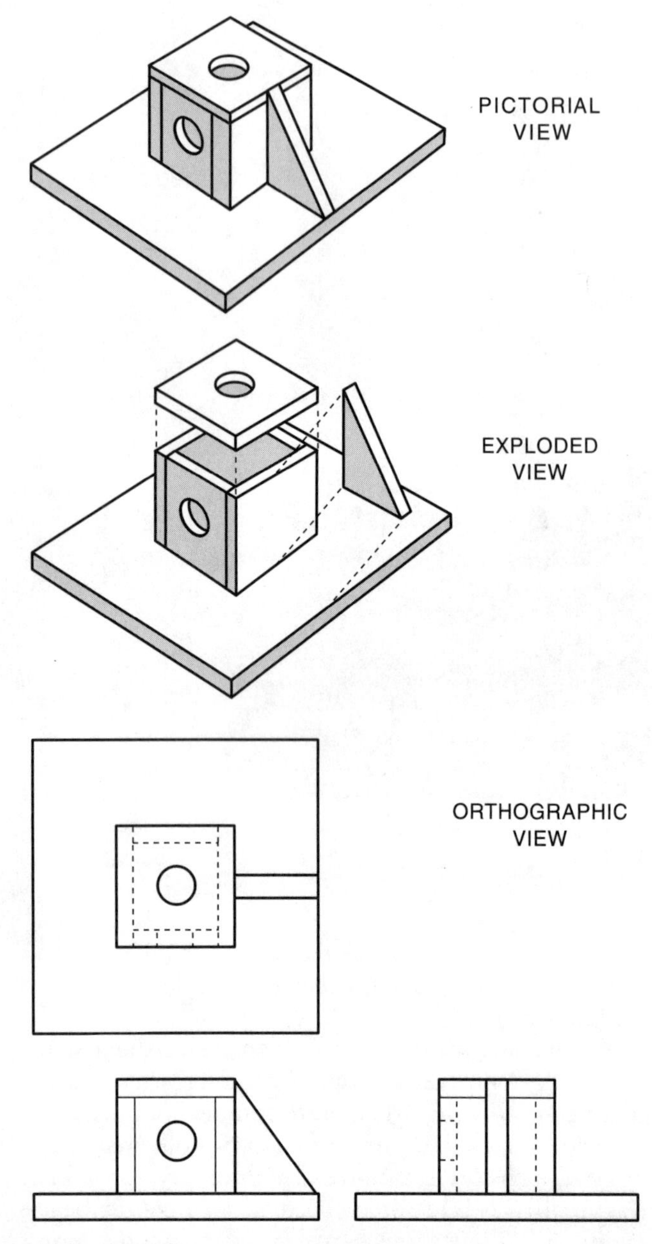

Figure 10-25

your assembly. Once this is done, all the other parts can be aligned to this one. Using a base component from which all other dimensioning is done helps to reduce location or dimensioning errors. A slight misalignment of one part, even within tolerances, can become compounded if one part is dimensioned from the next (Figure 10-29). Thus, using a base line or center line will result in a more accurate overall assembly.

You should identify each of the component parts and mark them before beginning the actual assembly. It may be helpful for you to hold the parts

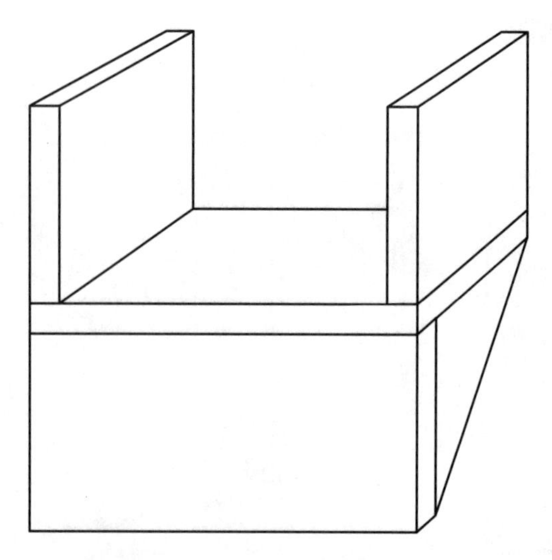

Figure 10-26 This weldment is assembled by placing all of the parts along the edge of other parts.

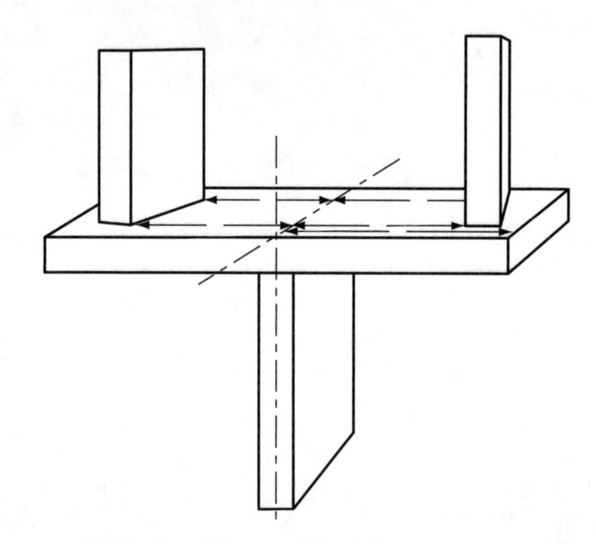

Figure 10-27 Parts aligned using centerlines.

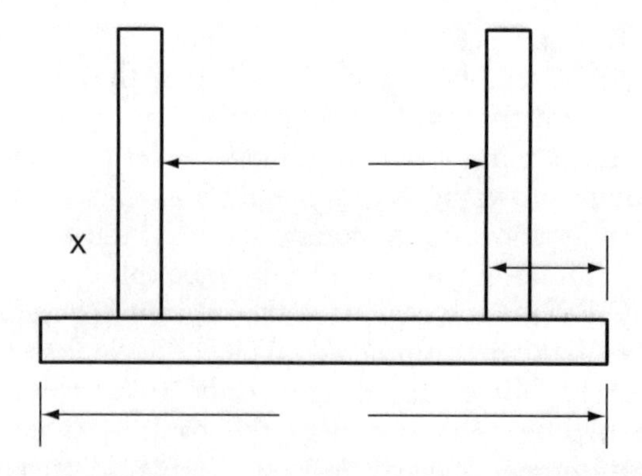

Figure 10-28 The dimension for distance *X* is omitted because it is not as critical as the other dimensions.

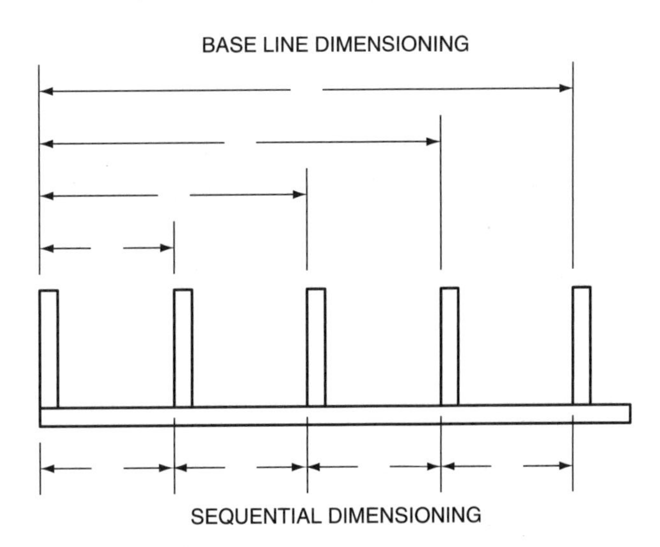

Figure 10-29 With sequential dimensioning, the accuracy for the location of the last part is affected by the accuracy of each preceding part's location.

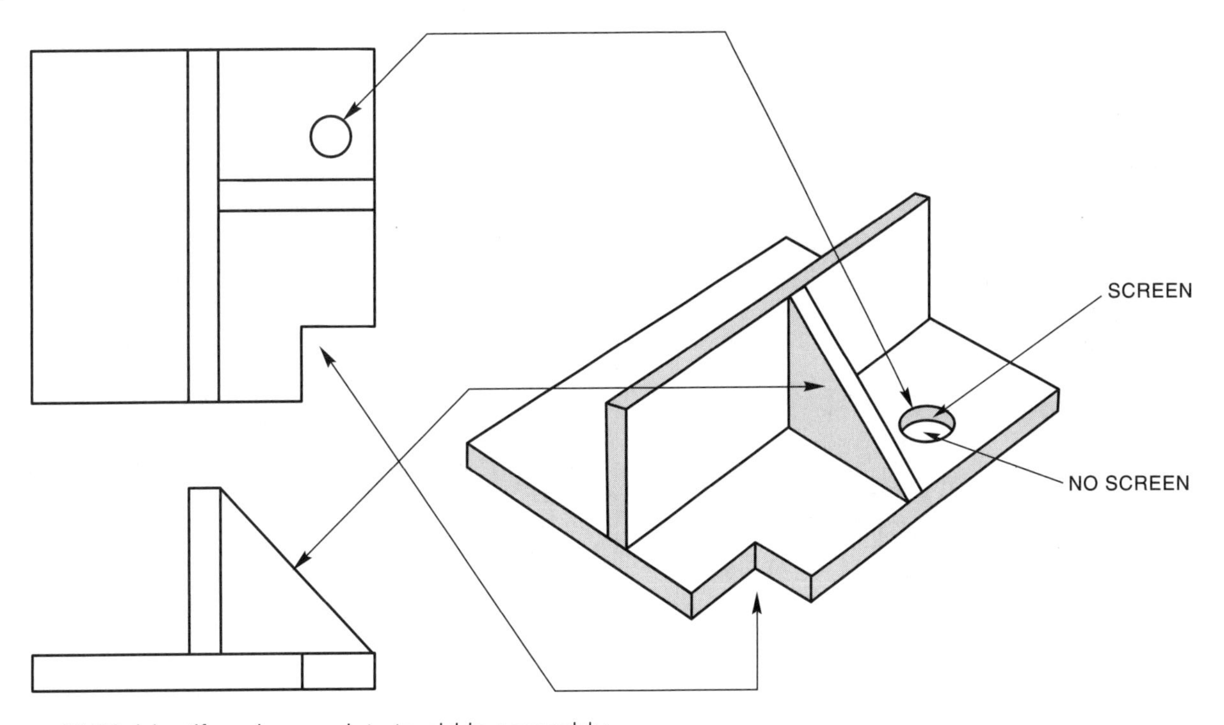

Figure 10-30 Identify unique points to aid in assembly.

together and orient them in the same direction as they are depicted in the drawings. Some parts are more easily identified than others because of their unique characteristics, such as notches, holes, or unusual shapes, that both help to identify parts and aid in locating them (Figure 10-30).

Lines may be drawn on the base to aid in the location of other parts. When a single line is used for this purpose, it is often a good idea to place an *X* on the side of the line that the part will be placed on. Another method is to draw parallel lines between which the part will be located (Figure 10-31). If the parts are being temporarily located on the part for later assembly, it may be helpful to draw small parallel location marks on both parts where they meet to aid in their relocation (Figure 10-32).

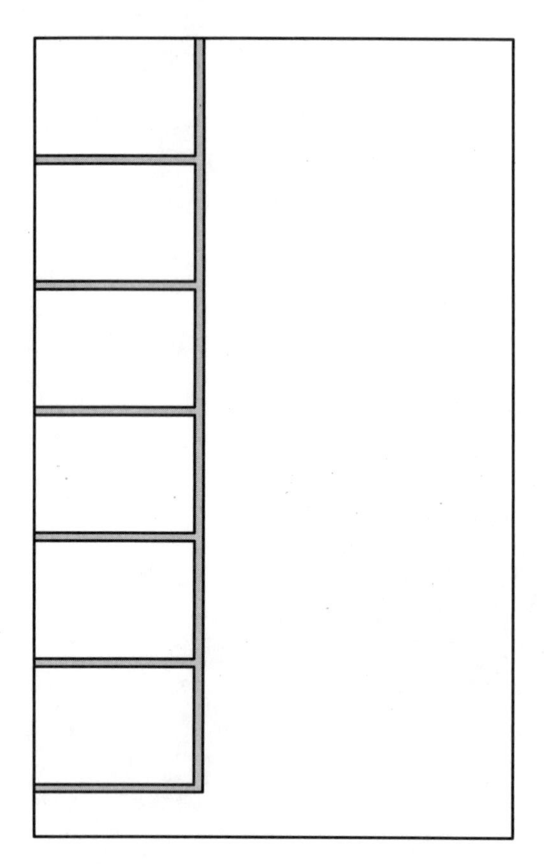

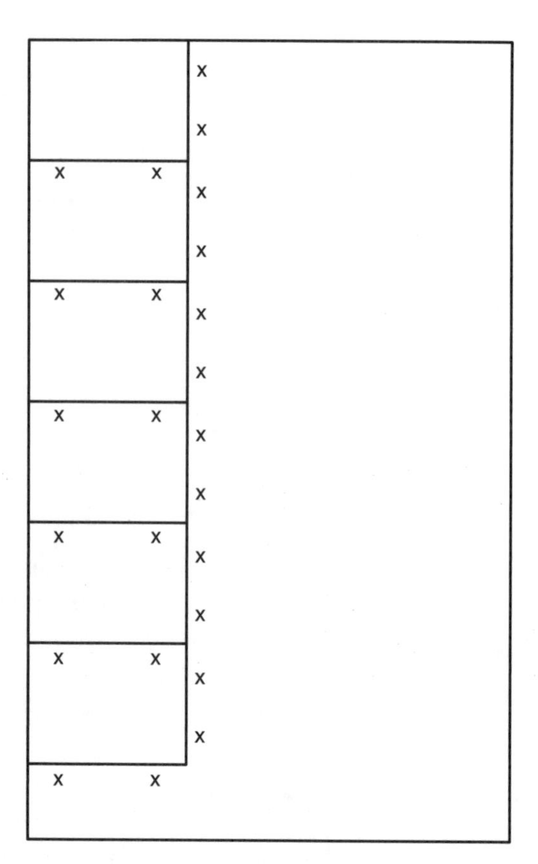

Figure 10-31 Kerf is made between the lines. Xs mark the side of the line on which the kerf is to be made.

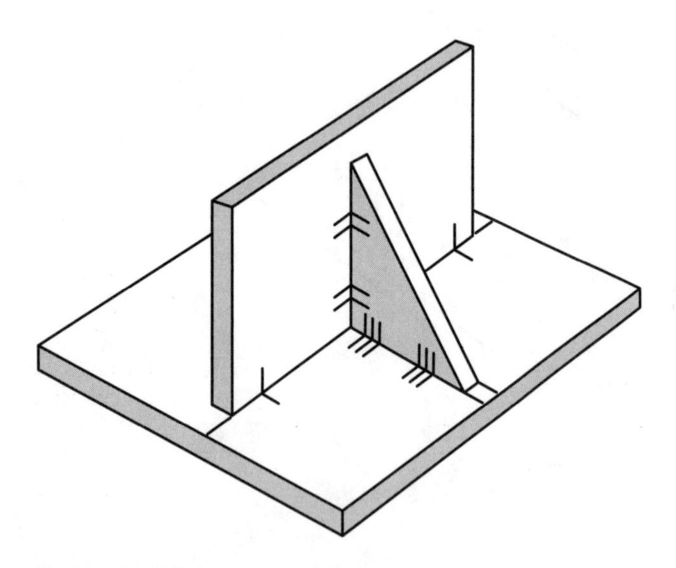

Figure 10-32 Lay out markings to help locate the parts for tack welding.

After the parts have been identified, marked, and located, they are ready for assembly. Parts can be held or clamped into place in preparation for tack welding. Holding parts in place by hand for tack welding is faster than clamping but often leads to errors because the parts may slip while the welder is preparing to tack weld. The more accurate the assembly requirements, the more critical is the clamping; in addition, gas tungsten arc welding often requires the use of both hands to make tack welds.

ASSEMBLY TOOLS

Clamps

A variety of types of clamps can be used to temporarily hold parts in place so that they can be tack welded:

- C-clamps are commonly used and come in a variety of sizes (Figure 10-33). Some C-clamps have been specially designed for welding, and may have a spatter cover over the screw; others have their screws made of spatter-resistant materials, such as copper alloys.

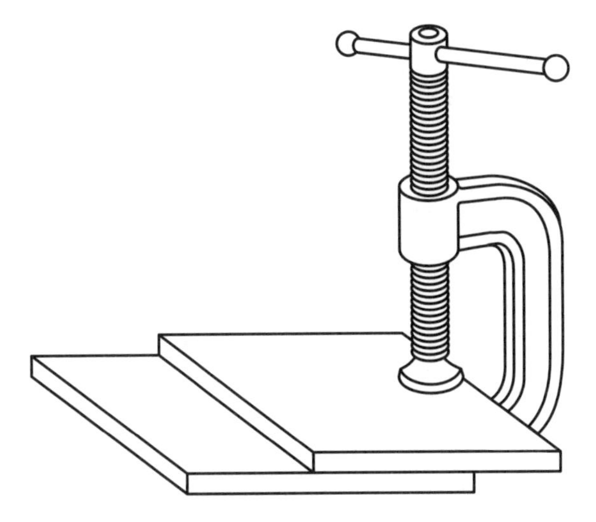

Figure 10-33 C-clamps hold plates for tack welding. (Courtesy of Mike Gellerman)

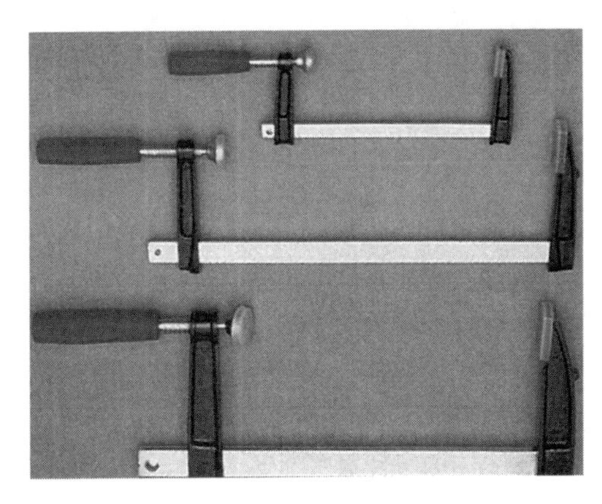

Figure 10-34 Bar clamps. (Courtesy of Woodworker's Supply Inc.)

- Bar clamps are useful for clamping larger parts. Bar clamps have a sliding lower jaw that can be snugged up against the part before tightening the screw clamping end (Figure 10-34). They are available in a variety of lengths.
- Pipe clamps are very similar to bar clamps. The advantage of pipe clamps is that the ends can be attached to a section of standard 1/2″ pipe. This allows for greater flexibility in length, and the pipe can easily be changed if it becomes damaged.
- Locking pliers are available in a range of sizes with a variety of jaw designs (Figure 10-35). The versatility and gripping strength make locking pliers very useful. Some locking pliers

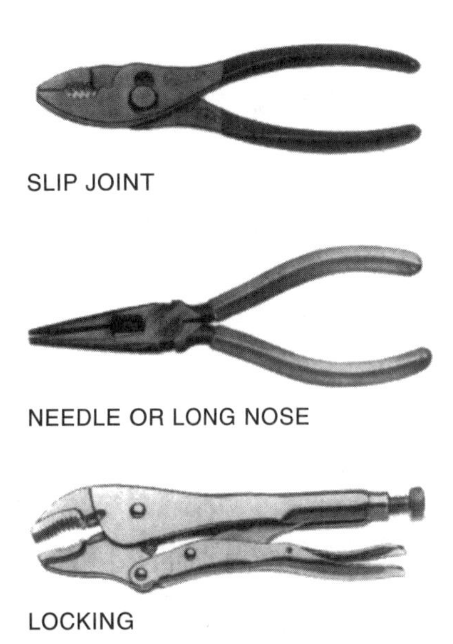

Figure 10-35 Three common types of pliers. (Courtesy of Stanley-Proto Industrial Tools, Covington, GA)

have a self-adjusting feature that allows them to be moved between different thicknesses without the need to readjust them.
- Cam-lock clamps are specialty clamps that are often used in conjunction with a jig or fixture. They can be preset, thus allowing for faster work (Figure 10-36).
- Specialty clamps such as these for pipe welding (see Figure 10-37), are available for many different types of jobs. Such specialty clamps make it possible to do faster and more accurate assembling.

Fixtures

Fixtures are devices that aid in the assembly and fabrication of weldments. Because of the time and expense associated with the fabrication of the fixture itself, they are typically used only when a large number for similar parts are to be fabricated or when they are required to obtain the necessary accuracy of the assembled parts. When used, they must be strong enough to support both the weights of the parts and weld stresses experienced during assembly and remain in tolerance. Some fixtures have manual or automatic clamping devices that aid in their use. Locating pins or other devices can ensure proper part location. A properly designed weld fixture allows adequate room for the welder

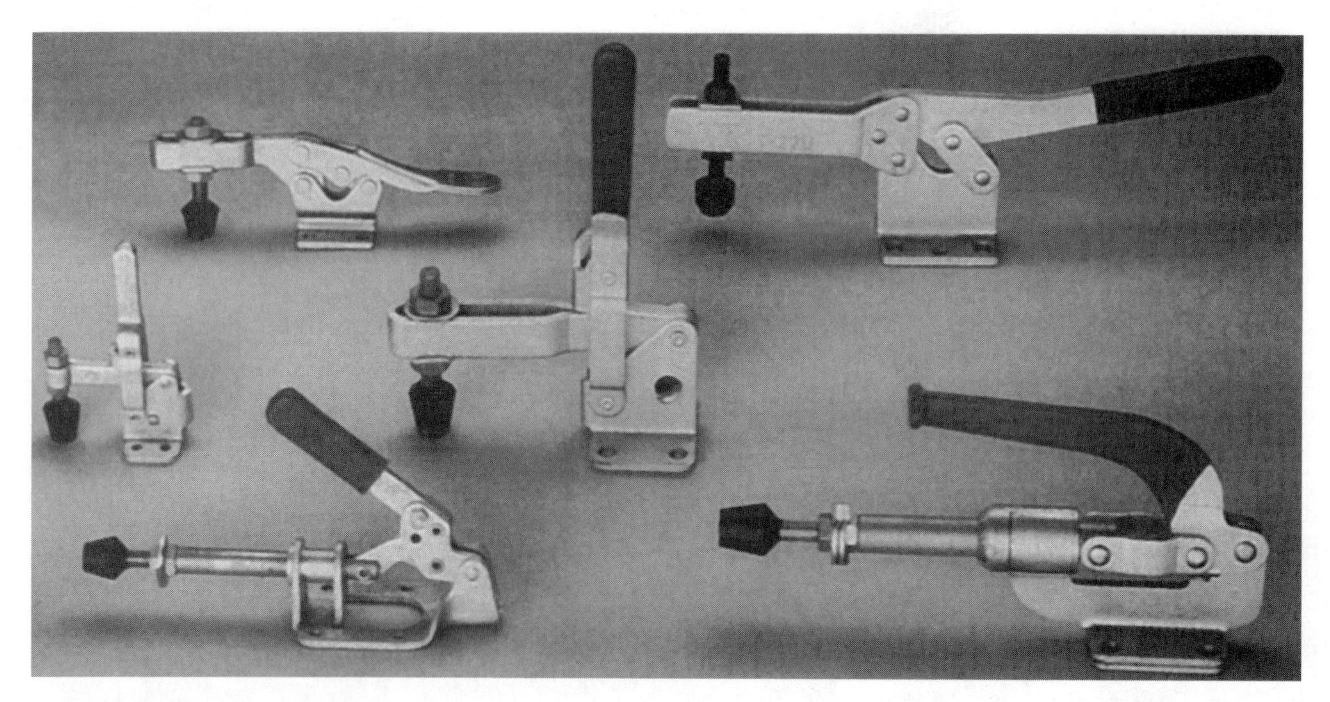

Figure 10-36 Toggle clamps. (Courtesy of Woodworker's Supply Inc.)

(A)

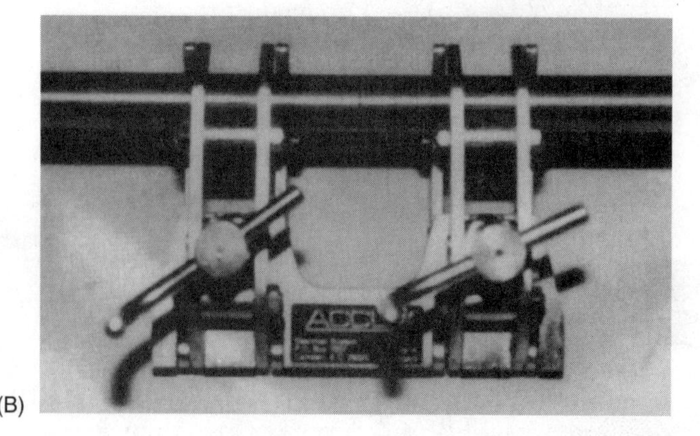

(B)

Figure 10-37 (A) Pipe alignment clamps. (B) Pipe clamps.

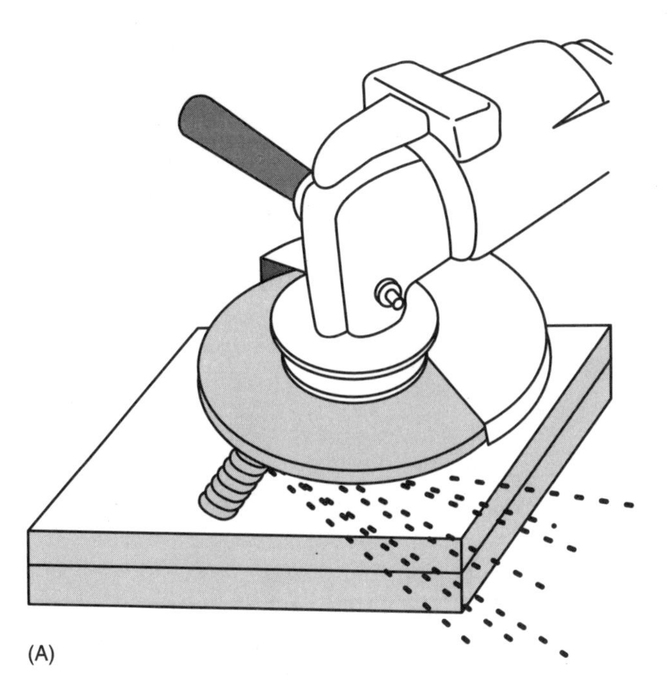

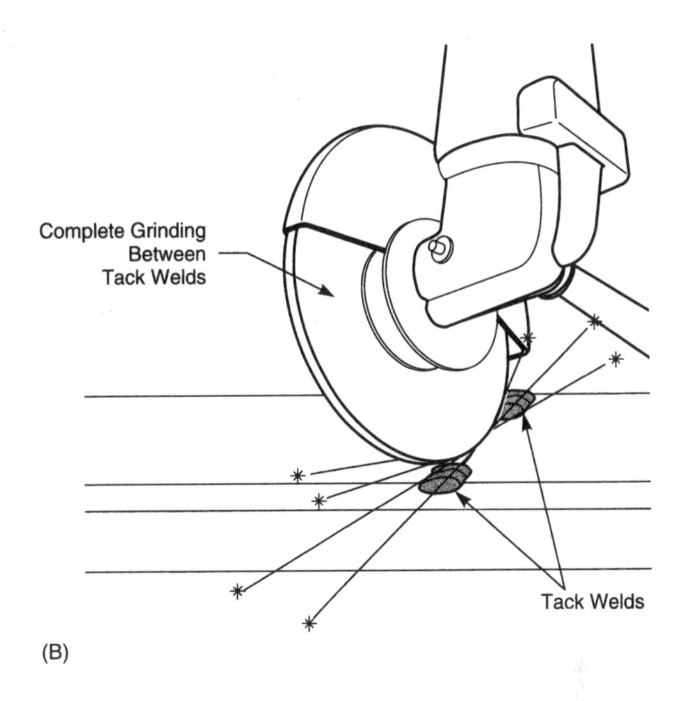

Figure 10-38

to easily perform the necessary tack welding. Some parts may be left in the fixture throughout the entire welding process in order to reduce distortion.

FITTING

Not all parts easily fit together exactly as they were designed. There may be slight imperfections, such as those caused by cutting errors or distortion that may prevent them from fitting into their exact location. Occasionally such problems can be resolved by grinding or filing the part to fit. Hand grinders can be used for some such fitting; however, you must be careful not to remove excessive material. Some metals, such as aluminum, require special grinding stones (Figure 10-38). On small or delicate components such as those fabricated from sheet metal, filing the parts to fit may be the only solution. Rubbing the filed face with a soapstone before it is used can help prevent metal chips, such as those from soft metals (for example, aluminum and brass) from becoming stuck in the file's surface. Some fitting requires that the parts be forced into alignment; one way to accomplish this is to make a small tack weld in the joint and then use a hammer and anvil to pound the parts into alignment (Figure 10-39). Small tack welds used in this manner will become part of the finished weld.

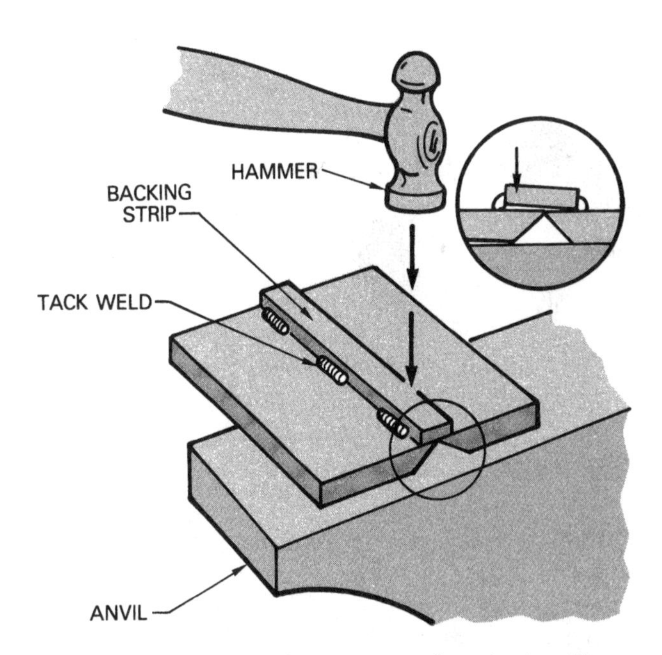

Figure 10-39 Using a hammer to align the backing strip and weld plates.

TACK WELDING

Tack welds are small welds that hold parts in place until they can be welded. Tack welds must be made in such a manner that they can be easily incorporated into the finished weld. They should be made using the same welding procedure and filler metal

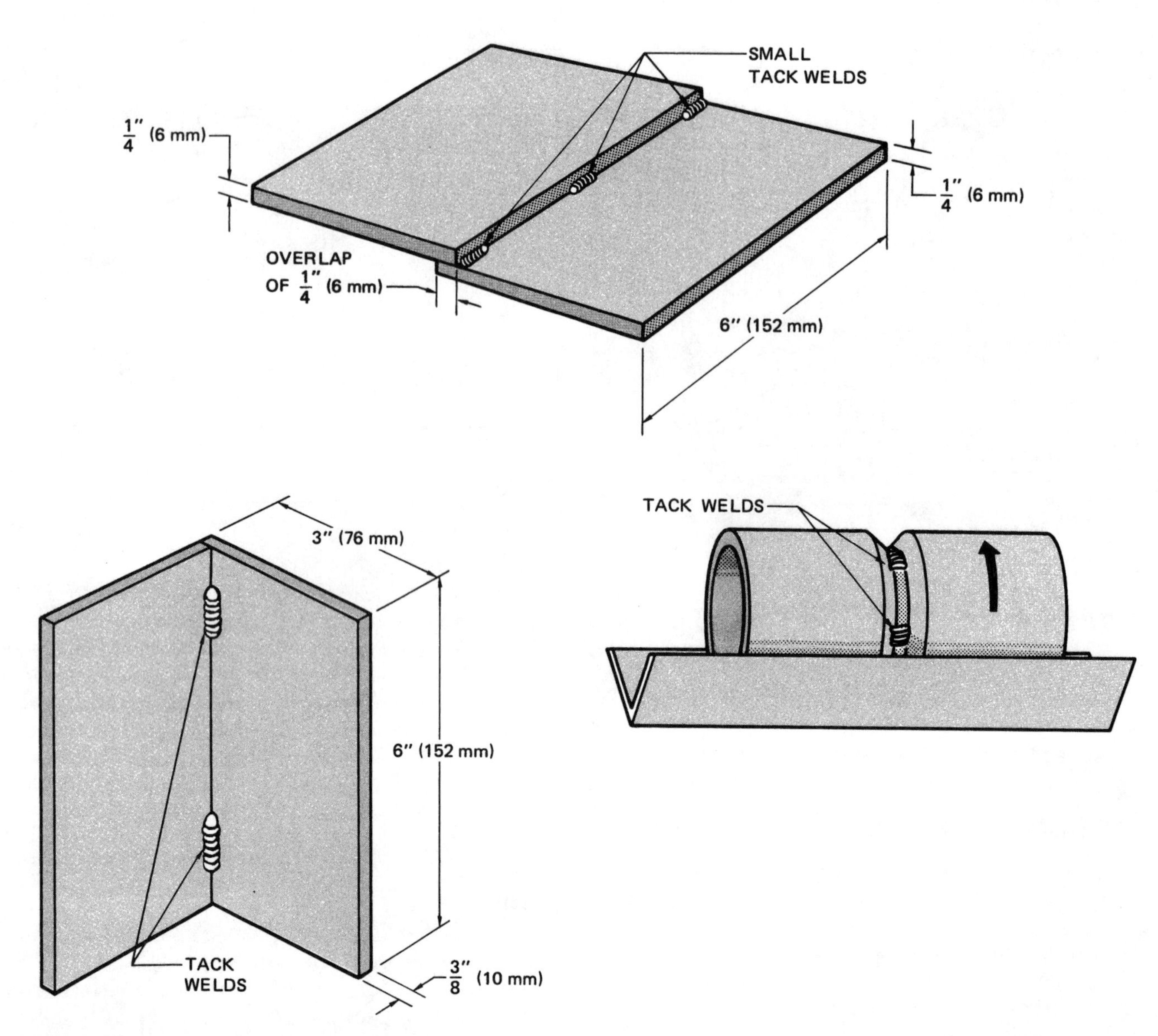

Figure 10-40 Make tack welds as small as possible.

that will be used for the finished weld. Tack welds must be large enough to withstand the forces of assembly and welding yet small enough to be easily incorporated into the finished weld without causing a discontinuity in its size or shape (Figure 10-40).

The size and location of tack welds will vary from weldment to weldment. Some thick sections may require a tack weld to be made only at each end of the joint; however, on some thin sections where distortion can become a problem, tack welds may need to be made closer together (Figure 10-41). Do not hesitate to add additional tack welds to a joint if, during the welding process, you notice that they are needed to prevent distortion.

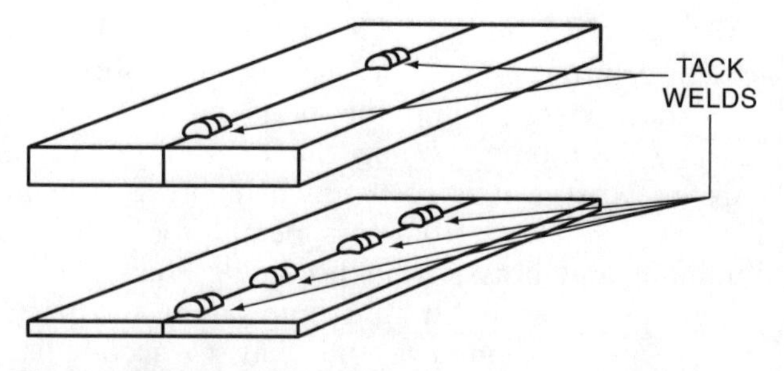

Figure 10-41

WELDING

The order and direction of welds can significantly affect weldment distortion. Generally welds should be staggered from one location on a part to another. This allows the welding heat and its related stresses to dissipate so that they do not cause distortions.

When starting a weld, be sure that the arc is struck in the joint only where the weld will be made. Arc strikes outside the joint can result in unsightly marks that must be removed and, under some code conditions, are considered defects.

Before you begin welding, it is a good idea to check your range and freedom of movement; this is especially important for gas tungsten arc welding because you must maintain a very accurate arc length throughout the weld. Any interference with the manipulation of the filler metal and torch can easily disrupt this arc distance. If the tungsten is accidently dipped into the molten weld pool, significant cleanup time may result.

New welders tend to place the practice plate squarely on the table; however, this may not be the most ideal position for you. With the power off and your welding hood up, make circular practice runs along the joint with the plate at various angles to your body (Figure 10-42). You may find that one angle enables you to follow the joint more easily than the others.

As you develop your GTA welding skills, it is a good idea to practice making welds in different positions. Not all field welds are produced on parts that are held in your most ideal welding position. Improving your skills in these out-of-position welding techniques will be beneficial to you.

If you are working on a weldment that is too large to fit into a welding booth, use portable welding screens to protect others in the welding area from the welding arc's light. Be sure to follow all safety procedures recommended by your welding shop.

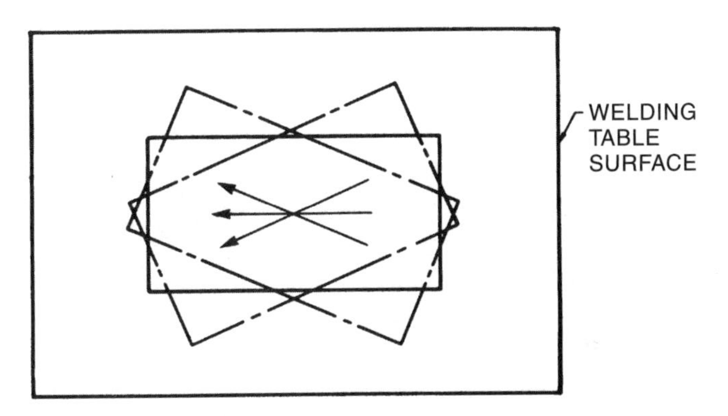

Figure 10-42 Change the plate angle to find the most comfortable welding position.

FINISHING

Depending on the size of the shop, the welder may be responsible for some or all of the finish work. This may include chipping, cleaning, or grinding the welds and applying paint or other protective surfaces.

Grinding of welds should be avoided if possible by properly sizing the weld as it is made. Grinding can add significant cost to the finished weldment. Sometimes it is necessary to grind for fitting purposes or for appearance, but even in these cases, it should be minimized.

CAUTION

When using a portable grinder, be sure that it is properly grounded, the sparks will not fall on others, cause damage, or start a fire, and maintain control to prevent the stone from catching and gouging you or the part.

CAUTION

Be sure that any stone or sandpaper is rated for a speed in revolutions per minute (rpm) that is equal to or greater than the speed of the grinder itself. Using a stone with a lower-rated rpm may cause it to fly apart explosively.

Most grinding is done with a hand angle grinder (Figure 10-43). These grinders can be used with a flat or cupped grinding stone or sandpaper. As the grinder is used, the stone will wear down. Once the grinding stone has worn down to the paper backing, you must discard it. Hold the grinder at an angle so that, if anything is thrown off of the stone or metal surface, it will strike neither you nor others in the area. Because of the speed at which the grinding stone turns, any such object can cause serious injury.

Hold the grinder securely and exert a constant pressure on the work. If the pressure is too great, the grinder motor will overheat and may burn out. If the pressure is too light, the grinder may bounce and possibly crack the grinding stone. Move the grinder in a smooth pattern along the weld. Watch the weld

Figure 10-43 Wire brushes and grinding stones used to clean up welds.

surface as it begins to take the desired shape and change your pattern as needed.

Painting and other finishes release fumes from volatile organic compounds (VOC) that are often regulated by local, state, and national governments. Special ventilation is required for most paints. A ventilation system will remove harmful fumes from the air before they are released back into the environment. Check with your local, state, or national regulating authority before using such products. Read and follow all manufacturers' instructions for the safe use of their products.

CAUTION

Most paints are flammable and must be stored well away from any welding.

REVIEW QUESTIONS

1. Define layout.
2. Define fabrication.
3. What is a chalk line used for?
4. How do you use a chalk line?
5. Where should a layout be started?
6. What factors can affect the laying out of a part?
7. What are two ways of locating the center point of a circle?
8. What tool can be used to lay out odd angles?
9. Why should a sturdy material be used to make templates?
10. What is a contour marker?
11. What does the term *weldment* refer to?
12. What is an assembly drawing?
13. What are bar clamps used for?
14. What is an advantage of pipe clamps?
15. When are cam-lock clamps used?
16. What are fixtures?
17. What are tack welds?
18. How large should a tack weld be?
19. What precautions should be taken when using a portable grinder?
20. What are VOCs?

CHAPTER 11

QUALIFIED AND CERTIFIED WELDERS

INTRODUCTION

Welding, in most cases, is one of the few professions that requires job applicants to demonstrate their skills even if they are already certified. Other jobs may have a written test or require a license, such as those for doctors, lawyers, and pilots. Welders, however, are often required to demonstrate both their knowledge and skills before they are hired because welding is unlike most other occupations in that it requires a high degree of eye–hand coordination.

A common way for welders to demonstrate their welding ability is for them to take a qualification or certification test. Welders who have passed such a test are referred to as qualified welders; if the proper written records are kept as proof of the test results, then they are referred to as certified welders. The basic difference between a qualified welder and a certified welder is that written records are kept for certified welders (Figure 11-1). Not all welding jobs require that the welder be certified. Some jobs require only that a basic welding test be passed.

Welder certification can be divided into two major areas. The first area covers the traditional welder certification that has been used for years. For this certification, welders take a welding test to demonstrate their welding skills for a specific process on a specific weld. This test may be taken to qualify for a welding assignment or as a requirement for employment.

The second and newer area of certification has been developed by the AWS. This certification has three levels. The first level is designed primarily for the new welder who needs to demonstrate Entry Level Weldering skills. The other levels cover advanced welders and expert welders. This chapter covers the traditional certification and the AWS QC10 Specification for Qualification and Certification for Entry Level Welder.

Code or Standard Requirements

The type, depth, angle, and location of the groove is usually determined by a code or standard that has been qualified for specific jobs. Organizations such as the AWS, the ASME, and the ABS issue such codes and specifications. The most common code or set of standards are the AWS D1.1 and the ASME Boiler and Pressure Vessel (BPV) Section IX. Some joint designs that have been tested and found to be reliable for the weldments for specific applications are called *prequalified*. Joint designs that have been modified and are no longer considered prequalified must be tested to ensure that they will be reliable.

WELDER AND WELDING OPERATOR QUALIFICATION TEST RECORD (WQR)

Welder or welding operator's name ______________________ Identification no. ______________

Welding process ____________ Manual ____________ Semiautomatic ____________ Machine ____________

Position ____

(Flat, horizontal, overhead, or vertical - if vertical, state whether up or down.)
In accordance with welding procedure specification no.____________________________

Material specification ____

Diameter and wall thickness (if pipe) - otherwise, joint thickness ____________________

Thickness range this qualifies____

FILLER METAL

Specification No. ______________ Classification ______________ F-number ______________

Describe filler metal (if not covered by AWS specification) ______________________

Is backing strip used? ____

Filler metal diameter and trade name ______________________ Flux for submerged arc or gas for gas metal arc or flux-cored arc welding ______________________

GUIDED BEND TEST RESULTS

Appearance ______________________ Weld Size ______________________

Type	Result	Type	Result

Test conducted by __________ Laboratory test no. __________

FILLET TEST RESULTS

Appearance ______________________ Fillet Size ______________________

Fracture test root penetration ____________________ Macroetch ____________________

(Describe the location, nature, and size of any crack or tearing of the specimen.)
Test conducted by ______________________ Laboratory test no.______________________

RADIOGRAPHIC TEST RESULTS

FILM IDENTIFICATION	RESULTS	REMARKS	FILM IDENTIFICATION	RESULTS	REMARKS

Test conducted by ______________________ Laboratory test no.______________________

We the undersigned, certifiy that the statements in this record are correct and that the welds were prepared and tested in accordance to these requriements.

Manufacturer or contractor ______________________

Authorized by ______________________

Date ______________________

Figure 11-1

QUALIFIED AND CERTIFIED WELDERS

Welder qualification and welder certification are often misunderstood. Sometimes it is assumed that a qualified or certified welder can weld anything. Being certified does not mean that a welder can weld everything, nor does it mean that every weld that is made is acceptable. It means that the welder has demonstrated the necessary skills and knowledge to make good welds. To ensure that a welder is consistently making welds that meet the standard, welds are inspected and tested. The more critical the welding, the more critical the inspection and the more extensive the testing of the welds.

All welding processes can be tested for qualification and certification. The testing can range from making spot welds with an electric resistant spot welder to making gas tungsten arc electron beam welds on aircraft. Being qualified or certified in one area of welding does not automatically mean that a welder can make quality welds in other areas. Most qualifications and certifications are restricted to a single welding process, position(s), and metal and thickness range.

Changes in any one of a number of essential variables can result in the need to recertify (Figure 11-2).

- Process — Welders can be certified in individual welding processes such as SMAW, GMAW, FCAW, GTAW, EBW, RSW and others. Therefore, a new test is required for each process.
- Material — The type of metal, such as steel, aluminum, stainless, or titanium being welded will require a change in the certification. Even

PROCEDURE QUALIFICATION RECORD (PQR)

Welding Qualification Record No: ________ WPS No: ________ Date: ________
Material specification ________ to ________
P-No. ________ Grade No. ________ to P-No. ________ Grade No. ________ Thickness and O.D. ________
Welding process: Manual ________ Automatic ________
Thickness Range ________

FILLER METAL

Specification No. ________ Classification ________ F-number ________
A-number ________ Filler Metal Size ________ Trade Name ________
Describe filler metal (if not covered by AWS specification) ________

FLUX OR ATMOSPHERE

Shielding Gas ________ Flow Rate ________ Purge ________
Flux Classification ________ Trade Name ________

WELDING VARIABLES

Joint Type ________ Position ________
Backing ________ Preheat ________
Passes and Size ________ Bead Type ________
No. of Arcs ________ Current ________
Ampere ________ Volts ________
Travel Speed ________ Oscillation ________
Interpass Temperature Range ________

WELD RESULTS

Appearance ________ Weld Size ________

Figure 11-2

GUIDED BEND TEST

SPECIMEN NO.	DIMENSIONS WIDTH/THICKNESS	AREA	ULTIMATE TOTAL LOAD, LB.	ULTIMATE UNIT STRESS, PSI.	CHARACTER OF FAILURE AND LOCATION

TENSILE TEST

TYPE	RESULT	TYPE	RESULT

Welder's Name ______________ Identification No. ______________ Laboratory Test No. ______________

By virtue of these tests meets welder performance requirements.

Test Conducted by ______________ Address ______________

per ______________ Date ______________

We certify that the statements in this record are correct and that the test welds were performed and tested in accordance to the WPS.

Manufacturer ______________

Signed by ______________

Date ______________

Figure 11-2 continued

a change in the alloy within a base metal type can require a change in certification.

- Thickness — Each certification is valid on a specific range of thickness of base metal. This range is dependent on the thickness of the metal used in the test. For example, if a 3/8″ (9.5 mm) plain carbon steel plate is used, then under some codes the welder would be qualified to make welds in plate thickness ranges from 3/16″ to 3/4″ (4.7 mm to 19 mm).
- Filler metal — Changes in the classification and size of the filler metal can require recertification.
- Shielding gas — If the process requires a shielding gas, then changes in gas type or mixture can affect the certification.
- Position — In most cases, if the weld test was taken in the flat, the certification is limited to flat and possible horizontal. If the test was taken in the vertical, this will usually allow the welder to work in the flat, horizontal, and vertical positions.
- Joint design — Changes in weld type, such as groove or fillet welds will require a new certification. Additionally, variations in the joint geometry such as groove type, groove angle, and number of passes can also require retesting.
- Welding current — In some cases, changing from AC to DC or changes such as pulsed power and high frequency can affect the certification.

Any welder qualification or certification process must include the specific welding skill level that is to be demonstrated. The detailed information for welding and testing is often given as part of a Welding Procedure Specification (WPS) or similar set of welding specifications or schedule (refer to Chapter 13, Figure 13-3). Such a specific set of written standards is needed so that everyone knows what skills are

WELDING PROCEDURE SPECIFICATION (WPS)

Welding Procedures Specifications No: ______________________ Date: ______________________

TITLE:

Welding ______________________ of ______________________ to ______________________

SCOPE:

This procedure is applicable for ______________________

within the range of ______________________ through ______________________

Welding may be performed in the following positions ______________________

BASE METAL:

The base metal shall conform to ______________________

Backing material specification ______________________

FILLER METAL:

The filler metal shall conform to AWS specification No. ______________________ from AWS

specification ______________________. This filler metal falls into F-number ______________________
and A-number ______________________

SHIELDING GAS:

The shielding gas, or gases, shall conform to the following compositions and purity:

____

JOINT DESIGN AND TOLERANCES:

PREPARATION OF BASE METAL:

ELECTRICAL CHARACTERISTICS:

The current shall be ______________________

The base metal shall be on the ______________________ side of the line.

PREHEAT:

BACKING GAS:

WELDING TECHNIQUE:

INTERPASS TEMPERATURE:

CLEANING:

INSPECTION:

REPAIR:

SKETCHES:

Figure 11-3

required. The WPS enables the welder to be prepared for the required welding test and the company to know what skills the welder has demonstrated. Varying from these strict limitations in the WPS usually requires that a different test be taken.

Welder Performance Qualification is the demonstration of a welder's ability to produce welds meeting very specific prescribed standards. The form that documents this test is called the Welding Qualification Test Record. The detailed written instructions that are to be followed by the welder are called the Welder Qualification Procedure. Welders who pass this are often referred to as Qualified Welders or just Qualified.

Welder Certification is the written verification that a welder has produced welds meeting a prescribed standard of welder performance. A welder who holds such a written verification is often referred to as Certified or as a Certified Welder.

An AWS Certified Welder is one who has complied with all of the provisions, requirements, and specifications of the AWS for this certification. Very specific requirements must be followed by any school or organization before they can offer this certification. Under the AWS program, the welder must pass both a closed-book exam in specific knowledge areas and a performance test. Written documentation including the welder's name and Social Security number along with test results must be sent to the AWS. These records are entered into the AWS National Registry for welders. The certification record expires after one year and at that time is automatically deleted from the registry.

REVIEW QUESTIONS

1. How does a welder's employment process differ from most other trades and professions?
2. What is the basic difference between a qualified welder and a certified welder?
3. What two major areas can welder certification be divided into?
4. What are the three levels of certification offered through the AWS program?
5. What types of information about welds is usually determined by a code or standard?
6. What are the two most commonly used codes?
7. Why are welders and welds inspected and tested?
8. Changes in any one of a number of essential variables can result in the need to recertify. What are these variables?
9. What is a WPS?
10. How long is an AWS welder certification good for?

CHAPTER 12

AWS ENTRY-LEVEL WELDER QUALIFICATION AND WELDER CERTIFICATION

INTRODUCTION

The AWS Entry Level Welders' qualification and certification program has a number of requirements not normally found in the traditional welder qualification and certification process. The additions to the AWS program have broadened the scope of the test. Areas such as practical knowledge have long been an assumed part of most certification programs. It has, however, not been a formal part of the process. Most companies assume that if a welder can produce code-quality welds, they understand enough of the technical aspects of welding.

Today a greater importance is placed on the technical knowledge of the process, code, and other aspects of the complete welding process. This change has been brought about by the greater complexity of many welding processes and an increased responsibility by the company and their welders to ensure the quality and reliability of weldments. Not only is it important that the weld be correctly performed, but it is also important that the welder know why it must be performed in a specific manner. This is all intended to increase accuracy and reduce rejection of welds.

PRACTICAL KNOWLEDGE

The practical knowledge requirement for welders is evaluated with a written exam that has two major components: safety and subject knowledge. The minimum overall grade for the safety portion is 90%; the minimum passing grade for the technical subject areas is 75%. The technical areas that are covered on the written exam are as follows:

- Welding and cutting
- Welding and cutting inspection and testing
- Welding and cutting terms and definitions
- Base and filler metal identification
- Common welding process variables
- Electrical fundamentals
- Drawing and welding symbol interpretation
- Fabrication principles and practices
- Safety

PERFORMANCE TEST

The performance test consists of more than just being able to produce acceptable welds. Areas covered under the practical test include the following:

- Layout
- Written procedures
- Written records
- Verbal instructions
- Safety
- Housekeeping
- Cutting out parts
- Fit up and assembly of parts
- Workmanship standards for welding
- Inspection

Layout

As part of the layout, you will be expected to interpret simple drawings and sketches, including welding symbols. This also includes making conversions between standard and S.I. units. Additionally, as part of the layout portion of the performance, you must complete a bill of material including all of the required materials and parts for the assigned fabrication.

The parts must be laid out to within a fractional tolerance of ±1/16″ (1.6 mm) with an angular tolerance of +10° or –5°. Be sure to leave an appropriate amount of space between parts for the torch kerf.

Written Procedures

Each drawing has a written list of specifications and notes that must be followed. Incorporated in these notes are references to specific welding procedure specifications (WPS) that also must be followed. Welding procedure specifications for mild steel, stainless steel, and aluminum have been included in the student lab manual. Additional information about welding procedure specifications can be found in the AWS EG2.0 publications.

Written Records

Written records must be kept by the welder for each welding performance test. These written records must include the following information:

- Welder's name
- Date
- Welding materials including filler metal, base metal, and shielding gas
- Welding current including amperage range, voltage, and polarity
- Metal type and thickness
- Filler metal diameter and type

Additional information may be included on the written record. All information must be neat and legible. These records must be turned in with the completed welding and will be considered as part of the overall evaluation of your skills. Similar records are often required by welding companies as a means of maintaining job-related records. Figure 12-1 shows

TIME CARD		
Name ______	Date ______	Job ______
Starting Time	Ending Time	Total Time
Sign		Total

(A)

Figure 12-1 (A) Useful forms for keeping track of weldments.

INSPECTION REPORT **JOB:** ____________

INSPECTION	PASS/FAIL	INSPECTOR'S INITIALS	DATE
Layout			
Cutout			
Assembly			
Welding			
Tack			
Interpass			
Finish			
Overall Rating			
Accuracy			
Appearance			

Welder ____________ Date ____________

(B)

Figure 12-1 continued

BILL OF MATERIALS

Name ____________ Date ____________ Job ____________

Part ID	Size Determination	SI Determination

Material Specification ____________

(C)

Figure 12-1 continued

the type of forms that may used for keeping and maintaining these records.

Verbal Instructions

From time to time on any job, additional instructions must be given to the welder verbally. These instructions are as important as those provided in writing. Many times in a welding shop minor modifications of a weld may be required perhaps to meet a specific customer's requirements or to enable the shop to take advantage of existing materials. In some cases critical information such as particular safety concerns may be given verbally. It is your responsibility to remember and follow all such verbal instructions.

Safety

Job safety is a major concern for everyone. Welding can present specific concerns that must be addressed by everyone working in a shop. Specific safety information pertaining to gas tungsten arc welding is included in Chapter 2, and additional information is available in *Safety for Welders* by Larry Jeffus and ANSI Z49.1. Safety is such an important area of concern that a score of 100% is required before you will be allowed to do any work in the welding lab.

Housekeeping

All welding processes produce some level of scrap or trash. Additionally, the work area can become cluttered with welding leads, hoses, torches, hand tools, power tools, clamps, jigs, and fixtures as well as other items. Welders are responsible for maintaining their work space in a clean and orderly manner.

On some welding jobs, a degree of housekeeping may be provided; however, the welder is usually held responsible for their specific work station's condition. Learning good housekeeping skills, therefore, is important.

Routine repairs of equipment may also fall under housekeeping chores. Most welders are expected to change their own welding hood lenses as needed and make minor repairs or adjustments to the GTA welding torch and work plant. Before making any such repairs or adjustments, your must first check with the manufacturer's literature for the equipment and with your instructor and make sure the power is turned off.

Cutting out Parts

The parts will be cut out using the manual oxyfuel gas cutting (OFC), machine oxyfuel gas cutting (OFC-track burner), air arc cutting (CAC-A), and manual plasma arc cutting (PAC) processes.

To cut out the various parts, you will need to make straight, angled, and circular cuts. In some cases it is up to you which method and process to use. On other cuts the method and process are specified on the drawing. Inspect all cut surfaces for flaws and defects and repair if necessary (Figure 12-2).

Fit up and Assembly of the Parts

Putting the parts of a weldment together in preparation for welding requires special skills. The more complex the weldment, the more difficult the assembling. Each part must be located and squared to other parts (Figure 12-3). To complicate this process, clamping may be necessary because the parts may not all be flat and straight.

As metal is cut, it may become distorted as a result of heat from the cutting process. Any such distortion or bend that may have been in the metal must be corrected during the fit-up process. Sometimes grinding is required to make a correct fit-up (Figure 12-4). At other times the parts can be fitted together correctly by using C-clamps or pliers (Figure 12-5). In more difficult cases, tack welds and cleats or dogs must be used to achieve the proper fit-up (Figure 12-6).

Workmanship Standards for Welding

Positioning of the weldment must be made so that the welds are made within 5° of the specified position.

All arc strikes must be within the groove. Arc strikes outside of the groove will be considered defects.

Tack welds must be small enough so that they do not interfere with the finished weld. They must, however, be large enough to withstand the shrinkage forces from the welds as they are being made. Sometimes it is a good idea to use several small tacks on the same joint to ensure that the parts are held in place (Figure 12-7).

The weld bead size is important. Beads must be sized in accordance with the WPS or the drawing for each specific weld. All weld bead starts and stops must be smooth (Figure 12-8).

The weld beads must be cleaned. All weld cleaning must be performed with the test plate in the welding position. A grinder may not be used to remove weld control problems.

Correct cut

Top edge square

Face smooth

Bottom square

Oxides, if any, easily removed

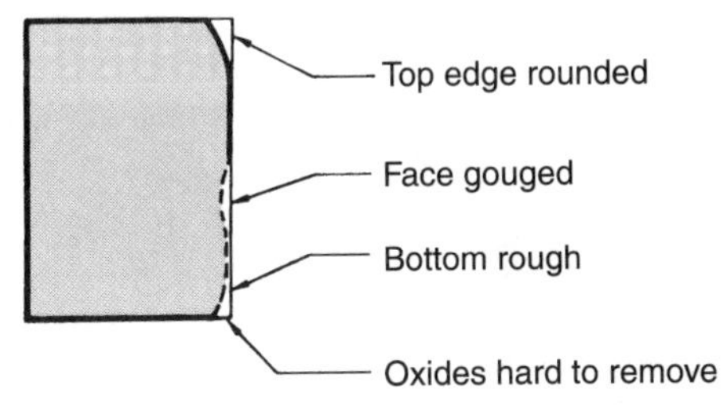

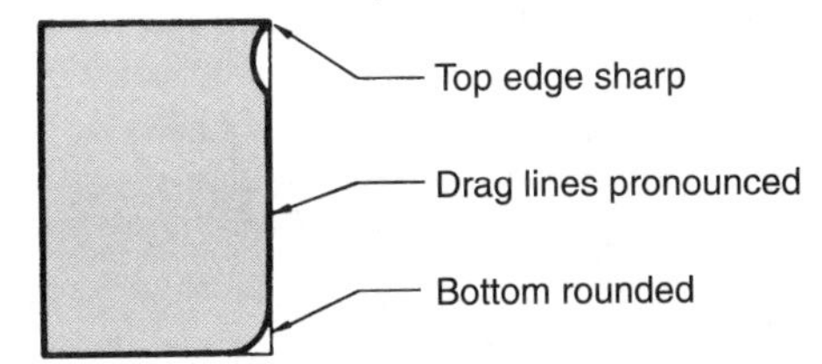

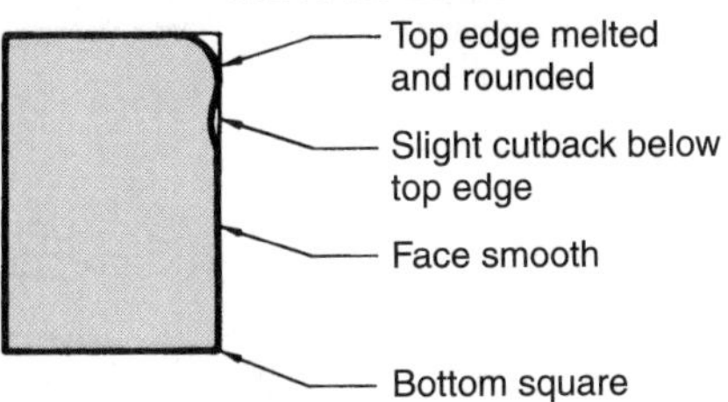

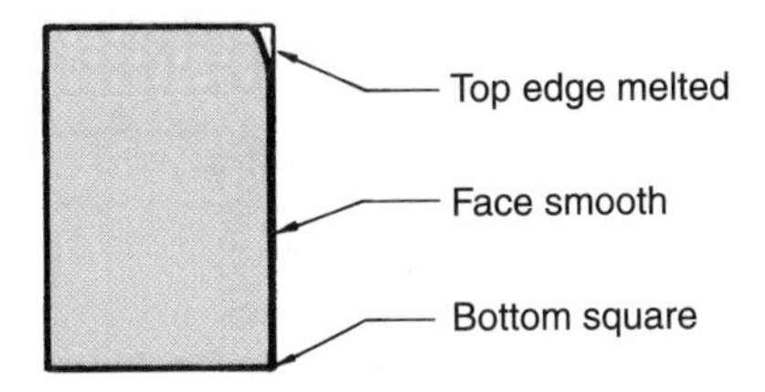

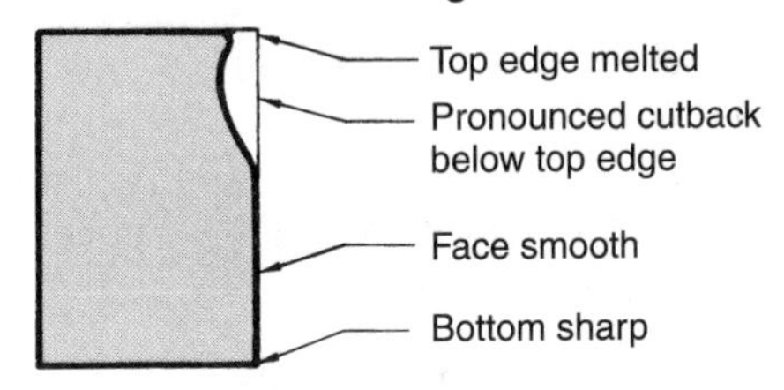

Figure 12-2 Flame-cut profiles and standards.

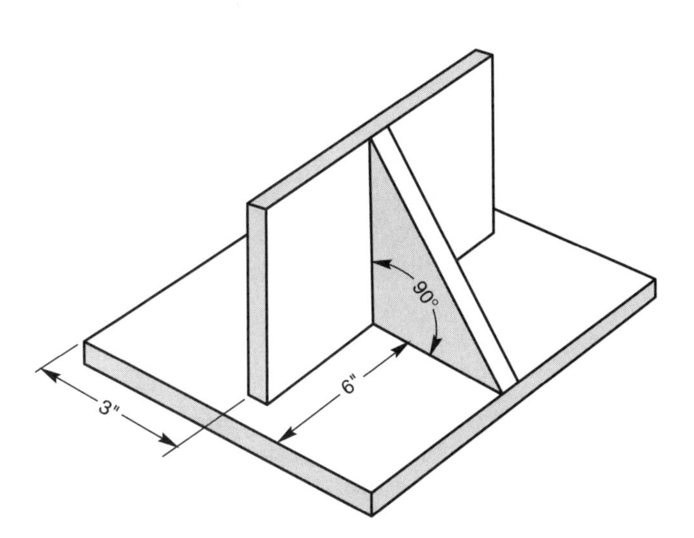

Figure 12-3 Locate and square parts to be welded.

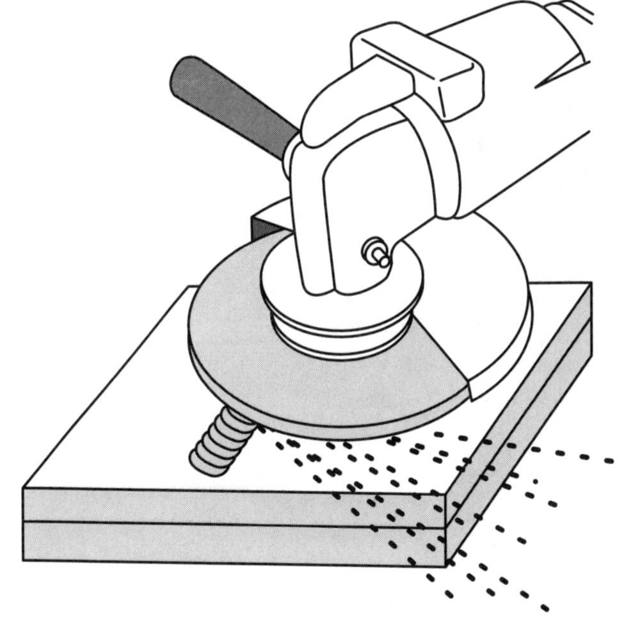

Figure 12-4 A portable grinder can be used to correct a cutting or fitting problem.

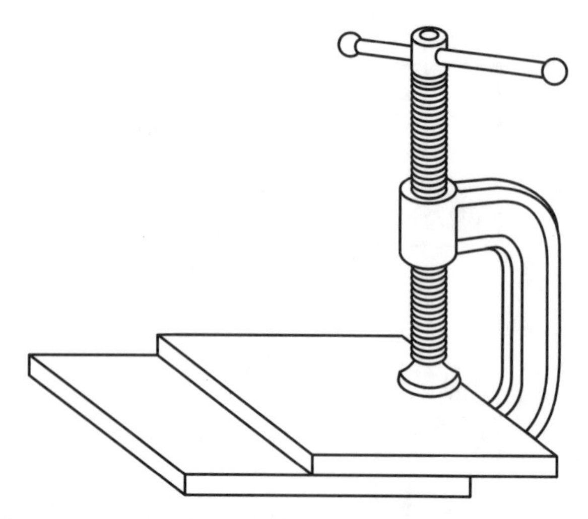

Figure 12-5 C-clamp being used to hold plates for tack welding.

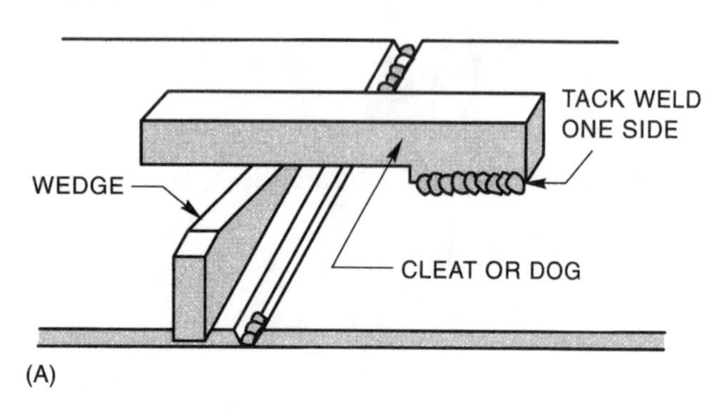

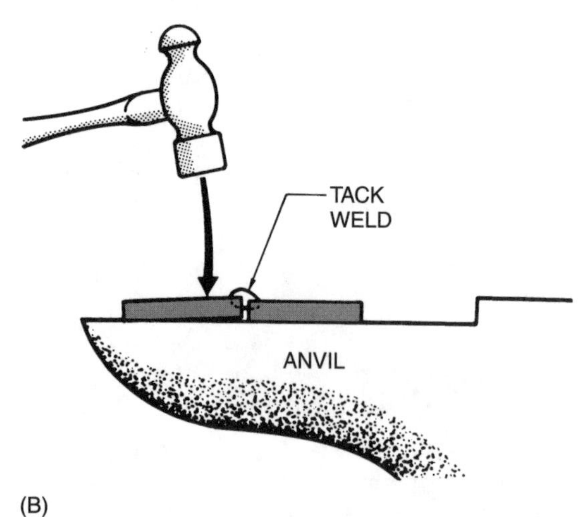

Figure 12-6 (A) Wedges and cleats can be used on heavier metal to pull the joint into alignment. (B) The plates can also be forced into alignment by striking them with a hammer.

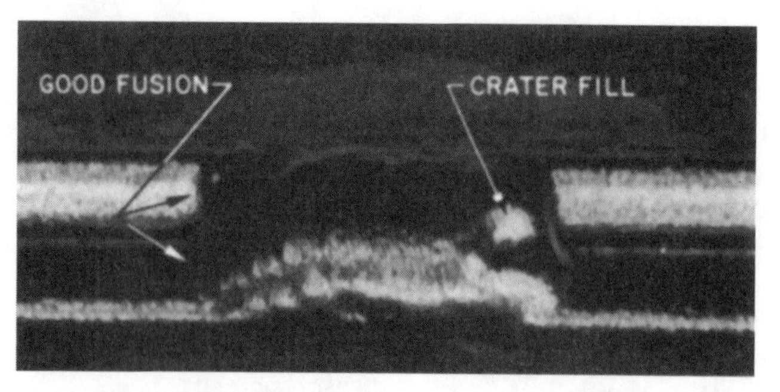

Figure 12-7 Tack weld. Note the good fusion at the start and the crater fill at the end.

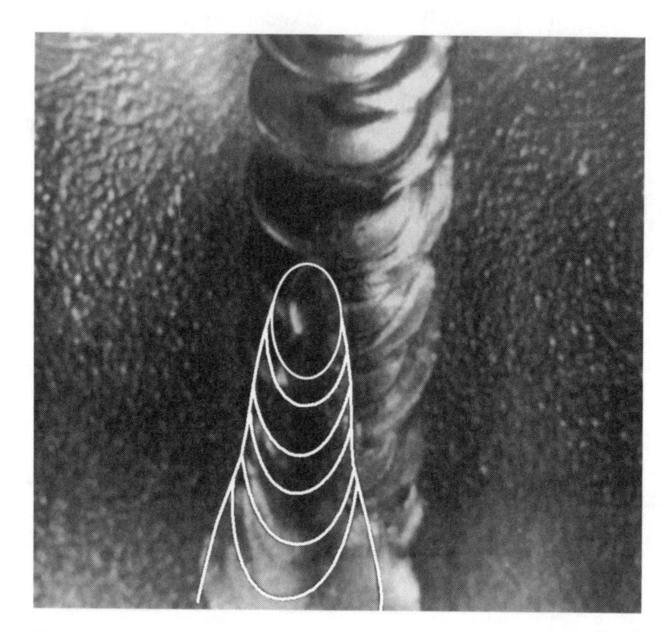

Figure 12-8 Taper down the weld when it is complete or when stopping to reposition. This will make the end smoother and ease restarting the next weld bead if necessary.

Weld Inspection

Each weld and/or weld pass is to be inspected visually for the following:

- There shall be no cracks or incomplete fusion.
- There shall be no incomplete joint penetration in groove welds except as permitted for partial joint penetration groove welds.
- The test supervisor shall examine the weld for acceptable appearance and shall be satisfied that the welder is skilled in using the process and procedure specified for the test.
- Undercut shall not exceed the lesser of 10% of the base metal thickness or 1/32″ (0.8 mm).

- Where visual examination is the only criterion for acceptance, all weld passes are subject to visual examination at the direction of the test supervisor.
- The frequency of porosity shall not exceed one in each 4″ (100 mm) of weld length, and the maximum diameter shall not exceed 3/32″ (2.4 mm).
- Welds shall be free from overlap.*

*American Welding Society AWS QC10-95 Specification for Qualification and Certification for Entry-Level Welders

WELDING SKILL DEVELOPMENT

The American Welding Society's *AWS EG2.0-95 Guide for the Training and Qualification of Welding Personnel Entry Level Welder* lists safety, related knowledge, and welding skills that must be mastered in a training program.

The amount of time required for each welder to develop the necessary skills to pass the American Welding Society Certified Welder program will vary greatly from individual to individual.

REVIEW QUESTIONS

1. How are the practical knowledge requirements for welders evaluated?
2. What is the minimum overall grade for the safety portion of the written exam before working in a shop is allowed?
3. What areas are covered under the practical test?
4. What information must be included in a welder's written records?
5. What type of verbal instructions may be given to a welder?
6. Who is responsible for maintaining the welder's work space and keeping it clean and orderly?
7. How can weldment parts be cut out for fabrication?
8. How can weld control problems be removed?

CHAPTER 13

WELDING SPECIFICATIONS

INTRODUCTION

As a result of years of experience and thousands of tests, specific welding criteria have been established that, when followed, have the greatest likelihood of producing quality welds. These welding specifications have ranges that allow welders to make changes within the specified range to suit their individual welding style.

When specific welding standards have been established for a particular weld, this material is provided in a document called *Welding Procedures Specifications (WPS)*. Welding procedure specifications should contain all of the necessary information required by the welder to produce a satisfactory weld.

WPS GENERAL INFORMATION

A WPS should contain information about the process material, filler metal, shielding gas, and other important material. Figure 13-1 is an example of a blank welding procedure specification form that can be used for processes other than gas tungsten arc welding; some blanks may not have information included in them.

1. The WPS (see Table 13-1) number may contain more information than simply a numeric identifier. Sometimes there may be a name or an abbreviation that would identify the welding school, shop, or organization that helped develop this specific document.
2. Date: This is the date that the WPS was accepted following testing.
3. Title: 3A The title includes information regarding the welding process that will be used and the types of material that will be joined. The process could be GTAW-pulsed or any other specific process identification.

   Blanks 3b and 3c are for the types of material, sheet, plate, pipe, or tubing to be welded. On some WPSs, plate-to-pipe welds, plate-to-tubing welds, tubing welds, or any other such combination could be included here.

TABLE 13-1

SPECIFICATION NUMBERS	
A5.10	Aluminum—bare electrodes and rods
A5.3	Aluminum—covered electrodes
A5.8	Brazing filler metal
A5.1	Steel, carbon, covered electrodes
A5.20	Steel, carbon, flux cored electrodes
A5.17	Steel-carbon, submerged arc wires and fluxes
A5.18	Steel-carbon, gas metal arc electrodes
A5.2	Steel—oxy-fuel gas welding
A5.5	Steel—low-alloy covered electrodes
A5.23	Steel—low-alloy electrodes and fluxes—submerged arc
A5.28	Steel—low-alloy filler metals for gas shielded arc welding
A5.29	Steel—low-alloy, flux cored electrodes

WELDING PROCEDURE SPECIFICATION (WPS)

Welding Procedures Specifications No: ______(1)______ Date: ______(2)______

TITLE: Welding ______(3a)______ of ______(3b)______ to ______(3c)______

SCOPE: This procedure is applicable for ______(4a)______

within the range of ______(4b)______ through ______(4c)______

Welding may be performed in the following positions ______(4d)______

BASE METAL: The base metal shall conform to ______(5a)______

Backing material specification ______(5b)______

FILLER METAL: The filler metal shall conform to AWS specification

No. ______(6a)______ from AWS specification ______(6b)______ . This filler metal falls into

F-number- ______(6c)______ and A-number ______(6d)______

ELECTRODE: The tungsten electrode shall conform to AWS

specification No. ______(7a)______ from AWS specification ______(7b)______ The tungsten diameter shall

be ______(7c)______ . The tungsten end shape shall be ______(7d)______

SHIELDING GAS: The shielding gas, or gases, shall conform to the following compositions and purity: ______(8)______

JOINT DESIGN AND TOLERANCES: ______(9)______

PREPARATION OF BASE METAL: ______(10)______

ELECTRICAL CHARACTERISTICS: The current shall be ______(11a)______

The base metal shall be on the ______(11b)______ side of the line.

METAL SPECIFICATIONS		GAS FLOW			NOZZLE SIZE	AMPERAGE
		RATES	PURGING TIMES			
THICKNESS	DIA. OF E70S-3*	CFM (L/MIN)	PREPURGING	POSTPURGING	IN. (MM)	MIN. MAX.
(12a)	(12b)	(12c)	(12d)	(12e)	(12f)	(12h)

OPERATING RANGE SPECIFICATIONS (12)

PREHEAT: ______(13)______

BACKING GAS: ______(14)______

SAFETY: ______(15)______

WELDING TECHNIQUE: (16)

TACK WELDS: ______(16a)______

SQUARE GROOVE AND FILLET WELDS: ______(16b)______

INTERPASS TEMPERATURE: ______(17)______

CLEANING: ______(18)______

INSPECTION: ______(19)______

REPAIR: ______(20)______

SKETCHES: ______(21)______

Figure 13-1 Welding Procedure Specification (WPS) form.

4. Scope: Included in 4a is the type of weld (for example, a groove weld, a fillet weld, or a plug weld). Blanks 4b and 4c are the thickness range. This could be for a single range such as 1/4 inch to 1/4 inch or a more inclusive range, such as 18 gauge to 10 gauge. Blanks 4d lists the welding positions that this WPS covers. Single or multiple positions can be included, such as 1G and 2G or 1G, 2G, 3G, and so on.
5. Base metal: Specific information regarding the makeup of the base material is included. General information such as mild steel, stainless steel, aluminum, titanium, anconal, or monel is included with the specific group or grouping of the base material. Blank 5b would include specifications for any backing material used for the joint, such as backing strips that might be used for a V-groove butt joint.
6. Blank 6a includes specific information regarding the AWS's identification number for the filler metal. Blank 6b is the specification documentation under which that specific filler metal can be found. Blank 6c is the F number group designation for the metal type (Table 13-2). Blank 6d is the A number, which is an additional identification number used for some metals.
7. Electrode: Blank 7a is the AWS's identification of a specific electrode classification (Table 13-3). Blank 7b is usually a 5.12, which is the AWS's specification for tungsten electrodes. Blank 7c is the maximum diameter tungsten that can be used with this WPS. Blank 7d gives the end shape specifications for the tungsten electrode: either tapered, rounded, or tapered with a balled end.
8. Shielding gas: The specific makeup of the shielding gas—whether it is 100% argon, 100% helium, or a mixture of gases—is included along with the grade, which is usually denoted as welding grade.
9. Joint design and tolerance: A cross-section of the specific welding joint, such as the one illustrated

TABLE 13-2

F NUMBERS

GROUP DESIGNATION	METAL TYPES	AWS ELECTRODE CLASSIFICATION
F1	Carbon Steel	EXX20, EXX24, EXX27, EXX28
F2	Carbon Steel	EXX12, EXX13, EXX14
F3	Carbon Steel	EXX10, EXX11
F4	Carbon Steel	EXX15, EXX16, EXX18
F5	Stainless Steel	EXXX15, EXXX16
F6	Stainless Steel	ERXXX
F22	Aluminum	ERXXXX

TABLE 13-3

TUNGSTEN ELECTRODE TYPES AND IDENTIFICATION

AWS CLASSIFICATION	TUNGSTEN COMPOSITION	TIP COLOR
EWP	Pure tungsten	Green
EWTh-1	1% thorium added	Yellow
EWTh-2	2% thorium added	Red
EWZr	1/4% to 1/2% zirconium added	Brown
EWCe-2	2% cerium added	
EWLa-1	1% lanthanum added	
EWG	Alloy not specified	

in Figure 13-2 is included. This drawing should include acceptable limits on groove and joint angles as well as root spacing, bagging material, and sizes.

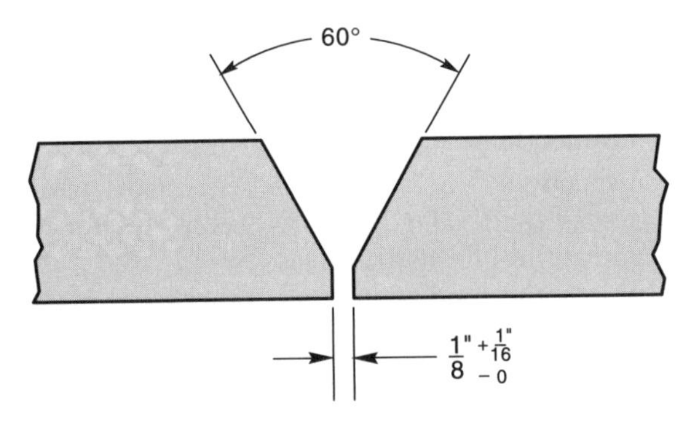

Figure 13-2

10. Preparation of base metal: Specific information regarding the cleaning and preparation of the base metal should be included. In most cases, for code-quality welds, the joint must be cleaned thoroughly in a minimum of 1 inch of the groove on both the inside and outside surfaces. Such details should be included.
11. Electrical characteristics: The type of welding current, direct current electrode negative, direct current electrode positive, for AC is specified here. Blank 11b lists the electrode polarity as either positive, negative, or nonapplicable (N/A) for alternating currents.
12. Operating range specifications: Table 13-4 is a detailed listing of the allowable ranges for electrode diameter gas flow rate pre- and postpurge times, nozzle sizes, and amperage for each of the thicknesses of metal covered under the scope of this WPS. This chart is used by first selecting the thickness of

TABLE 13-4

OPERATING RANGE SPECIFICATION						
METAL SPECIFICATIONS		GAS FLOW			NOZZLE SIZE IN. (MM)	AMPERAGE MIN. MAX.
		RATES	PURGING TIMES			
THICKNESS	DIA. OF E70S-3*	CFM (L/MIN)	PREPURGING	POSTPURGING		
18 ga.	1/16 in. (1.6 mm)	15 to 20 (7 to 9)	10 to 15 sec.	10 to 25 sec.	1/4 to 3/8 (6 to 10)	45 to 65
17 ga.	1/16 in. (1.6 mm)	15 to 20 (7 to 9)	10 to 15 sec.	10 to 25 sec.	1/4 to 3/8 (6 to 10)	45 to 70
16 ga.	1/16 in. (1.6 mm)	15 to 20 (7 to 9)	10 to 15 sec.	10 to 25 sec.	1/4 to 3/8 (6 to 10)	50 to 75
15 ga.	1/16 in. (1.6 mm)	15 to 20 (7 to 9)	10 to 15 sec.	10 to 25 sec.	1/4 to 3/8 (6 to 10)	55 to 80
14 ga.	3/32 in. (2.4 mm)	20 to 25 (9 to 12)	10 to 20 sec.	10 to 30 sec.	3/8 to 5/8 (10 to 16)	60 to 90
13 ga.	3/32 in. (2.4 mm)	20 to 25 (9 to 12)	10 to 20 sec.	10 to 30 sec.	3/8 to 5/8 (10 to 16)	60 to 100
12 ga.	3/32 in. (2.4 mm)	20 to 25 (9 to 12)	10 to 20 sec.	10 to 30 sec.	3/8 to 5/8 (10 to 16)	60 to 110
11 ga.	3/32 in. (2.4 mm)	20 to 25 (9 to 12)	10 to 20 sec.	10 to 30 sec.	3/8 to 5/8 (10 to 16)	65 to 120
10 ga.	3/32 in. (2.4 mm)	20 to 25 (9 to 12)	10 to 20 sec.	10 to 30 sec.	3/8 to 5/8 (10 to 16)	70 to 130

* Other E70S-X filler metal may be used.

the material to be welded and then following horizontally across the page to the right, where in each column specific ranges are given for each welding parameter to be used in making the weld.

13. Preheat: Unless a higher preheat is required for the particular type of metal to be welded, a minimum temperature of 50° F (10° C) is the lowest acceptable temperature allowed for welding.
14. Backing gas: On some critical welds, pipe welds, and welds made on reactive metals such as titanium, backing gas is used to surround the root face of the weld as it is being produced. This gas prevents contamination of the root face by the atmosphere. The specific type of gas, flow rate, purging procedures, and other detailed information would be included in this section.
15. Safety: General and specific safety procedures and requirements for the shop and job should be given here.
16. Welding techniques: Information pertaining to 16a tack welds such as size, location, and sequence of placement would be included as well as 16b, groove or fillet weld information such as number of passes, weld bead sequences, bead sizes, reinforcement, and stopping and starting techniques may be included here.
17. Inner pass temperature: This is the maximum allowable temperature that the base metal may reach during welding. Because long periods of elevated temperatures can significantly affect the mechanical properties of most metals and because elevated temperatures can adversely affect the weld bead's shape, it is important that this temperature be stated. Additionally, information regarding the method of cooling should be stated.
18. Cleaning: From time to time during the welding process, additional cleaning or recleaning of the weld may be required. In some cases, such as thick sections on mild steel, this cleaning could include some weld bead reshaping using grinders. Any such allowable operations must be outlined here.
19. Inspection: The method and extent of examination of the finished weld as well as tolerances of acceptability must be given.
20. Repair: If the weld fails during postwelding inspection, in many industrial applications specific repair procedures are allowed. Repair procedures may include methods of removing weld discontinuities and effects as well as allowable procedures for making such welding repairs. In many cases reference will be made to a completely separate welding procedure specification designed specifically for making such repairs.
21. Sketches: A series of sketches or complete mechanical drawings can be included. Drawings that are included or that accompany a WPS would be for a specific part or weldment that would be fabricated using this specific procedure. The same welding procedure specification can be used on any number of products as long as they fall within the scope of the welding procedure specifications.

WELDING PROCEDURE SPECIFICATION (WPS) FOR CARBON STEEL

Welding Procedures Specifications No: TEW-GTAW 1 Date: ______

TITLE:

Welding GTAW of sheet to sheet.

SCOPE:

This procedure is applicable for square groove and fillet welds within the range of 18 gauge through 10 gauge. Welding may be performed in the following positions 1G and 2F.

BASE METAL:

The base metal shall conform to Carbon Steel M-1 Group 1 Backing material specification NONE.

FILLER METAL:

The filler metal shall conform to AWS specification No. E70S-3 from AWS specification A5.18 . This filler metal falls into F-number F-6 and A-number A-1 .

ELECTRODE:

The tungsten electrode shall conform to AWS specification No. EWTh-2 from AWS specification A5.12 . The tungsten diameter shall be 1/8″ (3.2 mm) maximum . The tungsten end shape shall be Tapered at 2 to 3 times in length to its diameter .

SHIELDING GAS:

The shielding gas, or gases, shall conform to the following compositions and purity: Welding Grade Argon .

JOINT DESIGN AND TOLERANCES:

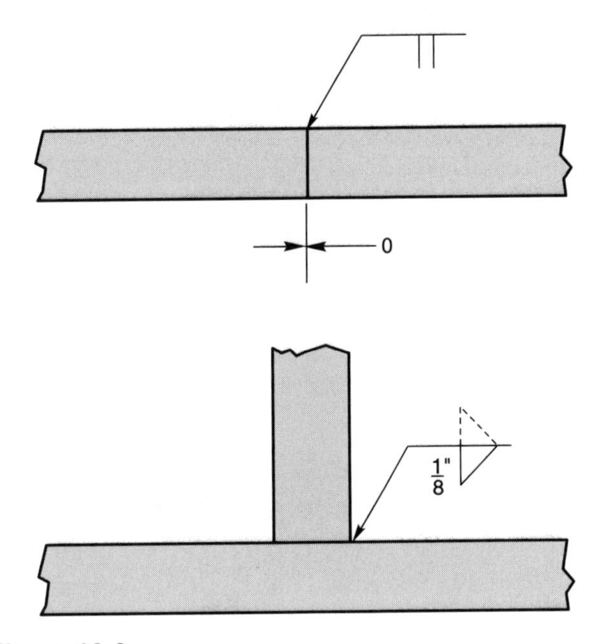

Figure 13-3

PREPARATION OF BASE METAL:

All hydrocarbons and other contaminations, such as cutting fluids, grease, oil, and primers, must be cleaned off of all parts and filler metals before welding. This cleaning can be done with any suitable solvents or detergents. The joint face, inside and outside plate surface within 1″ (25 mm) of the joint must be mechanically cleaned of slag, rust, and mill scale. Cleaning must be done with a wire brush or grinder down to bright metal.

ELECTRICAL CHARACTERISTICS:

The current shall be Direct Current Electrode Negative (DCEN). The base metal shall be on the positive side of the line.

OPERATING RANGE SPECIFICATIONS

METAL SPECIFICATIONS		GAS FLOW			NOZZLE SIZE IN. (MM)	AMPERAGE MIN. MAX.
THICKNESS	DIA. OF E70S-3*	RATES CFM (L/MIN)	PURGING TIMES PREPURGING	POSTPURGING		
18 ga.	1/16 in. (1.6 mm)	15 to 20 (7 to 9)	10 to 15 sec.	10 to 25 sec.	1/4 to 3/8 (6 to 10)	45 to 65
17 ga.	1/16 in. (1.6 mm)	15 to 20 (7 to 9)	10 to 15 sec.	10 to 25 sec.	1/4 to 3/8 (6 to 10)	45 to 70
16 ga.	1/16 in. (1.6 mm)	15 to 20 (7 to 9)	10 to 15 sec.	10 to 25 sec.	1/4 to 3/8 (6 to 10)	50 to 75
15 ga.	1/16 in. (1.6 mm)	15 to 20 (7 to 9)	10 to 15 sec.	10 to 25 sec.	1/4 to 3/8 (6 to 10)	55 to 80
14 ga.	3/32 in. (2.4 mm)	20 to 25 (9 to 12)	10 to 20 sec.	10 to 30 sec.	3/8 to 5/8 (10 to 16)	60 to 90
13 ga.	3/32 in. (2.4 mm)	20 to 25 (9 to 12)	10 to 20 sec.	10 to 30 sec.	3/8 to 5/8 (10 to 16)	60 to 100
12 ga.	3/32 in. (2.4 mm)	20 to 25 (9 to 12)	10 to 20 sec.	10 to 30 sec.	3/8 to 5/8 (10 to 16)	60 to 110
11 ga.	3/32 in. (2.4 mm)	20 to 25 (9 to 12)	10 to 20 sec.	10 to 30 sec.	3/8 to 5/8 (10 to 16)	65 to 120
10 ga.	3/32 in. (2.4 mm)	20 to 25 (9 to 12)	10 to 20 sec.	10 to 30 sec.	3/8 to 5/8 (10 to 16)	70 to 130

*Other E70S-X filler metal may be used.

PREHEAT:

The parts must be heated to a temperature higher than 50° F (10° C) before any welding is started.

BACKING GAS:

N/A

SAFETY:

Proper protective clothing and equipment must be used. The area must be free of all hazards that may affect the welder or others in the area. The welding machine, welding leads, work clamp, electrode holder, and other equipment must be in safe working order.

WELDING TECHNIQUE:

Tack Welds: With the parts securely clamped in place with the correct root gap, the tack welds are to be performed. Holding the electrode so that it is very close to the root face but not touching; slowly increase the current until the arc starts and a molten weld pool is formed. Add filler metal as required to maintain a slight convex weld face and a flat or slightly concave root face. When it is time to end the tack weld, lower the current slowly so that the molten weld pool can be tapered down in size. When all tack welds are complete, allow the parts to cool as needed before assembling the remaining parts. Repeat the tack welding procedure until the entire part is assembled.

Square Groove and Fillet Welds: Holding the electrode so that it is very close to the metal surface but not touching, slowly increase the current until the arc starts and a molten weld pool is formed. As the weld progresses, add filler metal as required to maintain a flat or slightly convex weld face. If it is necessary to stop the weld or to reposition yourself or if the weld is completed, the current must be lowered slowly so that the molten weld pool can be tapered down in size.

INTERPASS TEMPERATURE:

The plate should not be heated to a temperature higher than 120° F (49° C) during the welding process. After each weld pass is completed, allow the weld to cool but never to a temperature below 50° F (10° C). The weldment must not be quenched in water.

CLEANING:

Recleaning may be required if the parts or filler metal become contaminated or reoxidize to a degree that the weld quality will be affected. Reclean using the same procedure used for the original metal preparation. Any slag must cleaned off between passes.

INSPECTION:

Visually inspect the weld for uniformity and discontinuities. There shall be no cracks, no incomplete fusion, and no overlap. Undercut shall not exceed the lesser of 10% of the base metal thickness or 1/32″ (0.8 mm). The frequency of porosity shall not exceed one in each 4″ (100 mm) of weld length, and the maximum diameter shall not exceed 3/32″ (2.4 mm).

REPAIR:

No repairs of defects are allowed.

SKETCHES: (Figure 13-4, page 144)

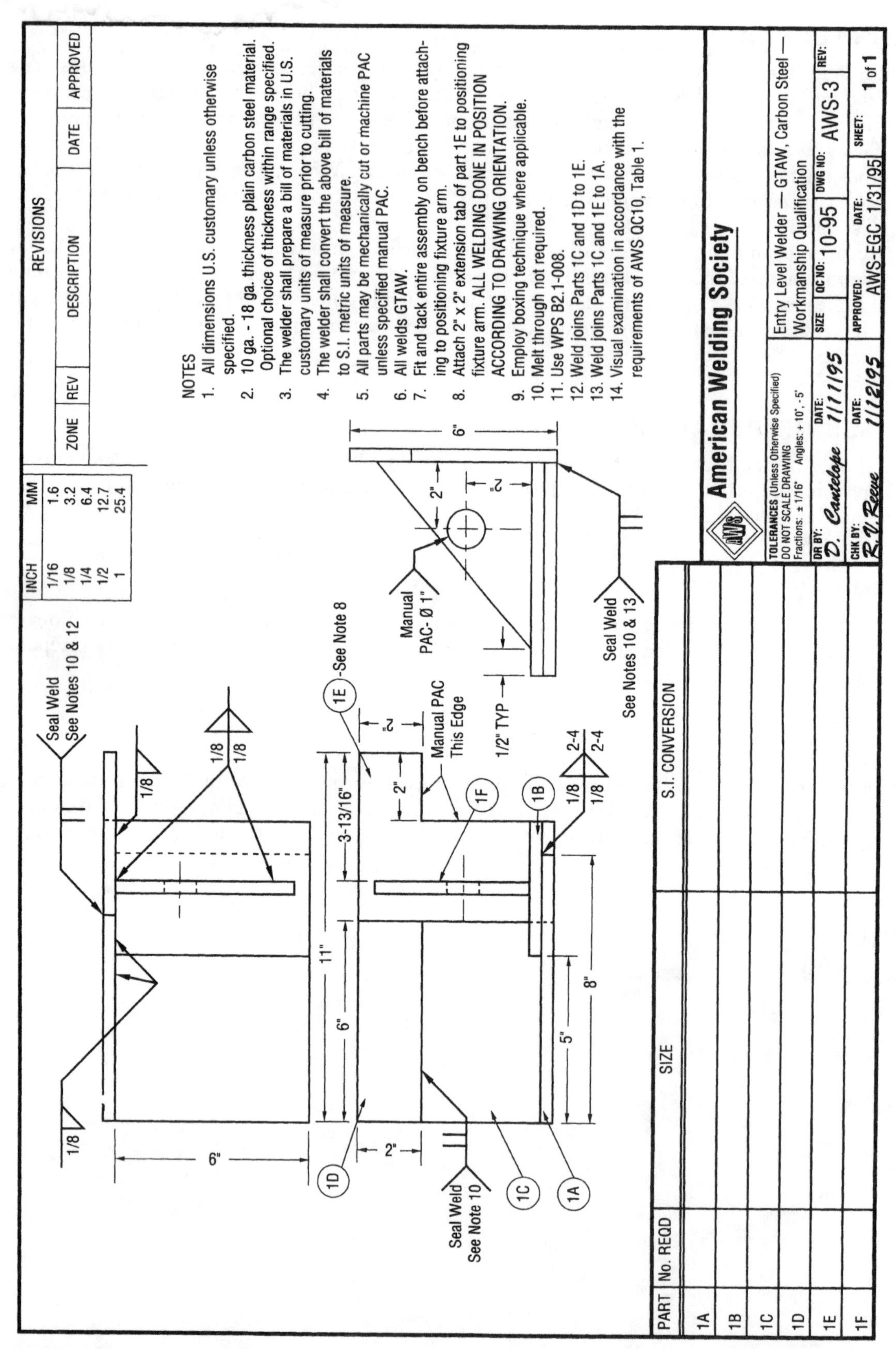

Figure 13-4 GTAW Plain Carbon Steel Workmanship Qualification Test. (Courtesy of the American Welding Society.)

WELDING PROCEDURE SPECIFICATION (WPS) FOR STAINLESS STEEL

Welding Procedures Specifications No: TEW-GTAW 2 Date: ____________

TITLE:

Welding GTAW of sheet to sheet.

SCOPE:

This procedure is applicable for square groove and fillet welds within the range of 18 gauge through 10 gauge. Welding may be performed in the following positions 1G and 2F.

BASE METAL:

The base metal shall conform to Austenitic Stainless Steel M-8 or P-8 Backing material specification NONE.

FILLER METAL:

The filler metal shall conform to AWS specification No. ER3XX from AWS specification A5.9. This filler metal falls into F-number F-6 and A-number A-8.

ELECTRODE:

The tungsten electrode shall conform to AWS specification No. EWTh-2 from AWS specification A5.12. The tungsten diameter shall be 1/8″ (3.2 mm) maximum. The tungsten end shape shall be Tapered at 2 to 3 times in length to its diameter.

SHIELDING GAS:

The shielding gas, or gases, shall conform to the following compositions and purity: Welding Grade Argon.

JOINT DESIGN AND TOLERANCES:

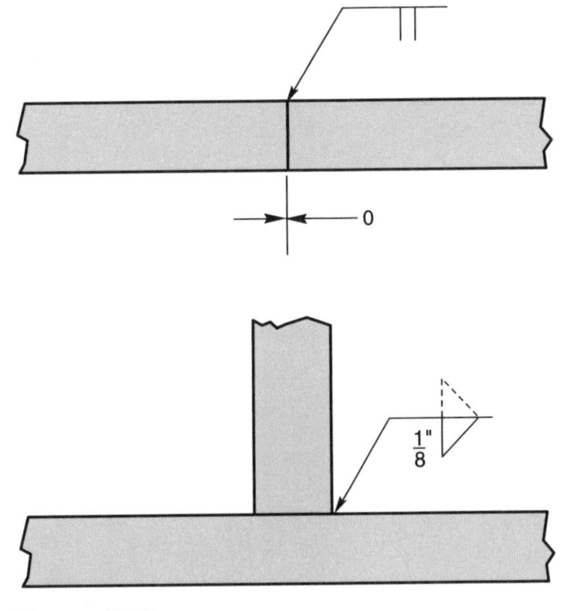

Figure 13-5

PREPARATION OF BASE METAL:

All hydrocarbons and other contaminations, such as cutting fluids, grease, oil and primers, must be cleaned off of all parts and filler metals before welding. This cleaning can be done with any suitable solvents or detergents. The joint face and inside and outside plate surface within 1 in. (25 mm) of the joint must be cleaned of slag, oxide and scale. Cleaning can be mechanical or chemical. Mechanical metal cleaning may be done by grinding, stainless steel wire brushing, scraping, machining, or filing. Chemical cleaning may be done by using acids, alkalies, solvents, or detergents. Cleaning must be done down to bright metal.

ELECTRICAL CHARACTERISTICS:

The current shall be Direct Current Electrode Negative (DCEN). The base metal shall be on the positive side of the line.

OPERATING RANGE SPECIFICATIONS

METAL SPECIFICATIONS		GAS FLOW			NOZZLE	
		RATES	PURGING TIMES		SIZE	AMPERAGE
THICKNESS	DIA. OF ER3XX*	CFM (L/MIN)	PREPURGING	POSTPURGING	IN. (MM)	MIN. MAX.
18 ga.	1/16 in. (1.6 mm)	15 to 20 (7 to 9)	10 to 15 sec.	10 to 25 sec.	1/4 to 3/8 (6 to 10)	35 to 60
17 ga.	1/16 in. (1.6 mm)	15 to 20 (7 to 9)	10 to 15 sec.	10 to 25 sec.	1/4 to 3/8 (6 to 10)	40 to 65
16 ga.	1/16 in. (1.6 mm)	15 to 20 (7 to 9)	10 to 15 sec.	10 to 25 sec.	1/4 to 3/8 (6 to 10)	40 to 75
15 ga.	1/16 in. (1.6 mm)	15 to 20 (7 to 9)	10 to 15 sec.	10 to 25 sec.	1/4 to 3/8 (6 to 10)	50 to 80
14 ga.	3/32 in. (2.4 mm)	20 to 25 (9 to 12)	10 to 20 sec.	10 to 30 sec.	3/8 to 5/8 (10 to 16)	50 to 90
13 ga.	3/32 in. (2.4 mm)	20 to 25 (9 to 12)	10 to 20 sec.	10 to 30 sec.	3/8 to 5/8 (10 to 16)	55 to 100
12 ga.	3/32 in. (2.4 mm)	20 to 25 (9 to 12)	10 to 20 sec.	10 to 30 sec.	3/8 to 5/8 (10 to 16)	60 to 110
11 ga.	3/32 in. (2.4 mm)	20 to 25 (9 to 12)	10 to 20 sec.	10 to 30 sec.	3/8 to 5/8 (10 to 16)	65 to 120
10 ga.	3/32 in. (2.4 mm)	20 to 25 (9 to 12)	10 to 20 sec.	10 to 30 sec.	3/8 to 5/8 (10 to 16)	70 to 130

*Any RE3XX stainless steel A5.9 filler metal may be used.

PREHEAT:

The parts must be heated to a temperature higher than 50° F (10° C) before any welding is started.

BACKING GAS:

N/A

SAFETY:

Proper protective clothing and equipment must be used. The area must be free of all hazards that may affect the welder or others in the area. The welding machine, welding leads, work clamp, electrode holder, and other equipment must be in safe working order.

WELDING TECHNIQUE:

Tack Welds: With the parts securely clamped in place with the correct root gap, the tack welds are to be performed. Holding the electrode so that it is very close to the root face but not touching, slowly increase the current until the arc starts and a molten weld pool is formed. Add filler metal as required to maintain a slight convex weld face and a flat or slightly concave root face. When it is time to end the tack weld, lower the current slowly so that the molten weld pool can be tapered down in size. When all tack welds are complete, allow the parts to cool as needed before assembling the remaining parts. Repeat the tack welding procedure until the entire part is assembled.

Square Groove and Fillet Welds: Holding the electrode so that it is very close to the metal surface but not touching, slowly increase the current until the arc starts and a molten weld pool is formed. As the weld progresses, add filler metal as required to maintain a flat or slightly convex weld face. If it is necessary to stop the weld or to reposition yourself or if the weld is completed, the current must be lowered slowly so that the molten weld pool can be tapered down in size.

INTERPASS TEMPERATURE:

The plate should not be heated to a temperature higher than 350° F (180° C) during the welding process. After each weld pass is completed, allow the weld to cool but never to a temperature below 50° F (10° C). The weldment must not be quenched in water.

CLEANING:

Recleaning may be required if the parts or filler metal become contaminated or reoxidize to a degree that the weld quality will be affected. Reclean using the same procedure used for the original metal preparation. Any slag must cleaned off between passes.

INSPECTION:

Visually inspect the weld for uniformity and discontinuities. There shall be no cracks, no incomplete fusion, and no overlap. Undercut shall not exceed the lesser of 10% of the base metal thickness or 1/32″ (0.8 mm). The frequency of porosity shall not exceed one in each 4″ (100 mm) of weld length, and the maximum diameter shall not exceed 3/32″ (2.4 mm).

REPAIR:

No repairs of defects are allowed.

SKETCHES: (Figure 13-6, page 148)

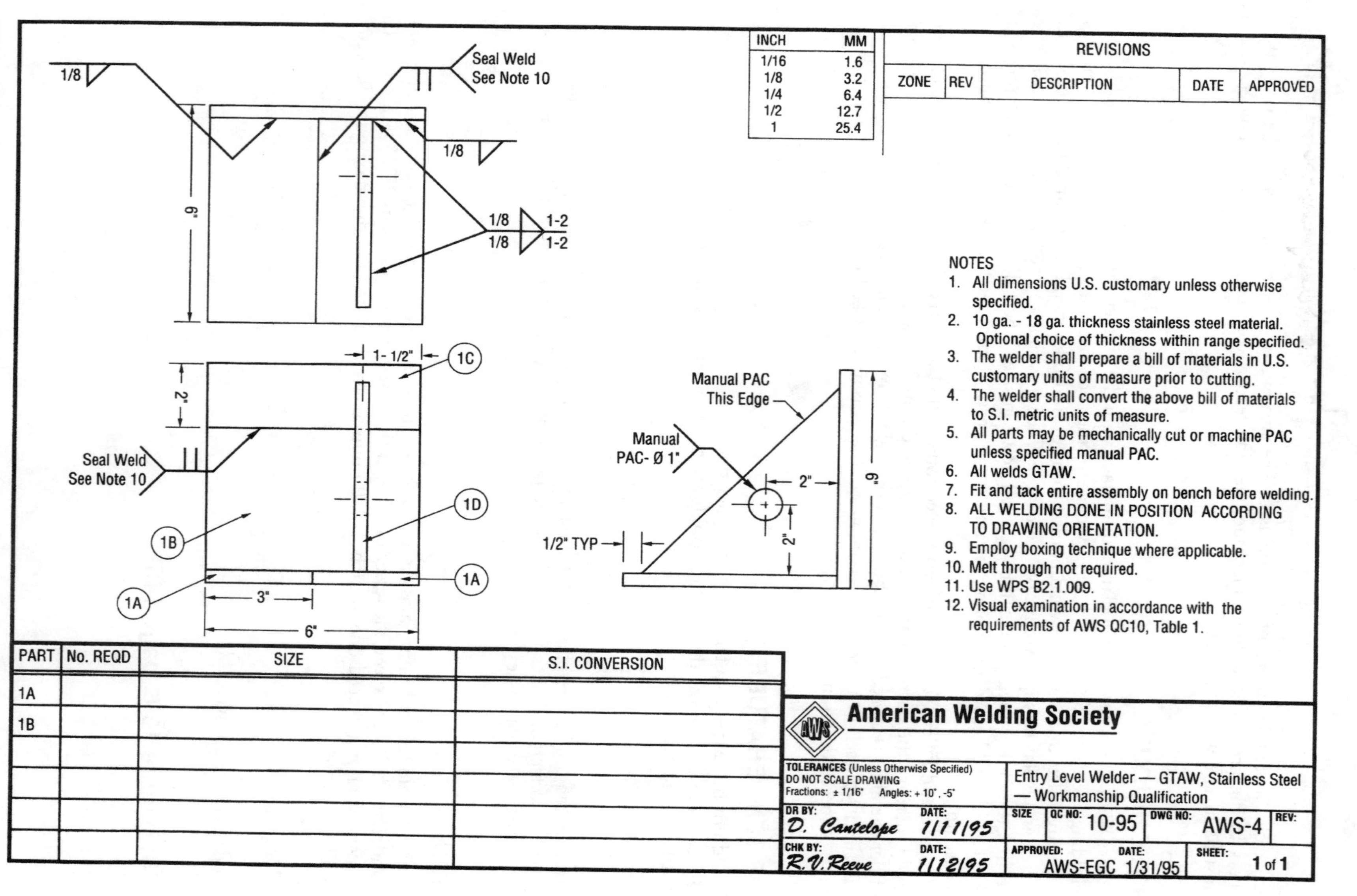

Figure 13-6 Stainless Steel Workmanship Qualification Test. (Courtesy of the American Welding Society.)

WELDING PROCEDURE SPECIFICATION (WPS) FOR ALUMINUM

Welding Procedures Specifications No: TEW-GTAW 3 Date: ____________

TITLE:

Welding GTAW of sheet to sheet.

SCOPE:

This procedure is applicable for square groove and fillet welds within the range of 18 gauge through 10 gauge. Welding may be performed in the following positions 1G and 2F.

BASE METAL:

The base metal shall conform to Aluminum M-22 or P-22 Backing material specification NONE.

FILLER METAL:

The filler metal shall conform to AWS specification No. ER4043 from AWS specification A5.10. This filler metal falls into F-number F-23 and A-number ________.

ELECTRODE:

The tungsten electrode shall conform to AWS specification No. EWP from AWS specification A5.12. The tungsten diameter shall be 1/8″ (3.2 mm) maximum. The tungsten end shape shall be Rounded.

SHIELDING GAS:

The shielding gas, or gases, shall conform to the following compositions and purity: Welding Grade Argon.

JOINT DESIGN AND TOLERANCES:

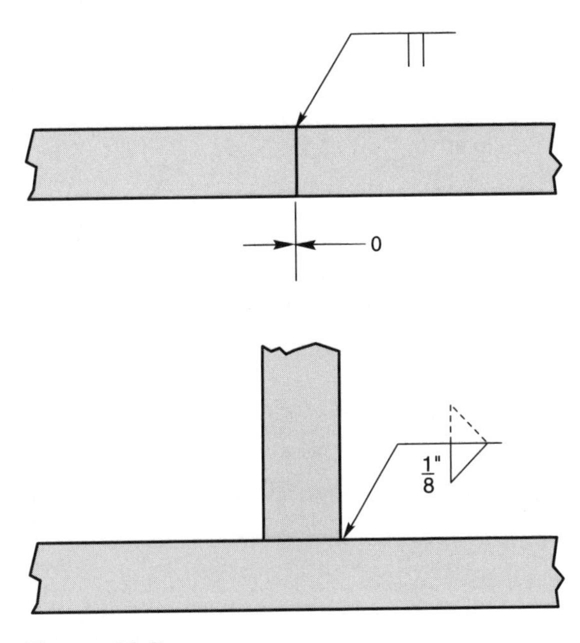

Figure 13-7

PREPARATION OF BASE METAL:

All hydrocarbons and other contaminations, such as cutting fluids, grease, oil, and primers, must be cleaned off of all parts and filler metals before welding. This cleaning can be done with any suitable solvents or detergents. The joint face, inside and outside plate surface within 1″ (25 mm) of the joint must be mechanically or chemically cleaned of oxides. Mechanical cleaning may be done by stainless steel wire brushing, scraping, machining, or filing. Chemical cleaning may be done by using acids, alkalies, solvents, or detergents. Because the oxide layer may reform quickly and affect the weld, welding should be started within 10 minutes of cleaning.

ELECTRICAL CHARACTERISTICS:

The current shall be Alternating Current High Frequency Stabilized (balanced wave preferably). The base metal shall be on the N/A side of the line.

OPERATING RANGE SPECIFICATIONS

METAL SPECIFICATIONS		GAS FLOW			NOZZLE SIZE	AMPERAGE
		RATES	PURGING TIMES			
THICKNESS	DIA. OF ER4043*	CFM (L/MIN)	PREPURGING	POSTPURGING	IN. (MM)	MIN. MAX.
18 ga.	3/32 in. (2.4 mm)	20 to 30 (9 to 14)	10 to 15 sec.	10 to 25 sec.	1/4 to 3/8 (6 to 10)	40 to 60
17 ga.	3/32 in. (2.4 mm)	20 to 30 (9 to 14)	10 to 15 sec.	10 to 25 sec.	1/4 to 3/8 (6 to 10)	50 to 70
16 ga.	3/32 in. (2.4 mm)	20 to 30 (9 to 14)	10 to 15 sec.	10 to 25 sec.	1/4 to 3/8 (6 to 10)	60 to 75
15 ga.	3/32 in. (2.4 mm)	20 to 30 (9 to 14)	10 to 15 sec.	10 to 25 sec.	1/4 to 3/8 (6 to 10)	65 to 85
14 ga.	3/32 in. (2.4 mm)	20 to 30 (9 to 14)	10 to 20 sec.	10 to 30 sec.	3/8 to 5/8 (10 to 16)	75 to 90
13 ga.	1/8 in. (3.2 mm)	25 to 40 (12 to 19)	10 to 20 sec.	10 to 30 sec.	3/8 to 5/8 (10 to 16)	85 to 100
12 ga.	1/8 in. (3.2 mm)	25 to 40 (12 to 19)	10 to 20 sec.	10 to 30 sec.	3/8 to 5/8 (10 to 16)	90 to 110
11 ga.	1/8 in. (3.2 mm)	25 to 40 (12 to 19)	10 to 20 sec.	10 to 30 sec.	3/8 to 5/8 (10 to 16)	100 to 115
10 ga.	1/8 in. (3.2 mm)	25 to 40 (12 to 19)	10 to 20 sec.	10 to 30 sec.	3/8 to 5/8 (10 to 16)	110 to 125

*Other aluminum A5.10 filler metal may be used if needed.

PREHEAT:

The parts must be heated to a temperature higher than 50° F (10° C) before any welding is started.

BACKING GAS:

N/A

SAFETY:

Proper protective clothing and equipment must be used. The area must be free of all hazards that may affect the welder or others in the area. The welding machine, welding leads, work clamp, electrode holder, and other equipment must be in safe working order.

WELDING TECHNIQUE:

The welder's hands or gloves must be clean and oil-free to prevent recontaminating the metal or filler rods.

Tack Welds: With the parts securely clamped in place with the correct root gap, the tack welds are to be performed. Holding the electrode so that it is very close to the root face but not touching; slowly increase the current until the arc starts and a molten weld pool is formed. Add filler metal as required to maintain a slight convex weld face and a flat or slightly concave root face. When it is time to end the tack weld, lower the current slowly so that the molten weld pool can be tapered down in size. When all tack welds are complete, allow the parts to cool as needed before assembling the remaining parts. Repeat the tack welding procedure until the entire part is assembled.

Square Groove and Fillet Welds: Holding the electrode so that it is very close to the metal surface but not touching, slowly increase the current until the arc starts and a molten weld pool is formed. As the weld progresses, add filler metal as required to maintain a flat or slightly convex weld face. If it is necessary to stop the weld or to reposition yourself or if the weld is completed, the current must be lowered slowly so that the molten weld pool can be tapered down in size.

INTERPASS TEMPERATURE:

The plate should not be heated to a temperature higher than 120° F (49° C) during the welding process. After each weld pass is completed, allow the weld to cool but never to a temperature below 50° F (10° C). The weldment must not be quenched in water.

CLEANING:

Recleaning may be required if the parts or filler metal become contaminated or reoxidize to a degree that the weld quality will be affected. Reclean using the same procedure used for the original metal preparation.

INSPECTION:

Visually inspect the weld for uniformity and discontinuities. There shall be no cracks, no incomplete fusion, and no overlap. Undercut shall not exceed the lesser of 10% of the base metal thickness or 1/32″ (0.8 mm). The frequency of porosity shall not exceed one in each 4″ (100 mm) of weld length, and the maximum diameter shall not exceed 3/32″ (2.4 mm).

REPAIR:

No repairs of defects are allowed.

SKETCHES: (Figure 13-8, page 152)

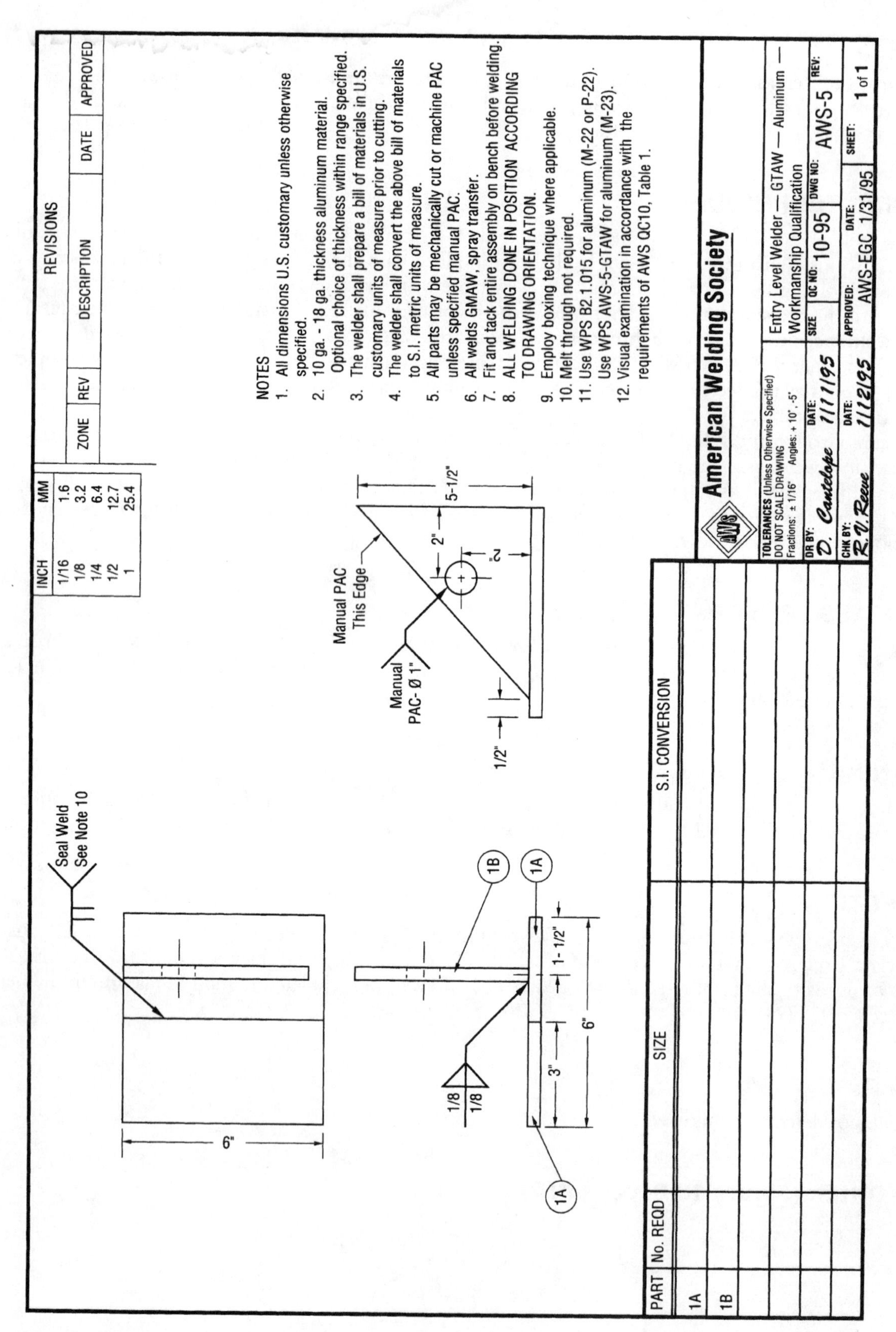

Figure 13-8 GTAW Aluminum Workmanship Qualification Test. (Courtesy of the American Welding Society.)

REVIEW QUESTIONS

1. What does WPS stand for?
2. What information does a WPS contain?
3. A welding procedure specification form title includes what information?
4. What specific information regarding the shielding gas makeup is required?
5. What should be included in the drawing of joint design and tolerances?
6. On some critical welds, pipe welds, and welds made on reactive metals such as titanium, what is backing gas used for?
7. What is the inner pass temperature?
8. In some cases such as thick sections on mild steel, cleaning could include ________________.
9. What is included in the inspection section of a WPS?
10. In a WPS what may repair procedures include?

CHAPTER 14

WELD TESTING AND INSPECTION

INTRODUCTION

A welder must know that a weld will meet a company's requirements and/or codes and standards. Weld testing and inspection ensure that welds meet the quality, reliability, and strength requirements of the weldments. The extent to which welders and products are subjected to testing and inspection depends on the requirements of the industry. Many welding products are used in non-critical applications, such as those in which the failure of the weld would result in only a minor inconvenience. (Contrasting with these are welds in critical industrial applications in which the failure of a weld could result in severe loss of property and life.) Noncritical welds occur in applications such as ornamental furniture, sculptures, jigs, fixtures, and other household or industrial products. Critical welds are welds produced in, for example, the nuclear power industry, oil refinery and chemical industries, and the aircraft, missile, spacecraft, and satellite industries.

The level and intensity of testing and inspection is greatly influenced by the intended purpose of the weld. The more intense the evaluation process, the more expensive the weld and weldment are to produce. Therefore, welds and weldments must meet the standards that ensure that the finished product is fit for service.

QUALITY CONTROL (QC)

The first step in establishing a quality control program is to determine the appropriate codes and standards.

The code or standard selected will then determine the type of testing. Two major classifications of testing products are *destructive* and *nondestructive* testing. Nondestructive testing can be performed on a product without affecting the product in any way. Destructive testing renders the weldment unusable. Nondestructive testing is often used to both qualify the welding procedure and ensure the quality of the weldment. During the establishment and verification of the welding procedure specifications, destructive testing may be used in conjunction with nondestructive examination. Destructive testing establishes the acceptable criteria for the nondestructive testing that will be applied to the products.

Destructive or mechanical testing is used as a baseline evaluation to establish welding procedure specifications. It may also be used randomly on completed products to ensure that the weld quality is being maintained.

DISCONTINUITIES AND DEFECTS

A discontinuity is an interruption of the typical structure of a weldment. It may be a lack of uniformity in the mechanical, metallurgical, or physical characteristics of the material or weldment. A discontinuity is not necessarily a defect.

A defect, according to AWS, is "a discontinuity or discontinuities, which by nature or accumulated effect (for example, total porosity or slag inclusion length) renders a part or product unable to meet minimum applicable acceptance standards or specifications."

In other words, many acceptable products may have welds that contain discontinuities. But no products may have welds that contain defects. The only difference between a discontinuity and a defect occurs when the discontinuity becomes so large (or when there are so many small discontinuities) that the weld is not acceptable under the standards of the code for that product. Some codes are more strict than others, which implies that the same weld might be acceptable under one code but not under another.

Ideally, a weld should not have any discontinuities, but that is practically impossible. The difference between what is acceptable, fit for service, and perfection is known as *tolerance*. In many industries, the tolerances for welds have been established and are available as codes or standards. Table 14-1 lists a few of the agencies that issue codes or standards. Each code or standard gives the tolerance that downgrades a discontinuity to a defect.

When evaluating a weld, it is important to note the type, size, and location of the discontinuity. Any one of these factors or all three can be the deciding factor that, based on the applicable code or standard, might change a discontinuity to a defect.

The twelve most common discontinuities are as follows:

- Porosity
- Inclusions
- Inadequate joint penetration
- Incomplete fusion
- Arc strikes
- Overlap
- Undercut
- Cracks
- Underfill
- Lamination
- Delaminations
- Lamellar tears

TABLE 14-1

MAJOR CODE ISSUING AGENCIES
American Bureau of Shipping
American Petroleum Institute
American Society of Mechanical Engineers
American Society for Testing and Materials
American Welding Society
British Welding Institute
United States Government

* A more complete listing of agencies with addresses is included in the Appendix.

Porosity

Porosity results from gas that was dissolved in the molten weld pool, forming bubbles that are trapped as the metal cools to become solid. The bubbles that make up porosity form within the weld metal, and for that reason they cannot be seen as they form. These gas pockets form in the same way that bubbles form in a carbonated drink as it warms up or like air dissolved in water that forms bubbles in the center of an ice cube.

Porosity forms either spherical (ball-shaped) or cylindrical (tube- or tunnel-shaped). Cylindrical porosity is called *wormhole*. The rounded edges reduce the stresses around them. Therefore, unless porosity is extensive, little or no loss in strength occurs.

Porosity is most often caused by improper welding techniques, contamination, or an improper chemical balance between the filler and base metals.

Improper welding techniques may result in shielding gas not properly protecting the molten weld pool. For example, drafts may blow the shielding gas coverage away from the molten weld pool, or inadequate prepurging may leave some atmosphere in the welding zone during start-up.

Often with gas tungsten arc welding, welders try to resolve a porosity problem by remelting the weld. Welders assume that this is possible because additional filler metal need not be added, thus preventing an apparently oversized weld from being produced. This practice is unacceptable because there is no way of removing the dissolved gases causing the porosity. Remelting simply results in these gases forming porosity below the surface or being dissolved in the weld metal. Such dissolved gases will result in significant reduction in the mechanical properties, strength, ductility, and hardness of the weld.

- Uniformly scattered porosity is most frequently caused by poor welding techniques or faulty materials (Figure 14-1).
- Clustered porosity is most often caused by improper starting and stopping techniques (Figure 14-1).
- Linear porosity is most frequently caused by contamination within the joint, root, or inter-bead boundaries (Figure 14-1).

- Piping porosity, or wormhole, is most often caused by contamination at the root (Figure 14-1). This porosity is unique because its formation depends on the gas escaping from the weld pool at the same rate as the pool is solidifying.

Inclusions

There are two major classifications of inclusions: metallic and nonmetallic. Nonmetallic inclusions are produced when such materials as slag become trapped in the weld metal (often between weld beads). Since GTA Welding does not use slag producing fluxes, nonmetallic inclusions are rare for GTA Welding.

Tungsten from the electrode is the most commonly found metallic inclusion. Tungsten can be transferred from the electrode to the weld metal as a result of erosion or inadvertently dipping the tungsten into the weld pool. Some degree of erosion will occur under the most ideal of welding conditions; therefore; acceptable limits for this type of inclusion are provided in all codes and standards. Tungsten can be inadvertently transferred as a result of the tungsten accidentally touching the molten weld pool.

Because of tungsten's high density, all tungsten inclusions can be found at the deepest point of the weld pool. On a single-pass weld, these inclusions will be near the root of the joint; on multiple-pass welds, tungsten inclusions will be at the bottom of the weld pass that they were deposited in. Because they are so deep, it is often difficult to remove them.

Inadequate Joint Penetration

Inadequate joint penetration occurs when the depth to which the weld penetrates the joint (Figure 14-2) is less than that needed to fuse through the plate or into the preceding weld. A defect usually results that could reduce the required cross-sectional area of the joint or become a source of stress concentration that leads to fatigue failure. The importance of such defects depends on the notch sensitivity of the metal and the factor of safety to which the weldment has been designed. Generally, if proper welding pro-

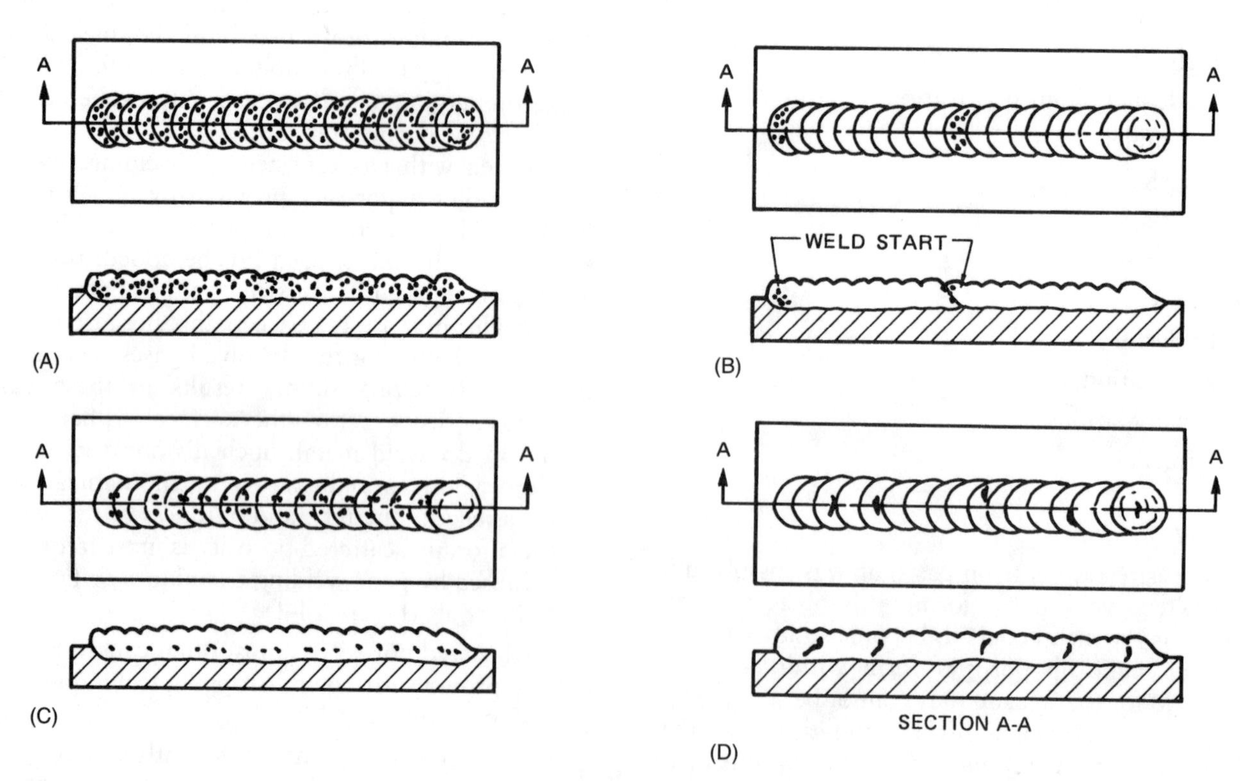

Figure 14-1 (A) Uniformly scattered porosities. (B) Clustered porosity. (C) Linear porosity. (D) Piping or wormhole porosity.

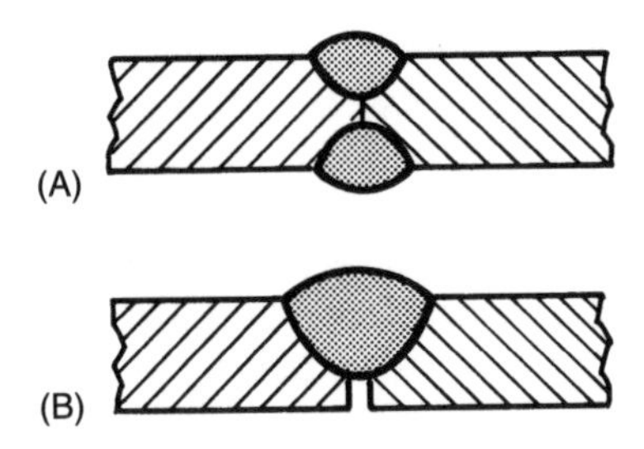

Figure 14-2 Inadequate joint penetration.

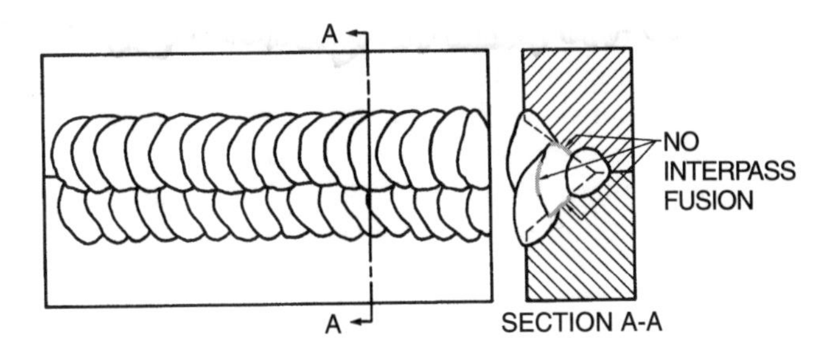

Figure 14-3 Incomplete fusion.

cedures were developed and followed, such defects will not occur.

The major causes of inadequate joint penetration are as follows:

- Improper welding technique — The most common cause is a misdirected arc. Also, the welding technique may require that both starting and run-out tabs be used so that the molten weld pool is well established before it reaches the joint. Sometimes a failure to back gouge the root sufficiently provides a deeper root face than allowed for.
- Not enough welding current — Metals that are thick or have a high thermal conductivity are often preheated so that the weld heat is not drawn away so quickly by the metal that it cannot penetrate the joint.
- Improper joint fit-up — This problem results when the weld joints are not prepared or fitted accurately. Too small a root gap or too large a root face will keep the weld from penetrating adequately.
- Improper joint design — When joints are accessible from both sides, back gouging is often used to ensure 100% root fusion.

Incomplete Fusion

Incomplete fusion is the lack of coalescence between the molten filler metal and previously deposited filler metal and/or the base metal (Figure 14-3). The lack of fusion between the filler metal and previously deposited weld metal is called *interpass cold lap*. The lack of fusion between the weld metal and the joint face is called *lack of sidewall fusion*. Both of these problems usually travel along all or most of the weld's length. This discontinuity is less detrimental to the weld's strength in service if it is near the center of the weld and is not open to the surface.

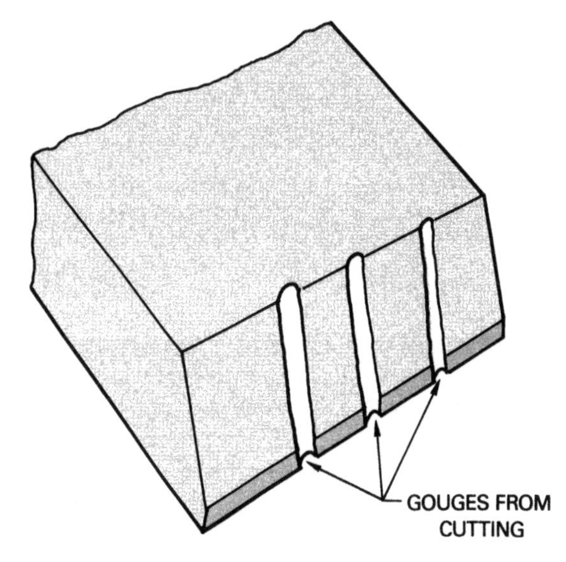

Figure 14-4 Remove gouges along the surface of the joint before welding.

Some major causes of lack of fusion are as follows:

- Inadequate agitation — Lack of weld agitation to break up oxide layers. The base metal or weld filler metal may melt, but a thin layer of oxide may prevent coalescence.
- Improper welding techniques — Poor manipulation, such as moving too fast.
- Improper edge preparation — Any notches or gouges in the edge of the weld joint must be removed. For example, if a flame-cut plate has notches along the cut, they could result in a lack of fusion in each notch (Figure 14-4).
- Improper joint design — Incomplete fusion may also result from inadequate heat to melt the base metal or inadequate space allowed by the joint designer for correct molten weld pool manipulation.

- Improper joint cleaning — Failure to clean oxides for the joint surfaces, resulting from the use of an oxy-fuel torch to cut the plate or failure to remove slag from a previous weld. Incomplete fusion can be found in welds produced by all major welding processes.

Arc Strikes

Arc strikes are small, localized points where surface melting occurred away from the joint. These spots may be caused by accidentally striking the arc in the wrong place and/or by faulty ground connections. Even though arc strikes can be ground smooth, they cannot be removed. These spots will always appear if an acid etch is used. They can also be localized hardness zones or the starting point for cracking. Arc strikes, even when ground flush for a guided bend, will open up to form small cracks or holes.

Overlap

Overlap occurs in fusion welds when weld deposits are larger than the joint is conditioned to accept. The weld metal then flows over a surface of the base metal without fusing to it (Figure 14-5). It generally occurs on the horizontal leg of a horizontal fillet weld under extreme conditions. It can also occur on both sides of flat-positioned capping passes.

Undercut

Undercut results when the arc plasma removes more metal from a joint face than is replaced by weld metal (Figure 14-6). It can result from excessive current.

Crater Cracks

Crater cracks are the tiny cracks that develop in the weld craters as the weld pool shrinks and solidifies (Figure 14-7). Low-melting materials are rejected toward the crater center while freezing. Because these materials are the last to freeze, they separate or are pulled apart as a result of the weld metal shrinking as it cools, leaving crater cracks. The high shrinkage stresses aggravate crack formation. Crater cracks can be minimized, if not prevented, by not interrupting the arc quickly at the end of a weld. This allows the arc to lengthen, the current to drop gradually, and the crater to fill and cool more slowly. GMAW equipment has a crater filling control that automatically and gradually reduces the wire feed speed at the end of a weld.

Underfill

Underfill on a groove weld occurs when the weld metal deposited is inadequate to bring the weld's face or root surfaces to a level equal to that of

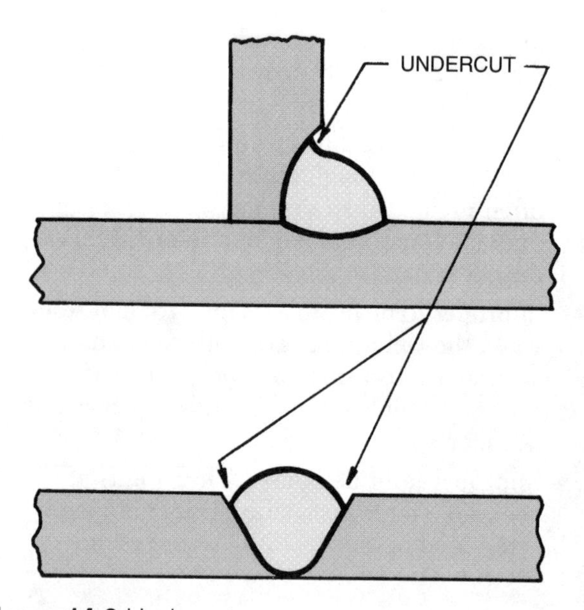

Figure 14-6 Undercut.

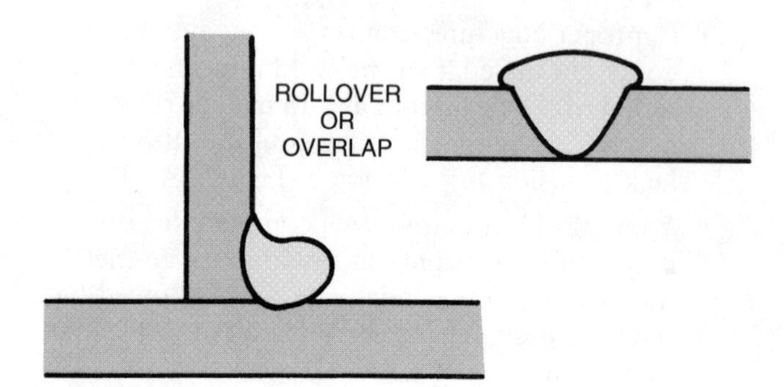

Figure 14-5 Rollover or overlap.

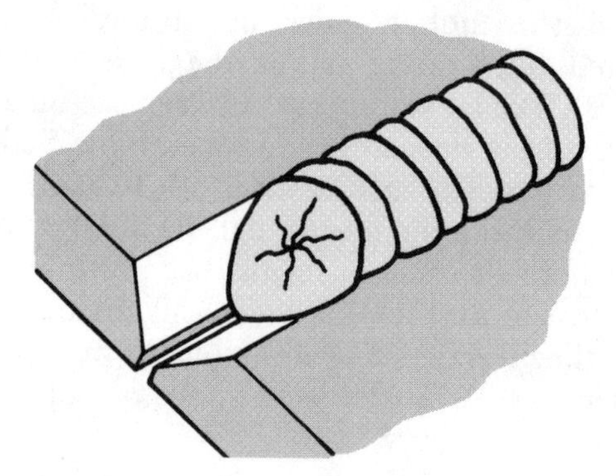

Figure 14-7 Crater or star cracks.

the original plane. For a fillet weld, underfill occurs when the weld deposit has an insufficient effective throat (Figure 14-8).

Plate-Generated Problems

Not all welding problems are caused by weld metal, the process, or the welder's lack of skill in depositing that metal. The material being fabricated can be at fault, too. Some problems result from internal plate defects that the welder cannot control. Others result from improper welding procedures that produce undesirable hard metallurgical structures in the heat-affected zone, as discussed in other chapters. The internal defects are the result of poor steel-making practices. Steel producers try to keep their steels as sound as possible, but the mistakes that occur are blamed, too frequently, on the welding operation.

Lamination

Lamination differs from lamellar tearing because it is more extensive and involves thicker layers of nonmetallic contaminants. Located toward the center of the plate (Figure 14-9), lamination is caused by insufficient cropping (removal of defects) of the pipe in ingots. The slag and oxidized steel in the pipe is rolled out with the steel, producing the lamination.

Delamination

When lamination intersects a joint being welded, some lamination may open up and become delaminated. Contamination of the weld metal may occur if the lamination contained large amounts of slag, mill scale, dirt, or other undesirable materials. Such contamination can cause wormhole porosity or lack-of-fusion defects.

The problems associated with delaminations are not easily corrected. An effective solution for thick plate is to weld over the lamination to seal it. A better solution is to replace the steel.

Lamellar Tears

These tears appear as cracks parallel to and under the steel surface. In general, they are not in the heat-affected zone and have a steplike configuration. They result from the thin layers of nonmetallic inclusions that lie beneath the plate surface and have very poor ductility. Although barely noticeable, these inclusions separate when severely stressed, producing laminated cracks. These cracks are evident if the plate edges are exposed (Figure 14-10).

A solution to the problem is to redesign the joints in order to impose the lowest possible strain throughout the plate thickness. This can be accomplished by making smaller welds so that each subsequent weld pass heat-treats the previous pass

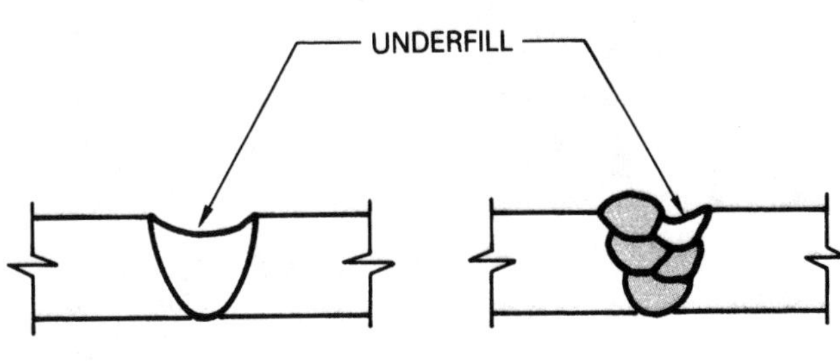

Figure 14-8

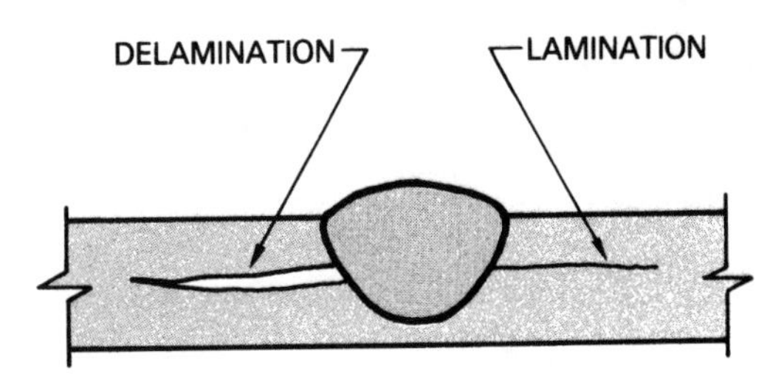

Figure 14-9

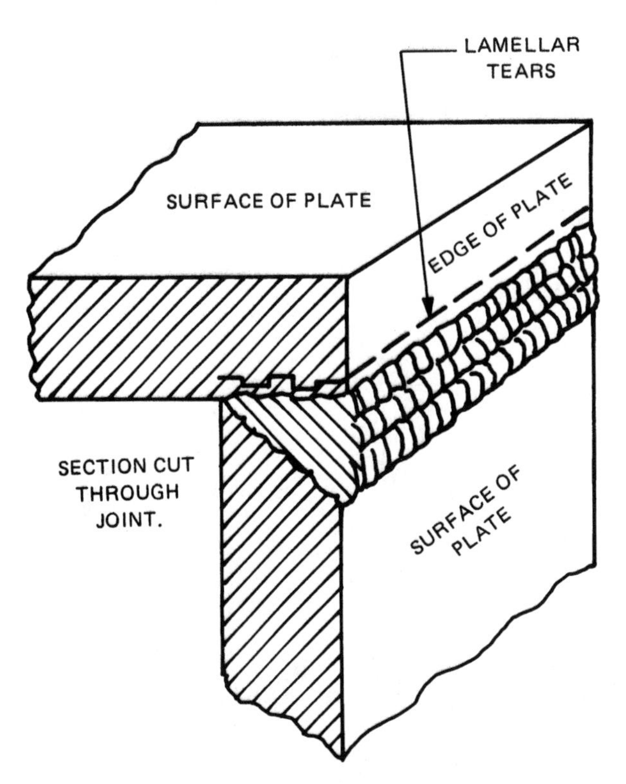

Figure 14-10 Example of lamellar tearing.

to reduce the total stress in the finished weld (Figure 14-11). The joint design can be changed to reduce the stress on the through thickness of the plate (Figure 14-12).

DESTRUCTIVE TESTING (DT)

Tensile Testing

Tensile tests are performed with specimens prepared as round bars or flat strips. The simple round bars are often used for testing only the weld metal, sometimes called "all weld-metal testing." Round specimens are cut from the center of the weld metal. The flat bars are often used to test both the weld and the surrounding metal. Bar size also depends on the size of the tensile testing equipment available (Figure 14-13).

Two flat specimens are commonly used for testing thinner sections of metal. When testing welds, the specimen should include the heat-affected zone and the base plate. If the weld metal is stronger than the plate, failure occurs in the plate; if the weld is weaker, failure occurs in the weld. This test, then, is open to interpretation.

After the weld section is machined to the specified dimensions, it is placed in the tensile testing machine and pulled apart. A specimen used to determine the strength of a welded butt joint for plate is shown in Figure 14-14.

Figure 14-11 Using multiple welds to reduce weld stresses.

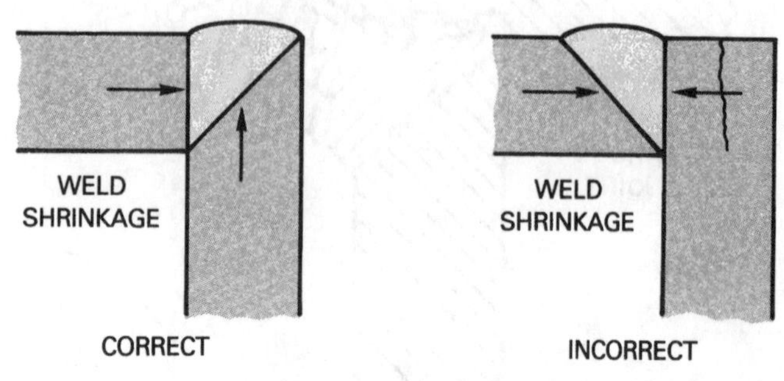

Figure 14-12 Correct joint design to reduce lamellar tears.

Fatigue Testing

Fatigue testing determines how well a weld can resist repeated fluctuating stresses or cyclic loading. The maximum value of the stresses is less than the tensile strength of the material.

In the fatigue test, the part is subjected to repeated changes in applied stress. This test may be performed in one of several ways, depending upon the type of service the tested part must withstand. The results are usually reported as the number of stress cycles that the part will resist without failure and the total stress used.

Nick-break Test

A specimen for this test is prepared as shown in Figure 14-15(A). The specimen is supported as shown in Figure 14-15(B). A force is then applied,

Figure 14-13 Typical tensile tester used for measuring the strength of welds (60,000-lb universal testing machines). (Courtesy of Tinius Olsen Testing Machine Co., Inc.)

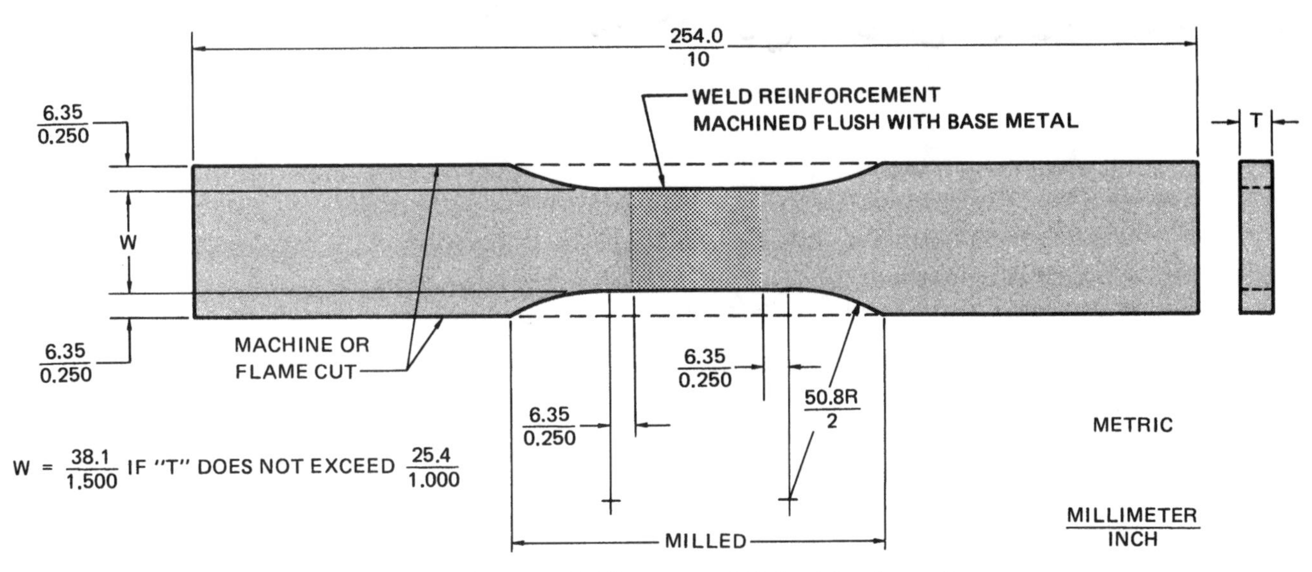

Figure 14-14 Tensile specimen for flat plate weld. (Courtesy of Hobart Brothers Company)

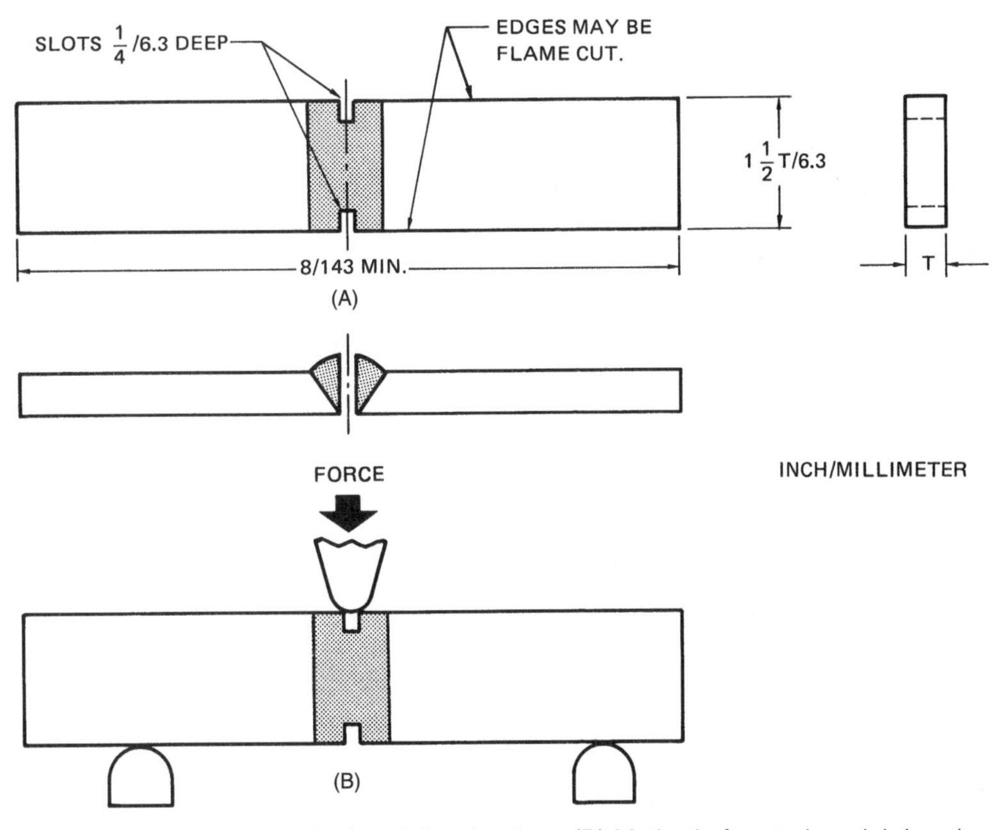

Figure 14-15 (A) Nick-break specimen for butt joints in plate. (B) Method of rupturing nick-break specimen. (Courtesy of Hobart Brothers Company)

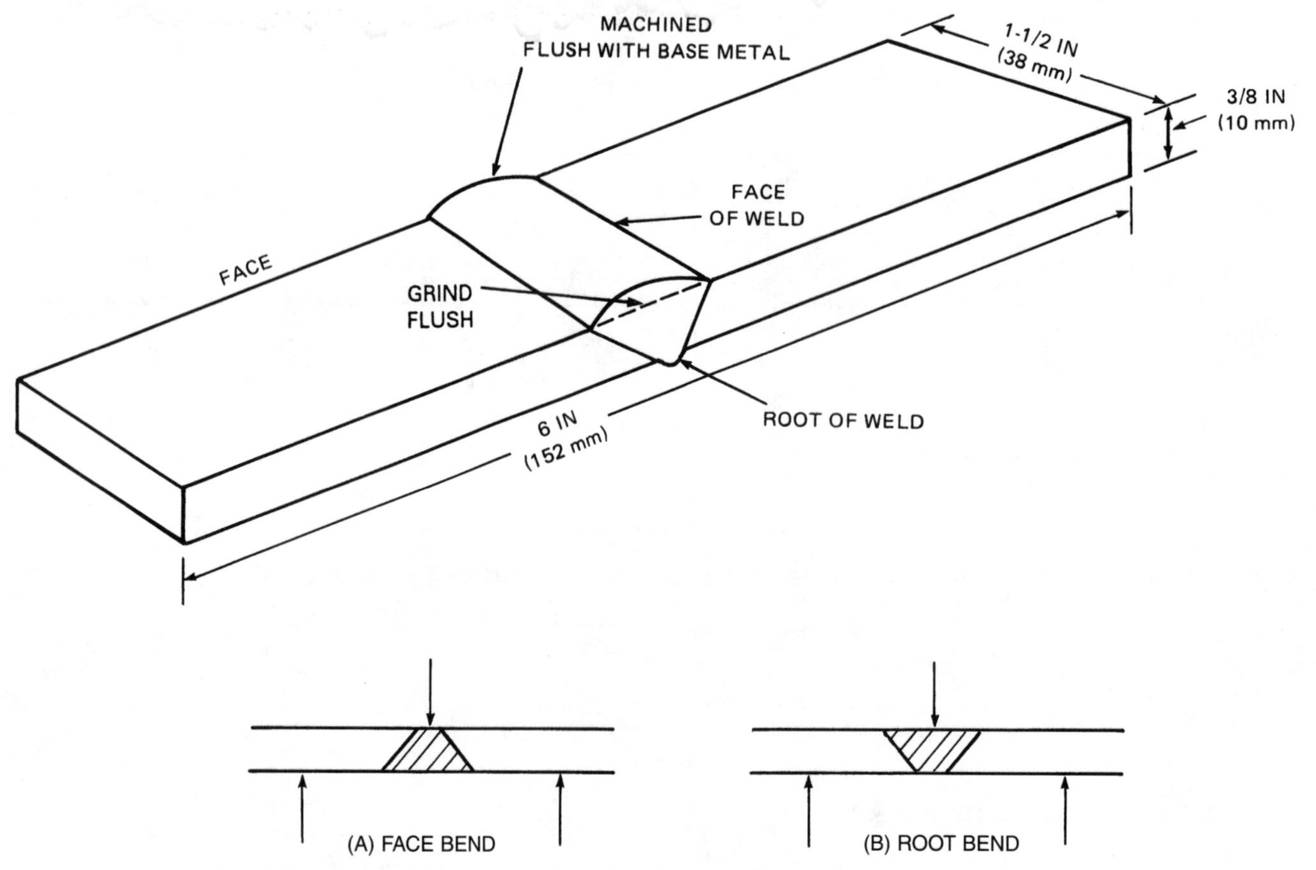

Figure 14-16 Root and face bend specimens for 3/8″ (10-mm) plate.

and the specimen is ruptured by one or more blows of a hammer. The force may be applied slowly or suddenly. Theoretically, the rate of application could affect how the specimen breaks, especially at a critical temperature. Generally, however, the appearance of the fractured surface does not correspond to the method of applying the force. The surfaces of the fracture should be checked for soundness of the weld.

Guided Bend Test

To test welded, grooved butt joints on metal that is 3/8″ (10 mm) thick or less, two specimens are prepared and tested—one face bend (Figure 14-16[A] and one root bend (Figure 14-16[B]). If the welds pass this test, the welder is qualified to make groove welds on plate having a thickness range from 3/8″ to 3/4″ (10 mm to 19 mm). These welds need to be machined as shown in Figure 14-17(A) and Figure 14-17(B). If these specimens pass, the welder will also be qualified to make fillet welds on materials of any (unlimited) thicknesses. For welded, grooved butt joints on metal 1″ (25 mm) thick, two side bend specimens are prepared and tested (Figure 14-17[C]). If the welds pass this test, the welder is qualified to weld on metals of unlimited thickness.

When the specimens are prepared, caution must be taken to ensure that all grinding marks run longitudinally to the specimen so that they do not cause stress cracking. In addition, the edges must be rounded to reduce cracking that tends to radiate from sharp edges.

Testing by Etching

Specimens to be tested by etching are etched for two purposes: (1) to determine the soundness of a weld or (2) to determine the location of a weld.

A test specimen is produced by cutting a portion from the welded joint so that a complete cross section is obtained. The face of the cut is then filed and polished with fine abrasive cloth. The specimen can then

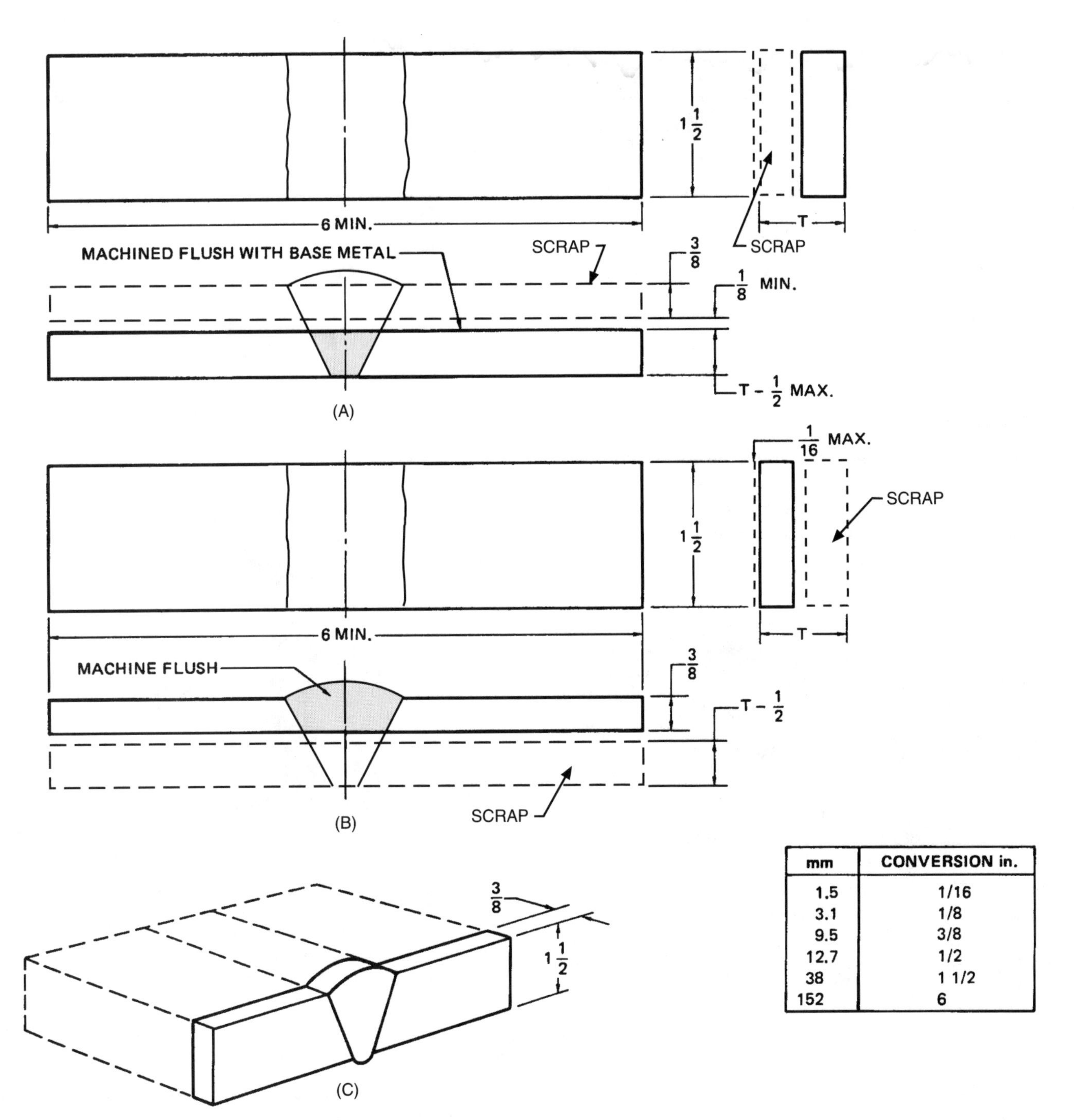

mm	CONVERSION in.
1.5	1/16
3.1	1/8
9.5	3/8
12.7	1/2
38	1 1/2
152	6

Figure 14-17 (A) Root bend specimen. (B) Specimen for face bend test. (C) Side bend specimen for plates thicker than 3/8″ (10 mm).

be placed in the etching solution. The etching solution or reagent makes the boundary between the weld metal and base metal visible, if the boundary is not already distinctly visible.

The most commonly used etching solutions are hydrochloric acid, ammonium persulphate, or nitric acid.

Hydrochloric Acid

Equal parts by volume of concentrated hydrochloric (muriatic) acid and water are mixed. The welds are immersed in the reagent at or near the boiling temperature. The acid usually enlarges gas pockets and dissolves slag inclusions, enlarging the resulting cavities.

CAUTION

When mixing the muriatic acid into the water, be sure to wear safety glasses and gloves to prevent injuries.

Ammonium Persulphate

A solution is prepared consisting of one part of ammonium persulphate (solid) to nine parts of water by weight. The surface of the weld is rubbed with cotton saturated with this reagent at room temperature.

Nitric Acid

A great deal of care should be exercised when using nitric acid because severe burns can result if it is used carelessly. One part of concentrated nitric acid is mixed with nine parts of water by volume.

CAUTION

When diluting, always pour the acid slowly into the water while continuously stirring the water. Careless handling of this material or pouring water into the acid can result in burns, excessive fuming, or explosion.

The reagent is applied to the surface of the weld with a glass stirring rod at room temperature. Nitric acid has the capacity to etch rapidly and should be used on polished surfaces only.

After etching the weld, rinse it in clear, hot water. Remove excess water and immerse the etched surface in ethyl alcohol and let it dry.

Impact Testing

A number of tests can be used to determine a weld's impact capability. One common test is the Izod test (Figure 14-18[A]), in which a notched specimen is struck by an anvil mounted on a pendulum. The energy in footpounds required to break the specimen is an indication of the impact resistance of the metal. This test compares the toughness of the weld metal with the base metal.

Another type of impact test is the Charpy test, which is similar to the Izod test. The difference is in the manner in which specimens are held. A typical impact tester is shown in Figure 14-18(B).

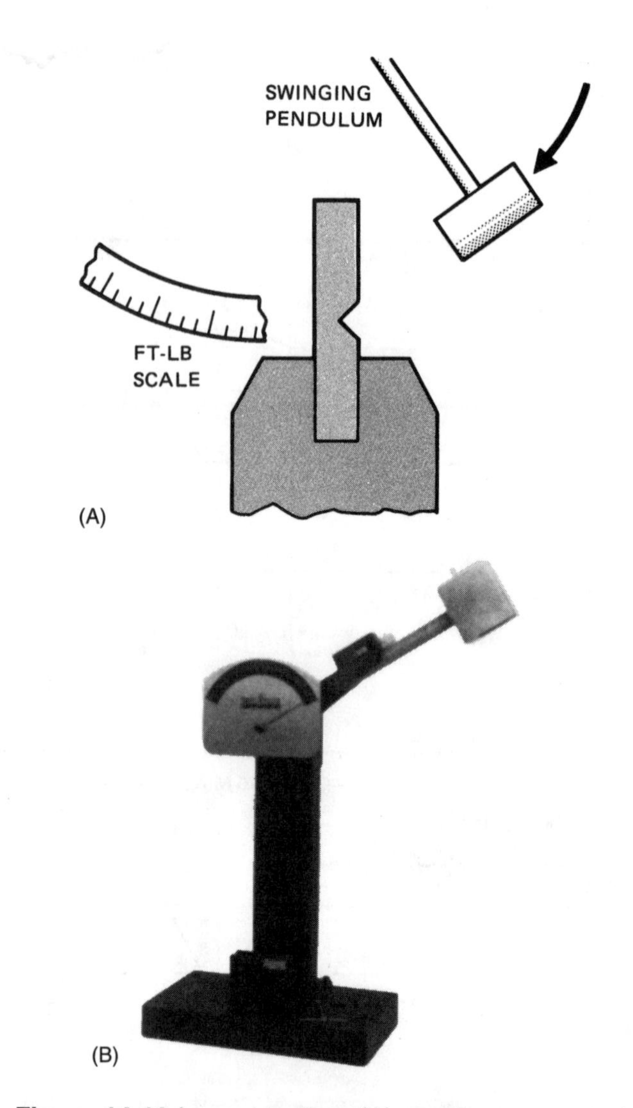

Figure 14-18 Impact testing: (A) specimen mounted for Izod impact toughness and (B) typical impact tester for measuring the toughness of metals. (Photo courtesy of Tinius Olsen Testing Machine Co., Inc.)

NONDESTRUCTIVE TESTING (NDT)

Nondestructive testing of welds tests materials for surface defects such as cracks, arc strikes, undercuts, and lack of penetration. Internal or subsurface defects can include tungsten inclusions, porosity, and unfused metal in the interior of the weld.

Visual Inspection (VT)

Visual inspection is the most frequently used nondestructive testing method. The majority of welds receive only visual inspection. In this method, if the weld looks good, it passes; if it looks bad, it is rejected. This procedure is often overlooked when more sophisticated nondestructive testing methods are used. However, it should not be overlooked.

Visual inspection can easily be used to check for fit-up, interpass acceptance, welder technique, and other variables that will affect the weld's quality. Minor problems can be identified and corrected before a weld is completed. This eliminates costly repairs or rejection.

Visual inspection should be used before any other nondestructive or mechanical tests to eliminate (reject) the obvious problem welds. Eliminating welds that have excessive surface discontinuities that will not pass the appropriate code or standards saves preparation time.

Penetrant Inspection (PT)

Penetrant inspection locates minute surface cracks and porosity. Two types of penetrants now in use are the color-contrast and the fluorescent versions. Color-contrast, often red, penetrants contain a colored dye that shows under ordinary white light. Fluorescent penetrants contain a more effective fluorescent dye that shows under black light.

Magnetic Particle Inspection (MT)

Magnetic particle inspection uses finely divided ferromagnetic particles (powder) to indicate defects on magnetic materials.

A magnetic field is induced in a part by passing an electric current through or around it. The magnetic field is always at right angles to the direction of current flow. Ferromagnetic powder registers an abrupt change in the resistance in the path of the magnetic field, such as would be caused by a crack lying at an angle to the direction of the magnetic poles at the crack. Finely divided ferromagnetic particles applied to the area will be attracted and outline the crack.

In Figure 14-19, the flow or discontinuity interrupting the magnetic field in a test part can be either longitudinal or circumferential. A different type of magnetization detects defects that run down the axis, as opposed to those occurring around the girth of a part.

Radiographic Inspection (RT)

Radiographic inspection detects flaws inside weldments. Instead of using visible light rays, the operator uses invisible, short-wavelength rays developed by X-ray machines, radioactive isotopes (gamma rays), and variations of these methods. These rays penetrate solid materials and reveal most flaws in a weldment on an X-ray film or fluorescent screen. Flaws are revealed on films as dark or light areas against a contrasting background after exposure and processing (Figure 14-20).

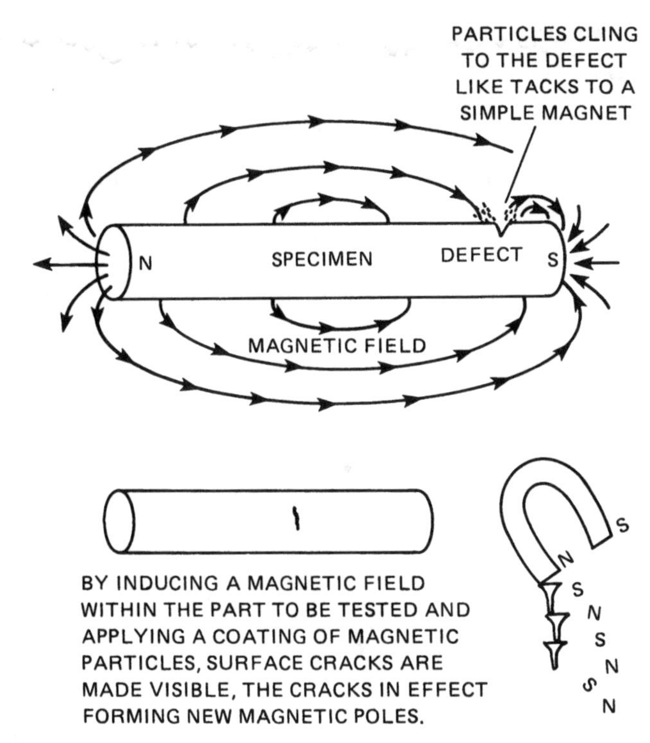

Figure 14-19 Magnetic deflection. (Adapted from Magnaflux Corporation)

The defect images in radiographs measure differences in how the X rays are absorbed as they penetrate the weld. The weld itself absorbs most X rays. If something less dense than the weld is present, such as a pore or lack of fusion defect, fewer X rays are absorbed, darkening the film. If something more dense is present, such as heavy ripples on the weld surface, more X rays will be absorbed, lightening the film.

Therefore, the foreign material's relative thickness (or lack of it) and differences in X-ray absorption determine the radiograph image's final shape and shading. Skilled readers of radiographs can interpret the significance of the light and dark regions by their shape and shading. The X-ray image is a shadow of the flaw. The farther the flaw is from the X-ray film, the fuzzier and larger the image appears. When X-raying thick material, flaws near the top surface may appear much larger than the same-sized flaw near the back surface. Those skilled at interpreting weld defects in radiographs must also be very knowledgeable about welding.

Ultrasonic Inspection (UT)

Ultrasonics is a fast and relatively low-cost nondestructive testing method that employs electronically produced, high-frequency sound waves (roughly 1/4 to 25 million cycles per second) that penetrate metals and many other materials at speeds of several thousand feet (meters) per second. A portable ultrasonic inspection unit is shown in Figure 14-21.

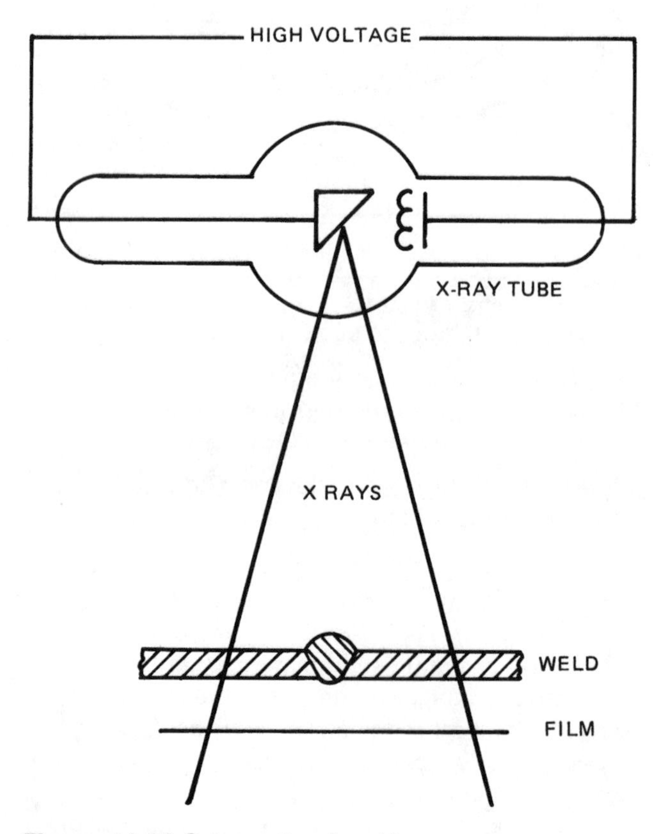

Figure 14-20 Schematic of an X-ray system.

The two types of ultrasonic equipment are pulse and resonance. The pulse-echo system, most often employed in the welding field, uses sound generated in short bursts or pulses. Because the high-frequency sound is at a relatively low power, it has little ability to travel through air so it must be conducted from the probe into the part through a medium such as oil or water.

Sound is directed into the part with a probe held in a preselected angle or direction so that flaws will reflect energy back to the probe. These ultrasonic devices operate very much like depth sounders or "fish finders." The speed of sound through a material is a known quantity. These devices measure the time required for a pulse to return from a reflective surface. Internal computers calculate the distance and present the information on a cathode ray tube, where an operator can interpret the results. The signals can be monitored electronically to operate alarms, print systems, or recording equipment. Sound not reflected by flaws continues into the part. If the angle is correct, the sound energy will be reflected back to the probe from the opposite side. Flaw size is determined by plotting the length, height, width, and shape using trigonometric formulas.

Leak Checking

Leak checking can be performed by filling the welded container with either a gas or liquid. Additional pressure may or may not be applied to the

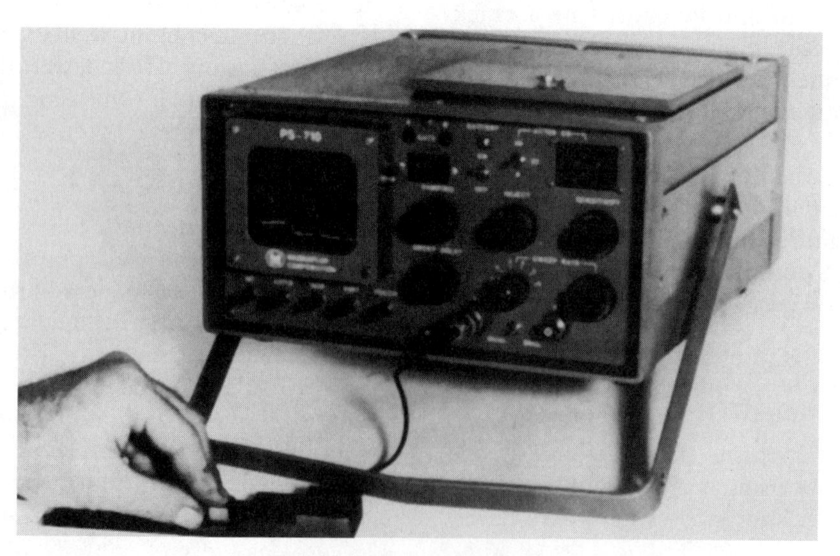

Figure 14-21 Portable ultrasonic inspection unit. (Courtesy of Magnaflux Corporation)

material in the weldment. Water is the most frequently used liquid, although sometimes a liquid with a lower viscosity is used. If gas is used, it may be either a gas that can be detected with an instrument when it escapes through a flaw in the weld, or it can be an air leak that is checked with bubbles.

Hardness Testing

Hardness is the resistance of metal to penetration and is an index of the wear-resistance and strength of the metal. Hardness tests can determine the relative hardness of the weld with the base metal. The two types of hardness-testing machines in common use are the Rockwell and the Brinell testers. The Rockwell hardness tester, Figure 14-22, uses a 120-degree diamond cone for hard metals and a 1/16″ (1.58-mm) hardened steel ball for softer metals. The method is based upon resistance-to-penetration measurement. The hardness is read directly from a dial on the tester. The depth of the impression is measured instead of the diameter. The tester has two scales for reading hardness known as the B-scale and the C-scale. The C-scale is used for harder metals, and the B-scale for softer metals.

Figure 14-22 Rockwell hardness tester. (Courtesy of Clark Instrument, Inc.)

The Brinell hardness tester measures the resistance of material to the penetration of a steel ball under constant pressure (about 3,000 kilograms) for a minimum of approximately 30 seconds. The diameter is measured microscopically, and the Brinell number is checked on a standard chart. Brinell hardness numbers are obtained by dividing the applied load by the area of the surface indentation.

REVIEW QUESTIONS

1. What are weld testing and inspection used for?
2. Where can noncritical welds be found?
3. What is the first step in establishing a quality control program?
4. What are the two major classifications of testing procedures?
5. What is a discontinuity?
6. What most often causes porosity?
7. What are the two major classifications of inclusions?
8. When does inadequate joint penetration occur?
9. What is incomplete fusion?
10. How may arc strikes be caused?
11. What are crater cracks?
12. What causes lamination in base metal?
13. What causes lamellar tears?
14. What is fatigue testing used for?
15. What are the two purposes of etching specimens?
16. What is the difference between a Charpy test and an Izod test?
17. What are the two types of penetrant inspection?
18. How does ultrasonic inspection operate?
19. What are the two types of ultrasonic equipment?
20. What are the two types of hardness-testing machines in common use?

APPENDIX

I. CONVERSION OF DECIMAL INCHES TO MILLIMETERS AND FRACTIONAL INCHES TO DECIMAL INCHES AND MILLIMETERS

II. CONVERSION FACTORS: U.S. CUSTOMARY (STANDARD) UNITS AND METRIC UNITS (SI)

III. WELDING CODES AND SPECIFICATIONS

I. CONVERSION OF DECIMAL INCHES TO MILLIMETERS AND FRACTIONAL INCHES TO DECIMAL INCHES AND MILLIMETERS

Inches dec	mm	Inches dec	mm
0.01	0.2540	0.51	12.9540
0.02	0.5080	0.52	13.2080
0.03	0.7620	0.53	13.4620
0.04	1.0160	0.54	13.7160
0.05	1.2700	0.55	13.9700
0.06	1.5240	0.56	14.2240
0.07	1.7780	0.57	14.4780
0.08	2.0320	0.58	14.7320
0.09	2.2860	0.59	14.9860
0.10	2.5400	0.60	15.2400
0.11	2.7940	0.61	15.4940
0.12	3.0480	0.62	15.7480
0.13	3.3020	0.63	16.0020
0.14	3.5560	0.64	16.2560
0.15	3.8100	0.65	16.5100
0.16	4.0640	0.66	16.7640
0.17	4.3180	0.67	17.0180
0.18	4.5720	0.68	17.2720
0.19	4.8260	0.69	17.5260
0.20	5.0800	0.70	17.7800
0.21	5.3340	0.71	18.0340
0.22	5.5880	0.72	18.2880
0.23	5.8420	0.73	18.5420
0.24	6.0960	0.74	18.7960
0.25	6.3500	0.75	19.0500
0.26	6.6040	0.76	19.3040
0.27	6.8580	0.77	19.5580
0.28	7.1120	0.78	19.8120
0.29	7.3660	0.79	20.0660
0.30	7.6200	0.80	20.3200
0.31	7.8740	0.81	20.5740
0.32	8.1280	0.82	20.8280
0.33	8.3820	0.83	21.0820
0.34	8.6360	0.84	21.3360
0.35	8.8900	0.85	21.5900
0.36	9.1440	0.86	21.8440
0.37	9.3980	0.87	22.0980
0.38	9.6520	0.88	22.3520
0.39	9.9060	0.89	22.6060
0.40	10.1600	0.90	22.8600
0.41	10.4140	0.91	23.1140
0.42	10.6680	0.92	23.3680
0.43	10.9220	0.93	23.6220
0.44	11.1760	0.94	23.8760
0.45	11.4300	0.95	24.1300
0.46	11.6840	0.96	24.3840
0.47	11.9380	0.97	24.6380
0.48	12.1920	0.98	24.8920
0.49	12.4460	0.99	25.1460
0.50	12.7000	1.00	25.4000

Inches		mm	Inches		mm
frac	dec		frac	dec	
1/64	0.015625	0.3969	33/64	0.515625	13.0969
1/32	0.031250	0.7938	17/32	0.531250	13.4938
3/64	0.046875	1.1906	35/64	0.546875	13.8906
1/16	0.062500	1.5875	9/16	0.562500	14.2875
5/64	0.078125	1.9844	37/64	0.578125	14.6844
3/32	0.093750	2.3812	19/32	0.593750	15.0812
7/64	0.109375	2.7781	39/64	0.609375	15.4781
1/8	0.125000	3.1750	5/8	0.625000	15.8750
9/64	0.140625	3.5719	41/64	0.640625	16.2719
5/32	0.156250	3.9688	21/32	0.656250	16.6688
11/64	0.171875	4.3656	43/64	0.671875	17.0656
3/16	0.187500	4.7625	11/16	0.687500	17.4625
13/64	0.203125	5.1594	45/64	0.703125	17.8594
7/32	0.218750	5.5562	23/32	0.718750	18.2562
15/64	0.234375	5.9531	47/64	0.734375	18.6531
1/4	0.250000	6.3500	3/4	0.750000	19.0500
17/64	0.265625	6.7469	49/64	0.765625	19.4469
9/32	0.281250	7.1438	25/32	0.781250	19.8437
19/64	0.296875	7.5406	51/64	0.796875	20.2406
5/16	0.312500	7.9375	13/16	0.812500	20.6375
21/64	0.328125	8.3344	53/64	0.828125	21.0344
11/32	0.343750	8.7312	27/32	0.843750	21.4312
23/64	0.359375	9.1281	55/64	0.859375	21.8281
3/8	0.375000	9.5250	7/8	0.875000	22.2250
25/64	0.390625	9.9219	57/64	0.890625	22.6219
13/32	0.406250	10.3188	29/32	0.906250	23.0188
27/64	0.421875	10.7156	59/64	0.921875	23.4156
7/16	0.437500	11.1125	15/16	0.937500	23.8125
29/64	0.453125	11.5094	61/64	0.953125	24.2094
15/32	0.468750	11.9062	31/32	0.968750	24.6062
31/64	0.484375	12.3031	62/64	0.984375	25.0031
1/2	0.500000	12.7000	1	1.000000	25.4000

For converting decimal-inches in "thousandths," move decimal point in both columns to left.

II. CONVERSION FACTORS: U.S. CUSTOMARY (STANDARD) UNITS AND METRIC UNITS (SI)

TEMPERATURE

Units

° F (each 1° change)	= 0.555° C (change)
° C (each 1° change)	= 1.8° F (change)
32° F (ice freezing)	= 0° Celsius
212° F (boiling water)	= 100° Celsius
–460° F (absolute zero)	= 0° Rankine
–273° C (absolute zero)	= 0° Kelvin

Conversions

° F to ° C _____ ° F – 32 = _____ × .555 = _____ ° C
° C to ° F _____ ° C × 1.8 = _____ + 32 = _____ ° F

LINEAR MEASUREMENT

Units

1 inch	= 25.4 millimeters
1 inch	= 2.54 centimeters
1 millimeter	= 0.0394 inch
1 centimeter	= 0.3937 inch
12 inches	= 1 foot
3 feet	= 1 yard
5280 feet	= 1 mile
10 millimeters	= 1 centimeter
10 centimeters	= 1 decimeter
10 decimeters	= 1 meter
1,000 meters	= 1 kilometer

Conversions

in. to mm _____ in. × 25.4 = _____ mm
in. to cm _____ in. × 2.54 = _____ cm
ft to mm _____ ft × 304.8 = _____ mm
ft to m _____ ft × 0.3048 = _____ m
mm to in. _____ mm × 0.0394 = _____ in.
cm to in. _____ cm × 0.3937 = _____ in.
mm to ft _____ mm × 0.00328 = _____ ft
m to ft _____ m × 32.8 = _____ ft

AREA MEASUREMENT

Units

1 sq in.	= 0.0069 sq ft
1 sq ft	= 144 sq in.
1 sq ft	= 0.111 sq yd
1 sq yd	= 9 sq ft
1 sq in.	= 645.16 sq mm
1 sq mm	= 0.00155 sq in.
1 sq cm	= 100 sq mm
1 sq m	= 1,000 sq cm

Conversions

sq in. to sq mm _____ sq in. × 645.16 = _____ sq mm
sq mm to sq in. _____ sq mm × 0.00155 = _____ sq in.

VOLUME MEASUREMENT

Units

1 cu in.	= 0.000578 cu ft
1 cu ft	= 1728 cu in.
1 cu ft	= 0.03704 cu yd
1 cu ft	= 28.32 L
1 cu ft	= 7.48 gal (U.S.)
1 gal (U.S.)	= 3.737 L
1 cu yd	= 27 cu ft
1 gal	= 0.1336 cu ft
1 cu in.	= 16.39 cu cm
1 L	= 1,000 cu cm

(*continued*)

II. CONVERSION FACTORS: U.S. CUSTOMARY (STANDARD) UNITS AND METRIC UNITS (SI) (continued)

1 L	= 61.02 cu in.
1 L	= 0.03531 cu ft
1 L	= 0.2642 gal (U.S.)
1 cu yd	= 0.769 cu m
1 cu m	= 1.3 cu yd

Conversions

cu in. to L ______ cu in. × 0.01638 = ______ L
L to cu in. ______ L × 61.02 = ______ cu in.
cu ft to L ______ cu ft × 28.32 = ______ L
L to cu ft ______ L × 0.03531 = ______ cu ft
L to gal ______ L × 0.2642 = ______ gal
gal to L ______ gal × 3.737 = ______ L

WEIGHT (MASS) MEASUREMENT

Units

1 oz	= 0.0625 lb
1 lb	= 16 oz
1 oz	= 28.35 g
1 g	= 0.03527 oz
1 lb	= 0.0005 ton
1 ton	= 2,000 lb
1 oz	= 0.283 kg
1 lb	= 0.4535 kg
1 kg	= 35.27 oz
1 kg	= 2.205 lb
1 kg	= 1,000 g

Conversions

lb to kg ______ lb × 0.4535 = ______ kg
kg to lb ______ kg × 2.205 = ______ lb
oz to g ______ oz × 0.03527 = ______ g
g to oz ______ g × 28.35 = ______ oz

PRESSURE and FORCE MEASUREMENTS

Units

1 psig	= 6.8948 kPa
1 kPa	= 0.145 psig
1 psig	= 0.000703 kg/sq mm
1 kg/sq mm	= 6894 psig
1 lb (force)	= 4.448 N
1 N (force)	= 0.2248 lb

Conversions

psig to kPa ______ psig × 6.8948 = ______ kPa
kPa to psig ______ kPa × 0.145 = ______ psig
lb to N ______ lb × 4.448 = ______ N
N to lb ______ N × 0.2248 = ______ psig

VELOCITY MEASUREMENTS

Units

1 in./sec	= 0.0833 ft/sec
1 ft/sec	= 12 in/sec
1 ft/min	= 720 in./sec
1 in./sec	= 0.4233 mm/sec
1 mm/sec	= 2.362 in./sec
1 cfm	= 0.4719 L/min
1 L/min	= 2.119 cfm

Conversions

ft/min to in./sec ______ ft/min × 720 = ______ in./sec
in./min to mm/sec ______ in./min × 0.4233 = ______ mm/sec
mm/sec to in./min ______ mm/sec × 2.362 = ______ in./min
cfm to L/min ______ cfm × 0.4719 = ______ L/min
L/min to cfm ______ L/min × 2.119 = ______ cfm

III. WELDING CODES AND SPECIFICATIONS

A *welding code* is a detailed listing of the rules or principles that are to be applied to a specific classification or type of product.

A *welding specification* is a detailed statement of the legal requirements for a specific classification or type of product. Products manufactured to code or specification requirements commonly must be inspected and tested to ensure compliance.

A number of agencies and organizations publish welding codes and specifications. The application of the particular code or specification to a weldment can be the result of one or more of the following requirements:

- Local, state, or federal government regulations
- Bonding or insuring company
- End user (customer) requirements
- Standard industrial practices

The three most popular codes are:

#1104, American Petroleum Institute—used for pipelines

Section IX, American Society of Mechanical Engineers—used for pressure vessels

D1.1, American Welding Society—used for bridges and buildings

The following organizations publish welding codes and/or specifications:

AASHT
American Association of State Highway and Transportation Officials
444 North Capitol Street, NW
Washington, DC 20001

AISC
American Institute of Steel Construction
1 East Wacker Drive
Chicago, IL 60601

ANSI
American National Standards Institute
11 W. 42nd Street
New York, NY 10036

API
American Petroleum Institute
2101 L Street, NW
Washington, DC 20005

AREA
American Railway Engineering Association
50 F. Street, NW
Washington, DC 20001

ASME
American Society of Mechanical Engineers
345 East 47th Street
New York, NY 10017

AWWA
American Water Works Association
6666 West Quincy Avenue
Denver, CO 80235

AWS
American Welding Society
550 NW LeJeune Road
Miami, FL 33126

AAR
Association of American Railroads
50 F. Street, NW
Washington, DC 20001

MIL
Department of Defense
Washington, DC 20301

SAE
Society of Automotive Engineers
400 Commonwealth Drive
Warrendale, PA 15096

GLOSSARY

AC high frequency (ACHF) Primary alternating welding current that has had a secondary high frequency, from 50,000 to 3,000,000 cycles per second, high 3,000 voltage, low amperage current induced on it in order to stabilize the primary welding arc.

acid A chemically active substance that is able to react with another material by releasing a proton. The reacion results in a chemical change in the material. Acids have a pH reading of less than 7.0.

air-cooling torch A gas tungsten arc welding torch that is able to withstand the heat of welding by releasing heat generated by the arc into the shielding gas and surrounding air.

alkaline A chemically active substance, also called basic, that is able to react with another material by taking a proton, which results in a chemical change in the material. Alkaline substances have a pH reading of more than 7.0.

alloy A metal with one or more elements added to it, resulting in a significant change in the metal's properties.

alternating current (AC) In alternating current the electrons flow in one direction, stop, and reverse their direction of flow. Each time this sequence is completed, it is referred to as a cycle. Alternating current power is supplied at either 50 or 60 cycles per second.

amperage (amps, A) A measurement of the rate of flow of electrons; amperage controls the size of the arc.

arc The flow of electrons across a gap.

argon A gaseous element produced as a byproduct of air reduction (oxygen produced by liquefying air). Argon is an inert gas that is heavier than air, colorless, and tasteless. When used as a shielding gas for GTA welding, argon provides both a stable arc and good weld pool protection.

atmosphere Gaseous envelope surrounding everything.

basic See alkaline.

butt-joint A joint between two members aligned approximately in the same plane.

carbide precipitation Migration of chromium from grains into grain boundaries in the chrome nickel stainless steels. Carbide precipitation can occur within an approximate temperature range of 800° F to 1500° F (450° C to 820° C).

chemically neutral A substance that is neither acid nor alkaline (basic). Neutral substances have a pH of 7.0.

code or standard Detailed listings of the rules or principles that are to be applied to a specific classification or type of work.

cold-rolled steel Forming of steel below its recrystallization temperature, which results in a harder metal with a very smooth surface.

conductive heat transfer Heat that transfers from molecule to molecule within a substance or from one substance to another.

contamination The accidental introduction of an undesirable substance such as dirt or oxygen into a material such as the weld or shielding gas.

convection heat transfer Heat that is transferred as the result of the flow or movement of a fluid such as air or water.

corner-joint A joint between two members located approximately at right angles to each other.

crater cracking A crack in the crater of a weld bead.

current flow The flow of electrons.

cylinder A portable container used for transportation and storage of a compressed gas.

DCEN (**direct current electrode negative**) Direct current that flows from the electrode to the work. Also called DCSP (direct current straight polarity).

DCEP (direct current electrode positive) Direct current that flows from the work to the electrode. Also called DCRP (direct current reverse polarity).

deoxidizer A material that, when added to molten metal, combines with oxygen in order to remove it from the completed weld metal.

direct current (DC) Flow of electrons in only one direction. Electrons can flow either from the workpiece or to the electrode.

direct current with negative electrode The arrangement of direct current arc welding leads on which the electrode is the negative pole and the workpiece is the positive pole of the welding arc.

direct current with positive electrode The arrangement of direct current arc welding leads on which the electrode is the positive pole and the workpiece is the negative pole of the welding.

electrical circuit An arrangement of a power source, a conductor, and a load through which a controlled flow of electricity can pass.

electrode A component of the electrical circuit that terminates at the arc, molten conductive slag, or base metal.

filler rod The metal or alloy to be added in making a welded, brazed, or soldered joint.

flowmeter A meter that measures the flow of shielding gases and allows a means to vary the flow as needed.

flow rate The rate at which a given volume of shielding gas is delivered to the weld zone.

flux A material used to hinder or prevent the formation of oxides and other undesirable substances in molten metal and on solid metal surfaces, and to dissolve or otherwise facilitate the removal of such substances.

frequency The rate at which AC cycles.

fuse Electrical circuit device designed to fail at a predetermined current level to protect the circuit or components from over-current.

gas tungsten arc welding (GTAW) An electric welding process that uses a nonconsumable tungsten electrode and a shielding gas.

heat A form of energy that brings about a change in the temperature of a substance. The quality of heat is measured in units such as British Thermal Units (BTU) or Joules (J).

heliarc welding Original registered trade name of the Linde Company used to identify the gas tungsten arc welding process.

helium An inert gas used as shielding gas. Helium has a higher electrical resistance than argon; therefore, its use will cause an increase in joint penetration of weld as well as allowing an increase in welding speed.

high frequency "Several thousand volts with high frequency is connected across the arc gap." See AC.

high frequency See alternating current high frequency (ACHF)

hot-rolled steel Forming of steel above its recrystallization temperature, which does not result in an increase in the hardness of the metal. Most hot-rolled steel has a thin layer of black iron oxide called "mill scale" tightly bonded to its surface.

inert gases A gas that will not combine chemically with materials. Inert gas is used in GTAW to shield tungsten and molten weld pool.

lap-joint A joint between two overlapping members.

mill scale The tightly bonded oxide layer that forms on the surface of hot-rolled steel when it is manufactured.

molten weld pool The liquid state of a weld prior to solidification as weld material.

nitrogen Chemically active gas that is common in the atmosphere.

non-consumable tungsten electrode An electrode that does not provide filler metal.

nozzle A device that directs shielding media.

oxygen Chemically active gas that is common in the atmosphere.

postpurge Once welding current has stopped in gas tungsten arc welding, this is the time during which the gas continues to flow to protect the molten pool and the tungsten electrode as cooling takes place to below a temperature at which they will not oxidize rapidly.

prepurge In gas tungsten arc welding, the time during which gas flows through the torch to clear out any air in the nozzle or surrounding the weld zone.

reverse polarity See direct current electrode positive.

shielding gas Protective gas used to prevent or reduce atmospheric contamination.

specifications A detailed statement of legal requirements for a specific classification or type of weld to be made on a specific product.

straight polarity See direct current electrode negative.

TIG a nonstandard term when used for gas tungsten arc welding.

tack weld A weld made to hold parts of a weldment in proper alignment until the final welds are made.

tee-joint A joint between two members located approximately at right angles to each other in the form of a T.

temperatures A unit of measure of the level of molecular activity within a substance. The two most common scales for this measurement are Fahrenheit and Celsius.

tungsten-alloy electrodes a nonconsumable metal electrode used in arc welding, arc cutting, and plasma spraying, made principally of tungsten.

voltage Electrical force or pressure that causes electrons to flow through a circuit.

walks the cup When a welder rests the torch on the workpiece and moves it back and forth along the joint.

water-cooling GTA welding torch that uses water to cool the welding torch head and power cable.

welding groove The joint penetration (depth of bevel plus the root penetration when specified).

weld pool The localized volume of molten metal in a weld prior to its solidification as weld metal.

welding helmet A device designed to be worn on the head to protect eyes, face, and neck from arc radiation, radiated heat, spatter, or other harmful matter expelled during arc welding, arc cutting, and thermal spraying.

workpiece The part that is welded, brazed, soldered, thermal cut, or thermal sprayed.

zero points Points in the middle of an AC cycle at which the flow is neither positive nor negative.

INDEX